PRACTICAL ELECTRONICS TROUBLESHOOTING

PRACTICAL ELECTRONICS TROUBLESHOOTING

James Perozzo

DELMAR PUBLISHERS INC.

Cover Photograph: Ted Kurihara

Delmar Staff
Administrative Editor: Gregory Spatz
Production Editor: Gerry East

For information, address
Delmar Publishers Inc.
2 Computer Drive West, Box 15-015
Albany, New York 12212

Printed in the United States of America
Published Simultaneously in Canada
by Nelson Canada
A Division of International Thomson Limited

10 9 8 7 6

Library of Congress Cataloging in Publication Data

Perozzo, James.
Practical electronics troubleshooting.

Includes index.
1. Electronic apparatus and appliances—Maintenance and repair. I. Title.
TK7870.2.P47 1985 621.3'028'8 84-28626
ISBN 0-8273-2433-2

NOTICE TO THE READER

Publisher does not warrant or guarantee any of the products described herein or perform any independent analysis in connection with any of the product information contained herein. Publisher does not assume, and expressly disclaims, any obligation to obtain and include information other than that provided to it by the manufacturer.

The reader is expressly warned to consider and adopt all safety precautions that might be indicated by the activities described herein and to avoid all potential hazards. By following the instructions contained herein, the reader willingly assumes all risks in connection with such instructions.

The publisher makes no representations or warranties of any kind, including but not limited to, the warranties of fitness for particular purpose or merchantability, nor are any such representations implied with respect to the material set forth herein, and the publisher takes no responsibility with respect to such material. The publisher shall not be liable for any special, consequential or exemplary damages resulting, in whole or in part, from the readers' use of, or reliance upon, this material.

Contents

Preface/xiii

chapter one **Some Necessary Basics**

Chapter Overview/1
Necessary Background/1
A Few Definitions/6
The Test Bench/10
Summary/12
Review Questions/12

chapter two **Kinds of Electronics Problems**

Chapter Overview/14
Complete Failures/14
Poor Performance/14
Tampered Equipment/15
Intermittents/15
Motorboating/18
Massive Traumas/18
Transients/21
Overheated Part Failures/22
Hum Problems/24
Distortion Problems/26
Multiple Problems/27
Microphonics/27
Noisy Controls/27
Operator-Induced Problems/28
The Tough Problems/28
Summary/29
Review Questions/29

chapter three **System Troubleshooting**

Chapter Overview/30
Investigate the Report/30

Finding the Bad Unit/31
Wire-Bundle and Coax-Cable Problems/33
Signal Levels as System Operation Indicators/36
Replacement of Suspected Units of the System/37
Verify Proper System Operation After the Replacement/38
Summary/38
Review Questions/39

chapter four **Live-Circuit Testing**

Chapter Overview/40
Safety and Cautionary Notes/40
Use of Circuit Card Extenders/42
Using the Schematic Diagram/42
If No Schematic Is Available/47
Summary/52
Review Questions/52

chapter five **DC Troubleshooting**

Chapter Overview/53
General DC Troubleshooting/53
DC Voltage Measurements to Find the Problem/65
A Most Important Concept/67
Opening the Equipment/70
Timely Shortcuts/70
Summary/71
Review Questions/71

chapter six **Fixing the Power Supply**

Chapter Overview/72
Finding the Open Fuse/72
Analyzing Fuse Failures/73
The Power-Supply Block Diagram/74
Applying Input Power to the Power Supply/76
Troubleshooting an Overloaded (Shorted) Power Supply/81
Troubleshooting an Inoperative Power Supply (No Output Voltage)/82
Troubleshooting the Power-Supply Regulator/83
Batteries as a Power Source/87
Summary/89
Review Questions/89

chapter seven How to Trace Signals

Chapter Overview/90
Troubleshooting Digital Circuits/90
Troubleshooting Analog Equipment/90
The Block Diagram—Finding the Bad Area/92
Using a Bench Power Supply/92
Narrowing the Problem to an Area/96
Narrowing the Problem to a Stage/97
Signal-Tracing Considerations/98
Tracing the Unwanted Signal/99
Summary/100
Review Questions/100

chapter eight Troubshooting and Signal Tracing in Low-Frequency Circuits

Chapter Overview/101
Source Instruments/101
Load Instruments/102
The Originating LF Circuit/105
The Processing LF Circuit/105
Tracing Signal Distortion/106
The Terminating LF Circuit/107
Summary/107
Review Questions/107

chapter nine Special Techniques for Discrete Semiconductors in Operation

Chapter Overview/108
Troubleshooting the PN Junction Diode in Operation/108
Troubleshooting the Zener Diode Circuit in Operation/109
Troubleshooting the Bipolar Transistor Circuit in Operation/109
Transistor and Resistor Arrays/118
Troubleshooting FET Circuits/118
Troubleshooting UJT Oscillators/122
Troubleshooting Thyristors (SCRs)/123
Summary/123
Review Questions/123

chapter ten Troubleshooting in Live Analog IC Circuits

Chapter Overview/125
Operational Amplifiers/125
Voltage Regulator ICs/132
The 555 Timer Chip/132
Consumer IC Chips/133
Analog Switches/134
Optical Isolators/135
Summary/138
Review Questions/138

chapter eleven Troubleshooting and Signal Tracing in RF Circuits

Chapter Overview/139
Why RF Circuits Are Different/139
Special Tools Needed/140
Source Instruments/141
Load Instruments/142
Signal Tracing in RF Circuits/150
Terminating RF Signals/153
Mixing RF Circuits/154
The RF Loop Circuit (Frequency Synthesizer)/154
Summary/156
Review Questions/156

chapter twelve Troubleshooting and Signal Tracing in Pulse Circuits

Chapter Overview/157
Terms Unique to Pulsed Waveforms/157
Source Instruments/159
Load Instruments/160
The Originating Pulse Circuit/164
The Processing Pulse Circuit/164
The Terminating Pulse Circuit/167
The Mixing Pulse Circuit/168
How Signals Change in Pulse Circuits/170
The Time Constant/172
Summary/179
Review Questions/179

chapter thirteen Digital Troubleshooting Techniques

Chapter Overview/180
First Checks/180

Digital IC References Are Absolutely Necessary/181
Digital Troubleshooting Basics/181
The Digital Schematic Diagram/187
Basic Digital Gates/188
Instruments for Digital Circuit Troubleshooting/194
Tracing Signals Through Digital Circuits/203
Summary/210
Review Questions/210

chapter fourteen **Troubleshooting Computer Circuitry**

Chapter Overview/212
Additional Knowledge Required/212
Types of Computer Problems/213
Hardware Problems/213
Let the Computer Test Itself/214
Additional Checks to Make/215
Static Stimulus Testing/216
Software Problems/217
Options and Hardware Setup Problems/219
Summary/219
Review Questions/220

chapter fifteen **Troubleshooting and Repairing Floppy Diskette Drives**

Chapter Overview/221
Cautions/221
Floppy Diskette Geometry/222
Testing and Aligning Diskette Drives/225
Summary/229
Review Questions/229

chapter sixteen **Dead-Circuit Testing**

Chapter Overview/230
Don't Ruin Your Ohmmeter!/230
Using the VOM on the Resistance Function/231
Using the DMM on the Resistance Function/233
Using a Solid-State Tester/234
Making the Checks/236
Finding Short Circuits on Printed Circuit Boards/238
Testing Component Resistance In-Circuit/239
Summary/242
Review Questions/242

chapter seventeen Replacing the Part and Analyzing the Failure

Chapter Overview/243
Some Mechanical Tips and Considerations/243
Point-to-Point Wiring Repairs
(Not Printed Circuits)/244
Printed-Circuit-Board Repairs/244
Soldering on Printed Circuit Boards/246
Where to Get Parts/246
Replacing Parts/247
Cleaning and Repairing Damaged Boards/249
Checking the Bad Component/249
Final Precautions/271
Summary/271
Review Questions/271

chapter eighteen Final Inspection and Return to Service

Chapter Overview/273
Aligning Equipment/273
Reassembling the Equipment/277
The Practical RF Decibel for Technicians/279
Final Tests/279
The Paperwork/284
Packing for Shipment/285
Summary/286
Review Questions/286

chapter nineteen Routine and Preventive Maintenance

Chapter Overview/287
Why Mess With It?/287
Using the Solid-State Tester to Predict Failures/288
TDR for Cable Maintenance and Troubleshooting/288
Mechanical Maintenance/290
Solvents for Cleaning/290
Weathered Electronics/290
Summary/293
Review Questions/293

chapter twenty Techniques for Live Vacuum-Tube Circuits

Chapter Overview/294
Special Tools/294

Safety Notes/295
Chassis Wiring Color Code/295
Typical Vacuum-Tube Voltages/296
Troubleshooting Vacuum-Tube Circuits/302
The Cathode-Ray Tube/305
Summary/307
Review Questions/307

chapter twenty-one **Basic Troubleshooting of Power-Line Circuits and Motors**

Chapter Overview/309
Single-Phase Power Distribution and Troubleshooting/309
Three-Phase Power Systems/310
A Special Circuit/311
DC Motor Basics/311
The Alternator/314
AC Motor Basics/315
Troubleshooting Motor Circuits/317
Using the Oscilloscope on AC Line Problems/318
Summary/318
Review Questions/319

chapter twenty-two **If the Right Part Isn't Available**

Chapter Overview/320
Parts Substitutions/320
Summary/328
Review Questions/328

appendix I **Resistor Color Codes/329**

appendix II **Electronics Schematic Symbols**

Common Electronics Symbols/331
Semiconductor Symbols/333
Digital Symbols/335
Logic Symbols/336
Common Vacuum-Tube Symbols/337

appendix III **Recommended Manufacturers of Test Equipment and Supplies/338**

appendix IV **BASIC Programs for Disk Drive Alignment on IBM and Compatible Computers/340**

Index/344

Preface

However sophisticated electronic equipment may become, it will still occasionally fail. This book will aid the new servicing technician in following a methodical course leading to the correction of the problem in a minimum amount of time, using both traditional techniques and some new approaches. The techniques presented here are real-world, actual ones that a practicing technician would use, techniques used to save time and make money. It is no longer necessary for a new technician to take years to learn troubleshooting techniques by trial and error.

These are *generic* troubleshooting techniques, procedures and tricks of the trade, from DC voltage troubleshooting to digital signal tracing in microcom puter circuits. There are many "Dos" and "Don'ts" to help the technician avoid costly errors. No specific field of electronics, such as consumer electronics or communications, is stressed.

No attempt is made to teach basic electronics or to explain in detail the operation of the test equipment except where appropriate. There are many other books available that explain circuit theory and the detailed use of test equipment, so only those points essential to practical troubleshooting receive emphasis. It is up to the technician to *read* and *know* the instruction books for both the equipment to be repaired (when available) and the test instruments he will be using. Once this is done, all that remains is to "tie them together"—the reason this book was written.

As a matter of practicality, mathematics is also held to a minimum. Only addition, subtraction, multiplication, and division are required to grasp the principles herein. Seldom does a real technician use a calculator while working. A basic foundation of electronic theory and circuits is necessary to understand electronic troubleshooting. The graduate of a vocational school basic electronics class or the holder of the FCC General Radiotelephone Certificate will easily follow the material presented.

I have presented the test equipment with which I am familiar, equipment I consider to be the best for the job at hand. The Fluke digital multimeter and the Huntron Tracker, for instance, are top-of-the-line instruments that perform exceedingly well, instruments that I can recommend through personal experience. If the reader uses other equipment, there will be some differences in the specifications, capabilities, and applications.

I wish to thank:

Mr. Bill Hunt of Huntron Instruments, Lynnwood, Washington, for encouragement and an education in semiconductor failure detection.

Mr. Chuck Newcombe and Mr. Tak Tsang of John Fluke Co., Everett, Washington for the latest update in digital multimeter instruments.

Mr. Rick Jorgenson and Ms. Elizabeth Dessuge of Dysan, Santa Clara, California, for their cooperation and assistance on computer disk drive alignment and practical troubleshooting tips.

Mr. Herbert Heller of the Bird Electronic Corp. for information and explanations of practical RF power measurements.

Mr. Scott Zelov of VIZ Manufacturing Co. for a chance to use the latest in affordable bench power supplies with built-in digital readout of voltage and current.

Mr. Frank Chipman of the Cooper Group for updating me on the latest Weller soldering equipment for sensitive circuitry.

ABOUT THE AUTHOR

Mr. Perozzo has 26 years' experience repairing electronic equipment of many different types. After 20 years' service as a technician and officer in the Coast Guard, he broadened his experience by working in marine and aviation electronics, microwave voice carriers, and computer repair. A native of the Pacific Northwest, his qualifications include an FCC General Radiotelephone Certificate, Amateur Extra Class License, FAA Avionics Certification, and an Electronics Teaching Certificate. Mr. Perozzo teaches electronics troubleshooting at Clover Park Vocational Technical Institute near Tacoma, Washington.

chapter one

Some Necessary Basics

CHAPTER OVERVIEW

There is no substitute for a sound background in electronic circuit theory. Such a background is essential to understand this book. There are some additional items that are essential, but that may not be covered in required electronics courses. These items are presented in this chapter. There is also clarification of some terms that might be misunderstood.

NECESSARY BACKGROUND

To accomplish electronics repairs with any reasonable amount of efficiency, it is necessary to know certain basic information. An excellent reference for the great many electronics terms used in this book is the *Modern Dictionary of Electronics,* available from Howard W. Sams Publishing Company. See Appendix III.

Converting Electronics Terms

It is important to be able to convert both ways from one term to another without having to write them down. Practice these until you are confident in your ability to do them:

- Converting between hertz, kilohertz, megahertz, and gigahertz.
- Converting between ohms, kilohms, and megohms.
- Converting between microfarads and picofarads.
- Converting between seconds, milliseconds, and microseconds.

There are rules for converting back and forth, but it seems that the rules are harder to remember than simply getting familiar with the terms.

Common Schematic Symbols

It will be necessary to recognize the common schematic symbols that you will see on schematics. See Appendix II for the ones you are most likely to use. Take special note of the symbols used for digital circuits, which are a different sort of schematic symbolism.

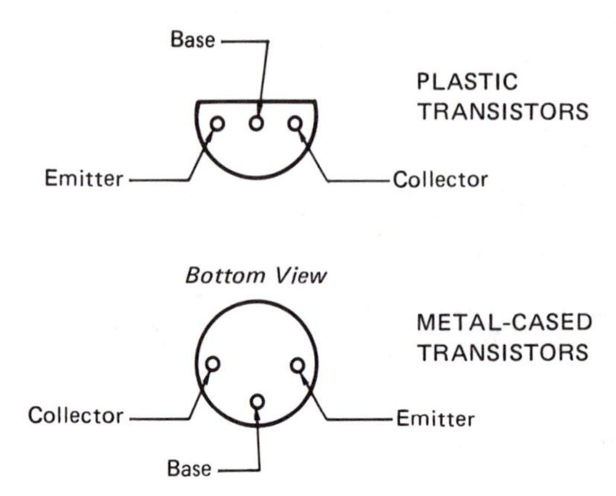

FIGURE 1-1 The most common connections to a bipolar transistor.

Common Transistor Lead Identification

Transistors are manufactured with just about every conceivable combination of lead layout. It is easier to keep lead identification straight in your mind if you have a beginning layout and then note differences, if any, in a specific lead layout.

Most transistors that have three leads in a circular pattern have the layout shown in Figure 1-1. It is not worthwhile to learn much more by memory since there are so many exceptions. Be sure to verify the lead identification of the specific transistor you are using.

Common Integrated Circuit Connections

Integrated circuit pins are numbered from one corner around the IC in a *clockwise* circular pattern when viewing the *bottom* of the IC. Of course if the IC is viewed from the *top,* the pins would be identified in a *counterclockwise* direction. Identification of the proper corner may be made by referring to Figure 1-2. Note that the ground and voltage supply pins are often at diagonal corners and situated so that if the IC is accidently reversed in the socket the IC will often be destroyed when power is applied. This is particularly the case when dealing with CMOS (complementary metal-oxide semiconductor) ICs.

Standard Circuits

There are a few simple circuits that should be memorized. They are encountered so often that the technician should be able to work around in them without a schematic. The circuits of Figure 1-3 are the ones that should be learned thoroughly. Other circuits may be obtained from actual equipment schematics.

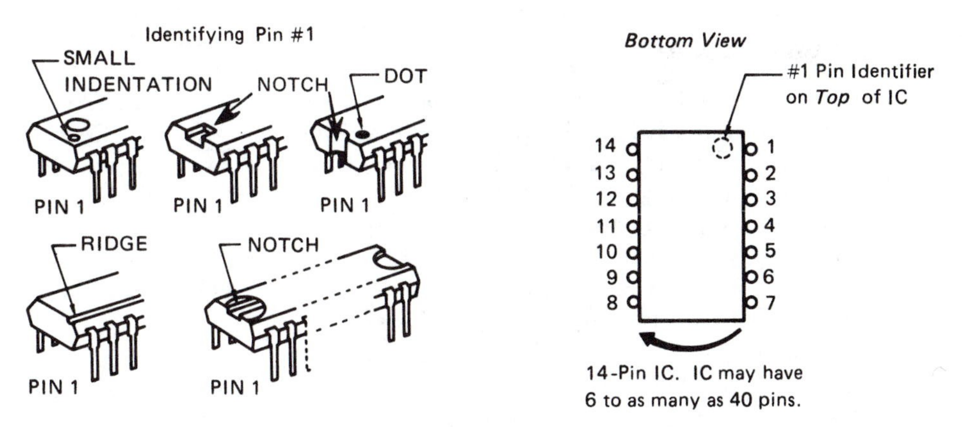

FIGURE 1-2 Identifying integrated circuit pin numbers. (Courtesy of Heath Co., Benton Harbor, MI)

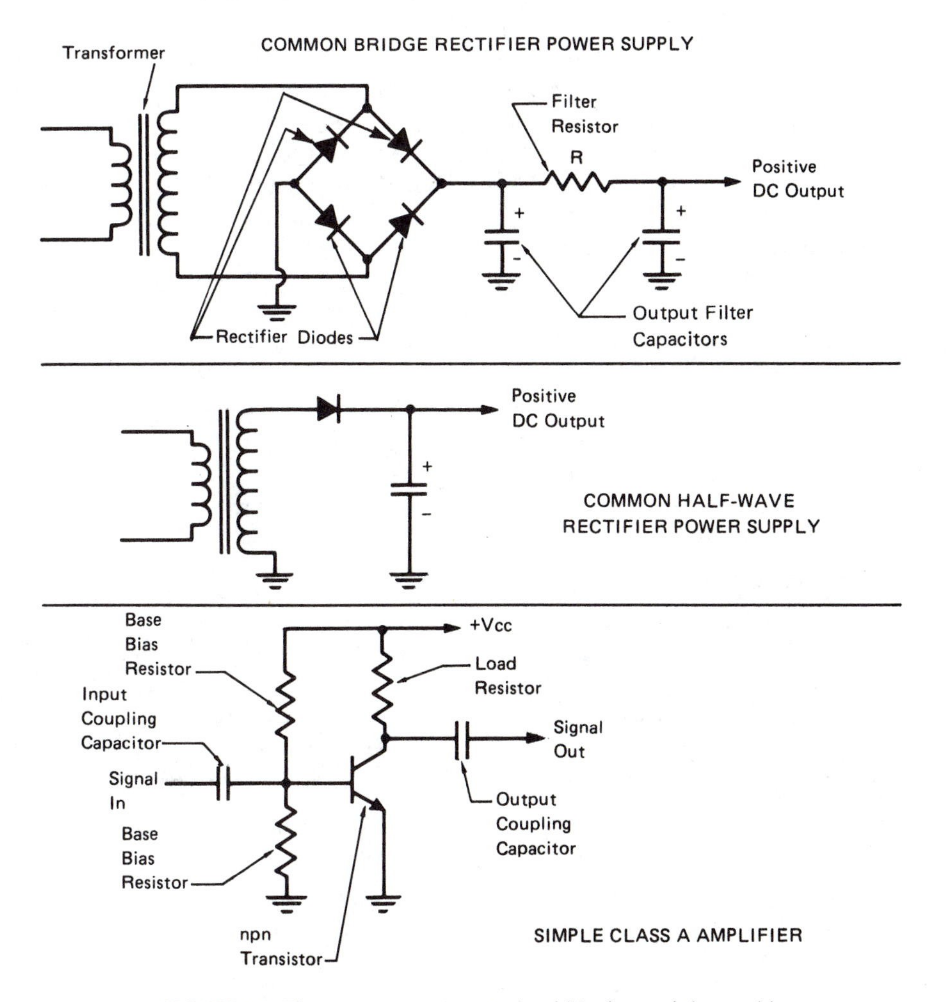

FIGURE 1-3 These common circuits should be learned thoroughly.

Resistor Color Code

It is important to be able to read resistors' values from their color code and to be able to distinquish them from one another. Knowing this code will keep you from wasting time by examining the wrong resistor. For example, knowing the code would immediately tell the experienced technician that the 10,000-ohm resistor being sought is not the one with yellow, violet, and orange bands.

There are two major color coding systems in use (Appendix I): the two-significant figure and the three-significant figure systems (see Figure 1-4). Learn them well; they are used frequently in troubleshooting.

If in doubt as to the color code value, do as any good technician would do—cheat! Use an ohmmeter. A digital ohmmeter will give a more accurate reading with less chance of misreading the value than an analog ohmmeter. On the high-resistance scales of either, keep your fingers off both connections at once to eliminate any error of reading through your body.

The same color code is often used with capacitors, too. The only problem with capacitors is that there are so many different coloring schemes in use. In one case the third color is a dot in one corner, in another it is a ring around the component, and in yet a third it is a spot of paint on the edge of the capacitor. Be careful when attempting to identify the values of capacitors using color code. The best method yet for telling the value of a capacitor is to find one with the value written on it in numbers.

Chassis wiring also sometimes uses color coding. Wires from a channel switch, for instance, might be wired such that channel one is brown, two is red, and so on. Watch for this, as it is a help at times in tracing wiring.

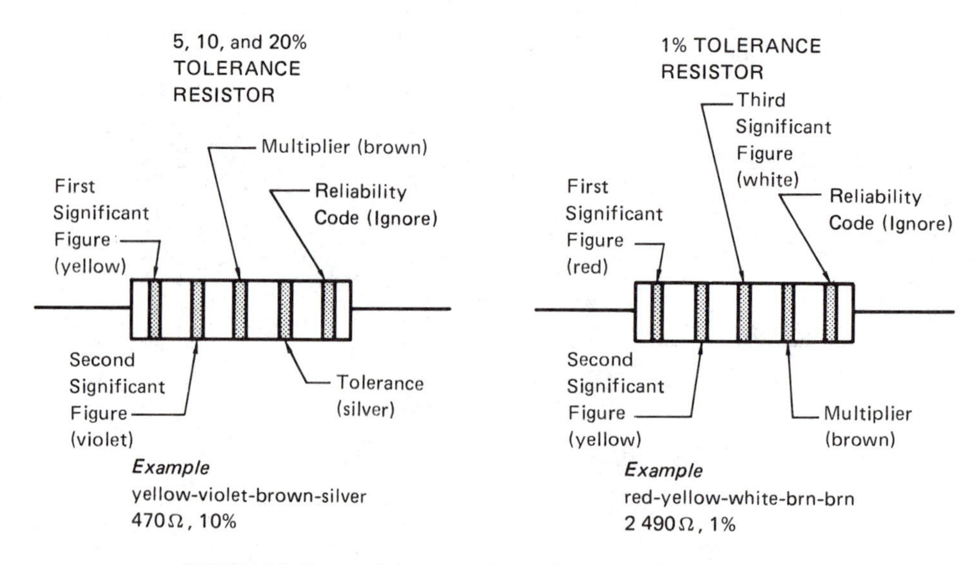

FIGURE 1-4 Two- and three-significant figure resistor color codes.

Wire Color Codes

When vacuum tubes were in general use, there was a color code for various different circuits in the equipment. With today's solid-state technology, all that remains of this old code is that black is still usually the circuit common and red is often the "hot" side of the supply, called Vcc or B+. Test leads of instruments using direct current (volt-ohm-milliameters, digital multimeters, power supplies) use black to indicate negative and red for positive.

Here is an important point to keep in mind for safety's sake, since it is just the opposite of conventional electronic wire color coding: 115-V AC power-line wiring uses black for the hot wire, white for the neutral, and green for earth ground.

A Feeling for Logic

When troubleshooting in digital logic circuits, the technician should keep in mind the terms *active high* and *active low.* Since logic circuits use a high or a low level, you could define either state as the desirable signal. To better clarify this point, consider the two inputs to a digital circuit shown in Figure 1-5. For a desired high output from the circuit the input must also be in the high or positive state. This is the most logical way to do things. These inputs would be called active high inputs.

The opposite case is possible; the two inputs must be *low* to give the desired AND output from the gate. This is negative logic and would be indicated by either a small circle at the input of the circuit symbol or by a small bar over any identifying letters next to the connection. This would be called an *active low* input.

Truth Tables

A technician who works on digital equipment must be very familiar with the truth tables for the digital building blocks that will be encountered. The common ones that must be thoroughly understood are AND, NAND, OR, NOR,

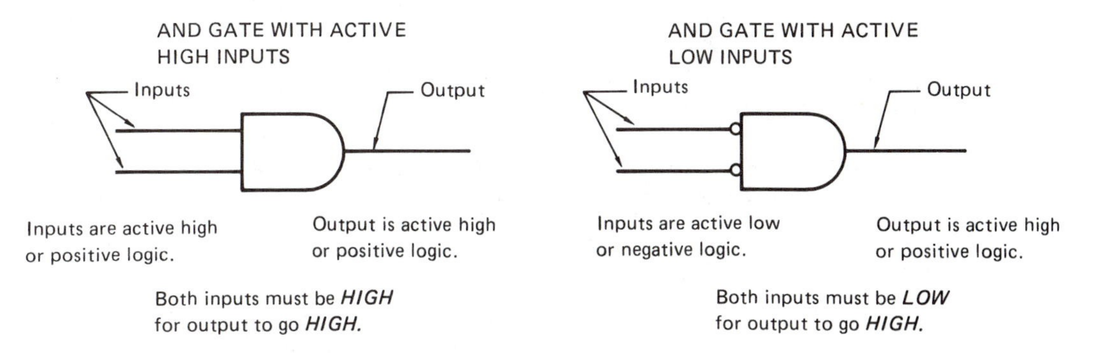

FIGURE 1-5 Sample logic symbols showing active high and active low signals.

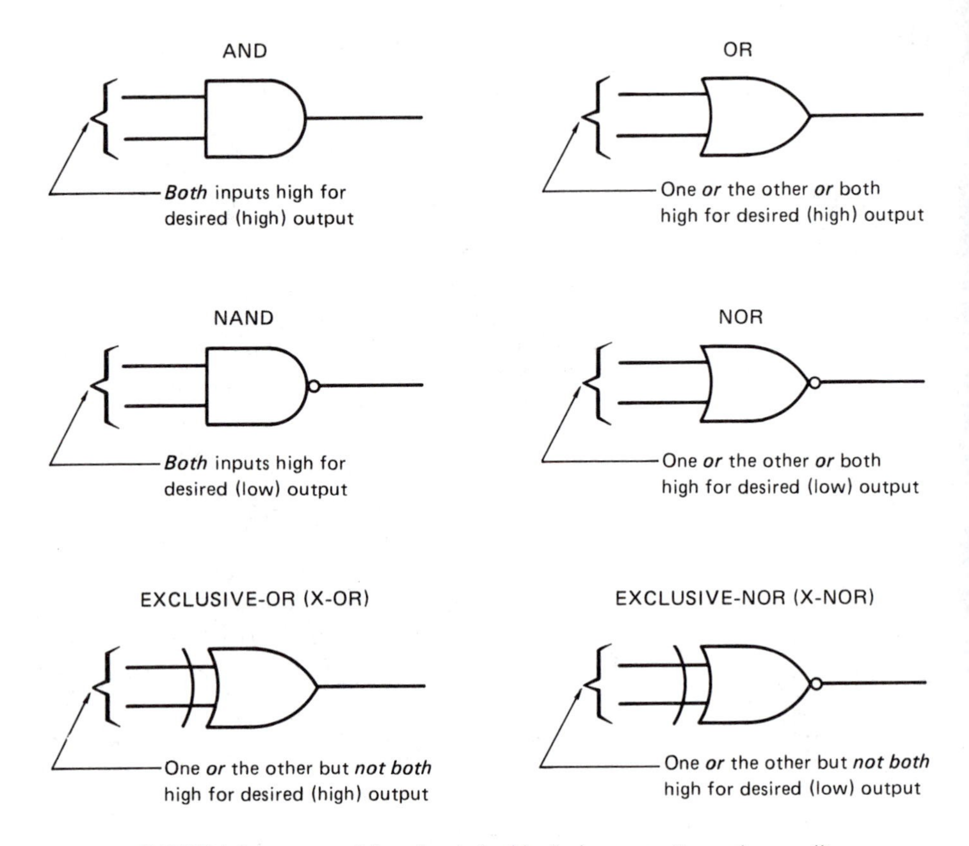

FIGURE 1-6 Input conditions for desired logical outputs. Know these well!

and EXCLUSIVE OR. It is not enough to memorize the tables themselves. Rather, the tables must be *understood* so well that the technician will be able to state without hesitation what the inputs of a given gate would have to be to obtain the desired output (see Figure 1-6). The tables themselves can often boil down to a simple statement. For instance, the truth table for an AND gate simply means that *both inputs must be high for the output to be high.*

A FEW DEFINITIONS

To clarify any possible misunderstanding, study the following terms and definitions for the purposes of this text.

Live and Dead Circuits

For the remainder of this book the term *live* shall simply mean that all operating voltages are present to operate the circuits under test. A *dead* circuit is one that has no external operating voltages connected to it whatsoever.

In-Circuit and Out-of-Circuit

If you can hold a single component in your hand, it is *out-of-circuit.* If one lead of the component is connected to the circuit but all other leads are completely disconnected, the component is still out-of-circuit. This is a common situation when testing components. Disconnect all component leads or leave no more than one connected. The component is then effectively out of the circuit for further testing.

An *in-circuit* component is fully installed and may show quite unexpected test results not due to defects, but due to other components connected to it.

Signal and No-Signal Testing

Depending upon the tests being made at the time, the technician may or may not require input signals to the circuits under test. Direct current voltage measurements do not require that the circuits being tested have signals flowing through them. As a matter of fact, the presence of a signal flowing through the circuit could complicate the troubleshooting by producing misleading meter indications. This is particularly true of RF (radio frequency) circuits. RF fields can cause some meters to strike their end peg violently, or digital types to have strange readings that have nothing to do with the actual values of DC voltage.

Classes of Operation

It is necessary to be familiar with the *classes of operation* of amplifiers. It is the biasing of these stages that determines their behavior in the circuit, and bias determination is one of the primary troubleshooting tools. The reader is referred to a basic electronics text for detailed review of Class A, B, and C amplifiers. Briefly, the *Class A* amplifier is biased such that the amplifier stage is only halfway turned on, allowing the stage to operate toward more or less conduction, thereby following the input signal faithfully. The *Class B* stage is biased right at the cutoff point and requires a signal of a particular polarity to turn it on and produce an output. An input signal of the opposite polarity will merely turn it more "off," producing no difference in the output circuit. The *Class C* stage requires an input signal of the proper polarity just like the Class B stage, but the initial turnoff bias on the Class C stage is so great that the incoming signal must be much larger to overcome this greater bias. Therefore, the Class C stage will require more amplitude of the input signal to turn it on. Of course, an opposite input signal polarity will serve only to keep the stage turned off.

These classes of operation will be used often in the topics following. Figure 1-7 summarizes these differences in classes of operations.

Circuit Common

Circuit common is the central point to which individual circuits connect. It is often but not always the external metal portion of the chassis. Common is that

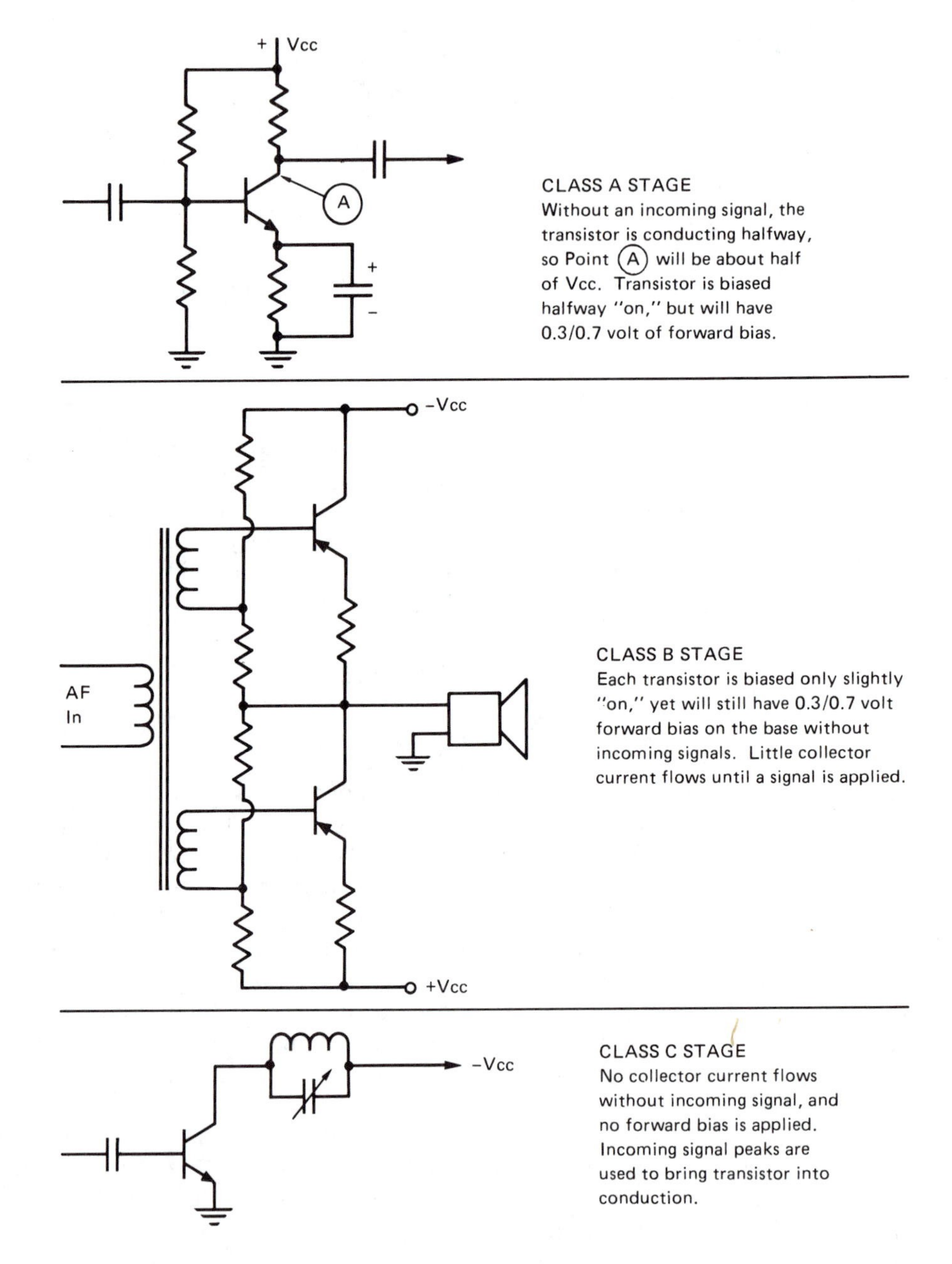

FIGURE 1-7 Classes of operation for amplifiers.

point to which the grounded or negative side of test equipment such as oscilloscopes and signal generators is usually connected. These instruments are often operated from the 117-V AC power line. Operation with their grounds connected to anything but the common bus of the circuit under test will invite problems such as damaging printed traces or causing other damage.

Circuits that use relays or photoisolators may have completely isolated circuitry from one end to the other of these components. One of the primary purposes of these components is to provide this isolation. Instruments that do not use or are completely isolated from (both DC and AC isolation) the power-line ground may be connected into most circuits at any point without incurring any damage due to grounding problems. Instruments meeting this requirement are the standard volt-ohm-milliameter (VOM) and the digital multimeter (DMM). When measuring voltages with the VOM or DMM, the black lead is the one usually connected to the circuit common.

Shorts and Opens

A *short* is an unintentional path of very low, essentially zero resistance, which in some instances can cause excessive current and seriously damage components, resulting in fuses blowing or more serious damage. In other cases, a short may simply eliminate a signal or DC voltage without any damage at all. Troubleshooting techniques sometimes take advantage of this fact. An *open* is a path of extremely high resistance that prevents any appreciable current flow. A switch in the "Off" position is an example of an open circuit.

Circuit Impedances

Throughout this book the terms *high impedance* and *low impedance* will be used. These are relative terms with no definite dividing line between them. They are rather like the terms high and low themselves. It depends on where you're standing at the time.

A high-impedance circuit will be "weak." It will have a high internal resistance, and its voltage will be easily changed by any external load. Figure 1-8 shows how an average VOM will give very inaccurate readings when connected

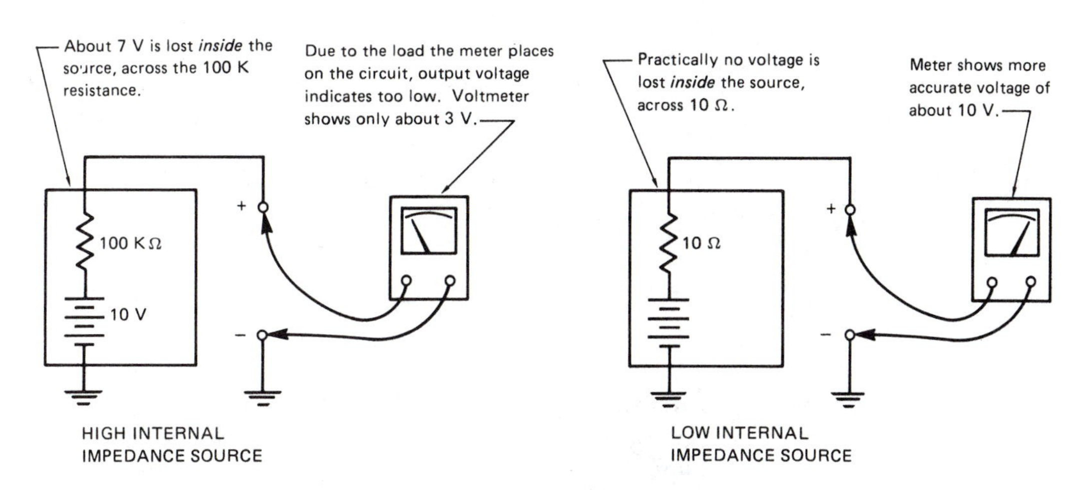

FIGURE 1-8 Loading effects of the volt-ohm-milliammeter (VOM).

to a high-impedance voltage source. Examples of high-impedance circuits are the inputs to vacuum tubes, FETs (field-effect transistors), and CMOS integrated circuits. A low-impedance circuit will be "strong." On such circuits the loading effect of a voltmeter will be negligible. Examples of low-impedance circuits are power supplies, bipolar transistor circuits, vacuum tube cathode circuits, and TTL (transistor-transistor logic) integrated circuits.

The main point to keep in mind is that high-impedance circuits are loaded down by external loads, such as test equipment test probes and leads. This will be pointed out in the text when appropriate.

Internal Resistance

A source of electrical power cannot be infinite. If its terminals are shorted together with a perfect short, there is no way that the resulting current can be infinite. What limits the current? The *internal resistance* of the source. Every source has an internal resistance. A 6-V flashlight battery has the same voltage as a 6-V car battery, but the the car battery can produce much more current under heavy loads than the lantern battery because the car battery has a much lower internal resistance. It is important to remember this term when troubleshooting voltage sources such as power supplies.

Mockup

A *mockup* is a series of connectors, signal generators of appropriate sorts, and power supplies that are required to bench test a printed circuit card or other equipment. In so doing, the mockup simulates the circuit's normal external operating conditions as closely as possible.

THE TEST BENCH

While much troubleshooting work is done in the field without the benefit of a fixed and convenient location, the bench technician needs basic furniture to be effective. Figure 1-9 is a suggested basic fixture that is sufficient for most bench testing and troubleshooting. Notice the slanted shelf above the bench surface. It should be at a convenient level and angle to make instruments approximately perpendicular to the technician's line of sight. By doing this errors due to viewing the instruments from the side (parallax errors) are minimized. Outlets for supplying AC power to the instruments should be available in the rear of the bench to avoid a tangle of line cords across the surface of the bench. It is also a good idea to provide a switch to turn off all of the outlets to equipment that is not needed at the end of each work shift. An outlet or two that is always on may be desirable for certain items such as battery chargers and power supplies needed for running equipment under test overnight.

A small vise mounted on the end of the bench or on a heavy metal plate is a valuable addition to the bench. Provide plenty of leg room under the bench and

FIGURE 1-9 Sample test bench arranged for operator comfort.

a place to put the feet either on the bench or on the stool used. The stool should provide back support for the technician to prevent back fatigue. A rotating stool is best for mobility.

Good lighting is important. When troubleshooting printed circuit boards, the technician must be able to see very tiny broken circuit traces. Fluorescent lighting is good. Insulating floor matting is highly recommended, particularly if the technician will be working on equipment operated from the AC power lines. A voltage rating of at least 1 kV is necessary.

Test equipment needed will depend upon the type of troubleshooting the technician will be doing. Table 1-1 is a suggested arsenal of equipment. It may seem that this is not much in the way of tools, but these tools alone will account for 95 percent or more of the working tools of the average bench or field technician. You don't need a toolbox full of tools to do a good job. As the technician gains experience, additional tools will be acquired as the specific job dictates.

SUMMARY

With the few definitions provided, a better understanding of some of the problems that the technician may encounter, a good bench, and applicable reference books to use, we are now ready to begin finding and repairing problems in electronic equipment.

TABLE 1-1 Suggested Troubleshooting Equipment

DC	LF	RF	PULSE	DIGITAL
Power Supply	Power Supply	Power Supply	Power Supply	Power Supply
VOM/DMM	Oscilloscope	Oscilloscope	Oscilloscope	Oscilloscope
Solid-State Tester	VOM/DMM	VOM/DMM	VOM/DMM	Solid-State Tester
	AF Oscillator	RF Probe	Storage Scope	Logic Probe
	Signal Injector	SWR Meter	Pulse Generator	Logic Clip
	AC Clamp-on Ammeter*	Dummy Load	Frequency Counter	Logic Comparator
		Power Meter	Function Generator	Current Tracer
		Frequency Counter		Logic Pulser
		Signal Injector		DMM
		RF Signal Generator		
		Spec. Analyzer		

**For working on certain 60-Hz power circuits*

Minimum standard tools required for most work:
Diagonal cutters, 6"
Long-nose pliers, 6"
¼" nut-driver; standard (−) and Phillips (+) screwdrivers
Temperature-controlled iron for PC-board work

REVIEW QUESTIONS

1. Convert 1050 kHz to Hz, MHz.
2. Draw schematic symbols for bipolar transistor, electrolytic capacitor, JFET transistor, NAND gate, SCR (silicon controlled rectifier).
3. Looking at the bottom of an IC, which way are the pins numbered around the component, clockwise or counterclockwise?

4. Draw a simple diagram of an AC power supply using a full-wave bridge rectifier and capacitor output filter.
5. Draw a simple transistor RC-coupled amplifier stage. Use four resistors, three capacitors, and the transistor.
6. What are the value and tolerance of a resistor having bands of red, yellow, orange, and silver?
7. What is the value of a precision resistor having bands of brown, black, black, and red?
8. How can you tell when a digital logic schematic diagram circuit has an *active low* input?
9. Is a circuit considered "dead" if a signal generator is still applying a small signal into the input, but the power supply for the circuit under test is disconnected?
10. Which will blow a fuse, an open circuit or a short circuit?
11. Can parallax error be either negative or positive?

chapter two

Kinds of Electronics Problems

CHAPTER OVERVIEW

This chapter familiarizes the technician with the wide variety of problems commonly seen on the test bench. For each type of equipment failure, the technician is directed to the proper procedures to use in evaluating and correcting the problem.

COMPLETE FAILURES

The complete failure is the easiest problem to correct. If everything is completely inoperative—panel lights, LEDs, meters, and all other indications—it might suggest that you forgot to pay the electric bill. Almost anyone could begin to troubleshoot this problem. Be sure the plug is in the wall, the fuses are good, and the circuit breakers are closed!

After this initial check, a technician would then check the power supply within the equipment that provides the rest of the circuitry with proper voltages.

Sometimes a poor connection at the wall socket is responsible. This can often be cured by spreading apart the tabs of the plug, by putting a small amount of "dogleg" in them, or by spreading apart the layers of some tabs (see Figure 2-1). See Chapter 4 to pursue this problem further.

POOR PERFORMANCE

Poor performance of equipment can be one of the more difficult problems to trace. It becomes more difficult as the poor performance approaches normal operation. For example, it is much more difficult to discover why a receiver is only slightly weak than why it is inoperative. Careful signal tracing and comparison with other equipment in good working order is the most effective way to handle poor performance. See Chapter 7 for signal tracing or Chapter 18 if alignment adjustments have been disturbed.

Poor performance may be due to a transistor amplifier failure. Often the transistor that is causing a poor gain problem is also sensitive to temperature changes. The gain of the transistor may vary greatly when heated. Apply a good signal at the lower limit of sensitivity of the receiver and touch an iron to each of the transistors in turn. Be sure to use a temperature-controlled iron to

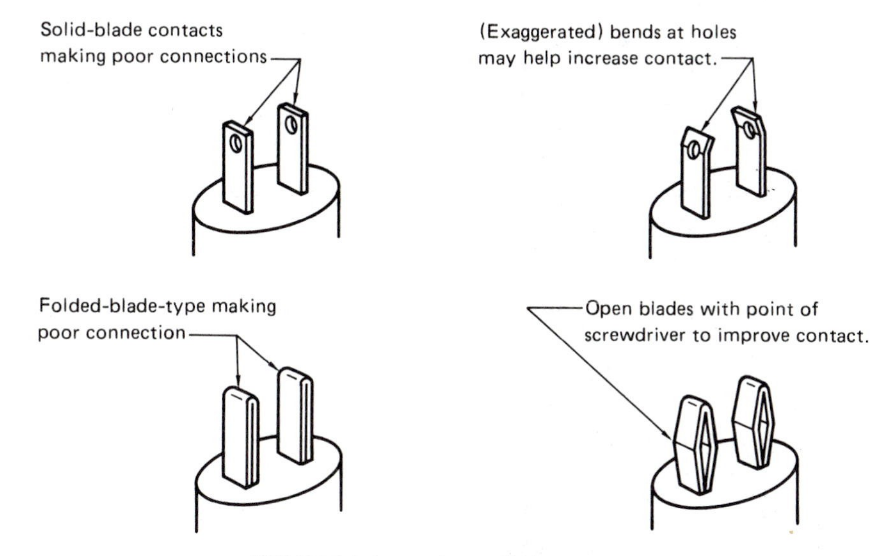

FIGURE 2-1 Improving wall plug contact.

prevent damaging good transistors, and don't heat them for more than about three seconds apiece. The gain may decrease drastically when the bad transistor is warmed up. Be especially careful using this procedure with plastic encapsulated transistors.

TAMPERED EQUIPMENT

Some people are just naturally curious. They like to look inside their equipment. There is no problem with that. The trouble starts when a curious person begins to think it might be possible to squeeze just a bit more power out of a transmitter or more audio from a receiver by "touching up" some of those interesting-looking little adjustments inside. The chances are very good that such efforts will be rewarded with the opposite effect. By the time a technician gets the equipment, it may be inoperative. Rare indeed is the person who will attach a note to their equipment confessing that they messed it up!

The servicing technician should watch for such tell-tale signs of tampering as chipped or cracked RF transformer slugs or broken paint seals on them or on other components. Such signs may indicate that alignment of the equipment will be needed to get it back to the manufacturer's specifications. See Chapter 18 for alignment instructions.

INTERMITTENTS

The key to repairing any kind of intermittent problem is to be able to *make the equipment stay inoperative long enough* to pinpoint the cause. There is little to be gained by working on an intermittent problem when the failed symptoms will not appear.

The Mechanical Intermittent

Have you ever seen a TV that quit working, only to respond favorably to a solid thump of the hand? This is called a mechanical intermittent because it responds to a mechanical thump. By very carefully noting the area of the equipment or the printed circuit board (PCB) that is most sensitive to mechanical stress applied, it is possible to narrow down the problem to the defective component in most cases. Take special note of the direction of flexing and the amount of force required when tracing the mechanical intermittent. As you get nearer to the source of the problem, the amount of force necessary to cause the symptoms should decrease.

The Thermal Intermittent

Characteristic of the thermal intermittent problem is that the problem usually occurs when the equipment is operated for some time. If turned off for a time, the problem is gone when the power is again applied. This thermal cycling can be simulated by the application of heat or cold in an effort to duplicate the problem. By selectively directing the heat or cold to smaller and smaller areas of the equipment, it should eventually be possible to isolate the defective component. Wide area cooling and heating is best done with a commercial freon-based aerosol product specifically manufactured for this purpose and with a heat gun or possibly a hair dryer. As the sensitive area is progressively reduced, one may revert to carefully using the tip of a small soldering iron to heat individual components. Use the iron with care on plastic parts.

The Erratic Intermittent

The erratic intermittent is perhaps the most difficult problem to repair. The symptoms seem somehow to disappear when the technician appears on the scene. The erratic intermittent is the source of many customer disputes. The customer will assume the technician can't find the problem because of incompetence, and the technician wonders if there really *is* a problem, because no unusual symptoms have appeared. Even if the technician believes the diagnosis of the customer, all the technician can do with this kind of intermittent is make an educated guess, replace a card within the equipment, or replace a likely cause of the problem as the operator describes it. The erratic equipment will not respond to changes in heat or cold, nor to thumping the equipment. It's just there sometimes and not at others. Suggestions to cure the erratic intermittent include operating the equipment at full ratings (as applicable) for long periods of time under a watchful eye. Cycle the equipment in an enclosure designed to subject equipment to higher-than-normal temperatures. Such a box can be made with a small heater resistor, circulating fan, and thermostat. See Figure 2-2 for one idea about how this might be constructed.

Another idea is to put the equipment in a refrigerator for a while to see if a drop in temperature causes the intermittent to appear. The erratic intermittent

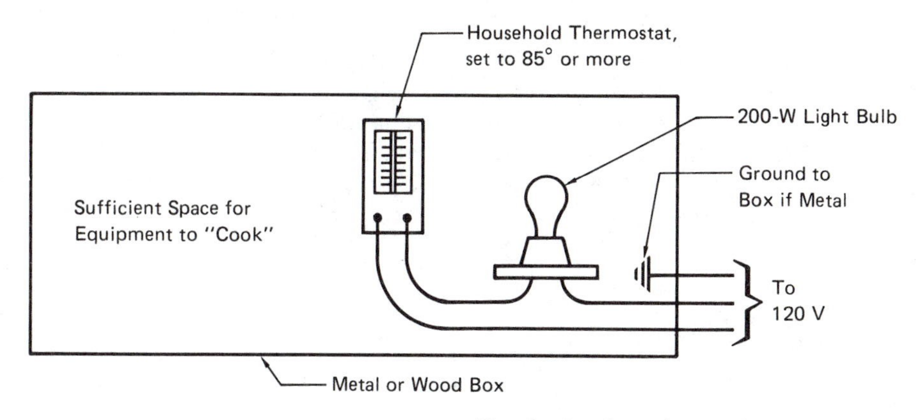

FIGURE 2-2 An idea for a "hot box".

may respond to twisting of circuit boards, pulling and pushing of wiring as it enters plugs and jacks, or the cleaning of plug contacts. Some plugs are prone to breaking contact between mating connectors. These contacts can sometimes be resprung so that their mating surfaces again fit tightly together. The methods of bending to restore normal tension in the contacts will be left to the ingenuity of the reader, as there are many different types of connectors. *Hint:* The common pencil eraser makes a very good abrasive in cleaning contacts, especially the edge connectors on printed circuit boards.

Printed circuit boards are subject to mechanical stresses where connector posts go through the board. Pushing and pulling on the wires that connect to these posts can easily cause a hairline break in the circuit where it should attach to the post. This is a particularly good place to visually inspect very carefully, using a magnifying glass if available. The same stresses are present when using some types of connectors that depend upon the printed circuit for mechanical mounting. These connectors are often not mounted in any other way, depending upon the bond between the circuit wiring and the board to absorb all of the stress of plugging and unplugging wiring to the connector. Again, this is a good place to look carefully for breaks. Stressing the connector back and forth gently can sometimes make a break easier to see.

The final attempt at making an erratic intermittent appear may be the application of higher than normal supply voltages. This must be done with the greatest of care, so as not to cause more damage. Some circuitry, such as TTL ICs, will not tolerate overvoltage without failing. A rule of thumb is to subject the equipment to no more than a 20-percent overvoltage.

Vacuum-tube grid circuits, FET gates, and CMOS ICs of the analog and digital types are subject to erratic behavior if one of their high-impedance inputs is left disconnected for any reason. A break in a circuit board can produce such an open, which will behave very erratically. Be aware of this unique possibility when dealing with erratic intermittents in high-impedance circuits.

In digital circuits a voltage level that is between the on and off levels can cause intermittent symptoms that are erratic in nature. Such levels may be detected with an oscilloscope or a logic probe.

If an intermittent is general, i.e., everything seems to be affected, then the power supply is a likely suspect to receive detailed investigation. See Chapter 6 for the procedures to follow in troubleshooting the power supply.

If the erratic intermittent will appear long enough, troubleshooting can progress as outlined in Chapter 4.

MOTORBOATING

Motorboating is the "plup-plup-plup" sound that sometimes occurs when the batteries get weak in a radio or the electrolytic capacitors in the power supply filter dry out and cease to filter. If the electrolytics are very bad, the oscillation can be a squeal instead of a low-frequency plopping sound.

The technical reason for the oscillation is that the final amplifier stage of the audio amplifier is loading the power supply down and the filter or batteries are unable to maintain the proper voltage for the very short period of time required until the audio waveform changes polarity. When the power supply cannot maintain the voltage, there is a feedback path for audio frequencies through the power supply that can affect earlier stages, resulting in oscillations. Oscillations are particularly likely at the higher volume-control settings. The cure for this problem is to replace the filter capacitors that are not doing their job, or to replace the batteries that have developed too much internal resistance.

MASSIVE TRAUMAS

Massive traumas are those incidents that cause more than just one or two components to go bad.

Fire or Smoke Damage

A close visual inspection of the equipment will provide a pretty sure indication of whether it is worth trying to put the equipment back into service. If plastic parts of the equipment are melted or burned off, it is best to let the insurance pay off the equipment with no further effort wasted on it. If there is no evidence of high heat, then the equipment may be only smoke damaged and possibly worth an effort to return it to service. The approach to use is to begin testing the equipment by applying power and watching for any unusual indications. Proceed to Chapter 6 on power supplies, and begin testing the equipment as outlined there.

Primary DC Polarity Reversal

For the purposes of this section, I will assume that DC primary refers to 12 volts DC, though it could be 24, 48, or any voltage specified by the manufacturer.

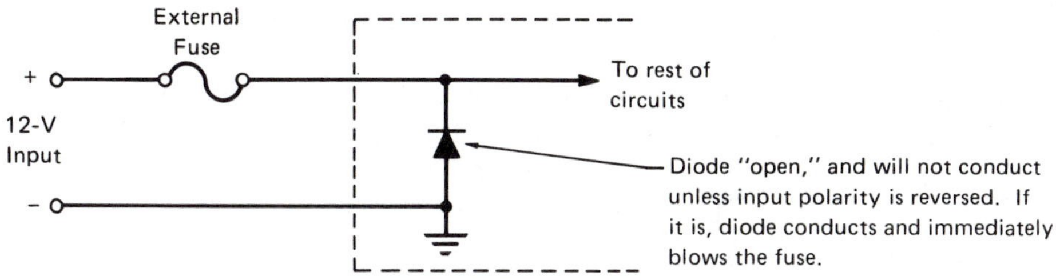

FIGURE 2-3 Sample polarity protection circuit.

Accidental reversal of the 12-volt input to equipment will probably not harm it *if* the circuit shown in Figure 2-3 is incorporated. This circuit can also be added to DC-operated equipment if there is any chance of accidental polarity reversal.

If this protection circuit is defeated by the installation of a fuse of too large a rating, it is possible to burn open the protection diode and thereby subject the entire equipment to the uninhibited rush of reversed current. Extensive damage can be expected in this case.

Circuits without the above protection may suffer greatly from application of reversed polarity. The following components are most susceptible to damage from reversed polarity: electrolytic capacitors, diodes, transistors of any type, regulator ICs, and all other ICs of any type. Each and every one must be tested for damage. Simple testing of shorted and opened components is not sufficient, as there may be internal damage that will not show in DC voltage testing. After mass parts replacement, signal tracing through the circuitry will be the best way to verify operation stage by stage. Be sure to begin with the assumption that the primary input will be shorted. See Chapter 4 to begin troubleshooting.

Dropped Equipment

Solid-state equipment is amazingly resistant to damage from dropping so far as the semiconductors are concerned. Vacuum tubes do not fare so well. If the equipment is not completely smashed and obviously junk, a simple drop and resultant inoperation is not necessarily as bad as one might think.

Possible problems are that the heavier components are pulled loose from the circuit board, wiring is pulled loose, or the printed circuit board may develop open circuits. If there is no extensive mechanical damage, such as binding of the dial mechanism or other mechanical devices, the equipment is treated as a normal failure and troubleshooting is done in a routine manner. Chapter 4 is the place to begin.

Water Immersion

The first thing to do with equipment that has been submerged is to rinse off the damaging salts and contaminants as soon as possible. If the immersion was in salt water, then the equipment should be thoroughly rinsed by dunking and agitating in fresh water, preferably warm. If the original accident involved fresh water or if the equipment has been rinsed clean of salt, then it should be submerged in a commercial liquid specifically made for such purpose. One such product is called CRC. See Appendix III for availability.

Several factors will determine the amount of damage sustained and the feasibility of returning the equipment to service:

1. Did the equipment have power applied during immersion?
2. How long was the equipment submerged?
3. What kind of water was it?
4. What does a close visual inspection reveal?
5. What was the equipment originally worth?
6. With depreciation, what is it worth now?
7. Will any future failures as a result of the immersion be tolerable? (How critical is this equipment—for instance, will it be used in aviation, where reliability is of utmost importance?)

The answers to these questions will aid in the final decision as to whether to attempt repair or to scrap the equipment without further investment of a technician's time.

Let us consider each of these items in detail. The application of power during an immersion will cause current flow between portions of the circuit that normally do not have it. The resulting corrosion can have a disastrous effect. Current flow within liquids causes the movement of metal molecules from one point to another in great quantities. This is the principle of electroplating and can render equipment useless by erasing printed circuit patterns completely and applying resistances where there should be open circuits. This effect is increased many times if the submergence is for a long time or in salt water. Pure water on the other hand is a fairly good insulator and will not cause the extensive damage that salt water will. A close visual inspection may show corrosion damage or other problems such as rusted transistor leads. They can rust completely off the board in severe cases. Remember that this kind of deterioration may be taking place under the case of a transistor where you cannot see it. The owner of the equipment will have to make a decision as to whether or not to proceed with repairs based upon the technician's findings and any requirements placed upon the owner by the insurance company. It is wise to set water damaged equipment aside after further damage is halted and await the decision of the owner and the insurance company. They will take into account the original value of the equipment and its value before the immersion. These decisions plus the acceptability of further unreliability will determine whether or not it is practical to pursue repairs. If the decision is made to put the equipment back into service, see Chapter 4 for the starting point.

Application of Voltage in Wrong Place

When a technician is using a probe or any metal tool within equipment while it is in operation, there is a risk of shorting a pair of IC pins or printed circuit traces together. Often this does no damage, but once in a while the wrong combination comes up and in a few milliseconds a great deal of damage is done. As a simple precautionary measure, the technician should not use anything to probe into live circuitry other than a test probe designed for the job. A good probe will be needle-sharp at the tip and very small in diameter. Ideally, it should be small enough that it will not short IC pins together if placed between them.

If the equipment suddenly emits a strange noise, smokes, or otherwise ceases operation when the technician "slips," there is a very good chance that a big job lies ahead just to get back to the original problem. A slip of a tool like this can cause a string of many ICs to die an instant death. TTL ICs, for instance, will not tolerate 12 V DC applied to them without permanent damage. A slip from input to output pins of a regulator, for instance, could accomplish this in the twinkling of an eye. Be careful!

Investigation into the problems caused by such an accident should begin by noting exactly where the improper voltage was applied. Carefully note where those points are on the schematic diagram and what components would most likely be damaged. These components will usually be ICs, regulators, or electrolytic capacitors. Other components are much more tolerant of such accidents. If by chance the accidental voltage applied is reversed from normal, one can expect about the same results. Troubleshooting should then proceed to correct or verify as good those components possibly damaged by the accident. This may mean a stage-by-stage signal tracing. If so, see Chapter 4 to proceed with the repair.

Lightning Damage

Equipment on the receiving end of a lightning strike or near-strike should generally be written off as an act of God and replacements procured. The damage caused by lightning has to be seen to be believed. Seldom can the many hours of technician time be justified in repairing such equipment. Partially operative equipment from the vicinity of a strike may be considered as covered in "Transients," below.

TRANSIENTS

Transients that cause failures of electronic equipment are actually voltage "spikes" or surges of the primary power source. These surges can cause the instantaneous voltage to increase to a point that can cause failures of filter capacitors, semiconductors, insulation in transformers, and components farther into the circuitry from the power source.

Transients are common in mobile installations. The normal operation of the starter motor on a vehicle can cause great excursions of the battery voltage, sometimes sufficient to damage semiconductor equipment in particular. For this reason, any equipment, such as a radio, that contains semiconductors should be turned off during cranking of the engine, thereby isolating them from the potentially dangerous transients.

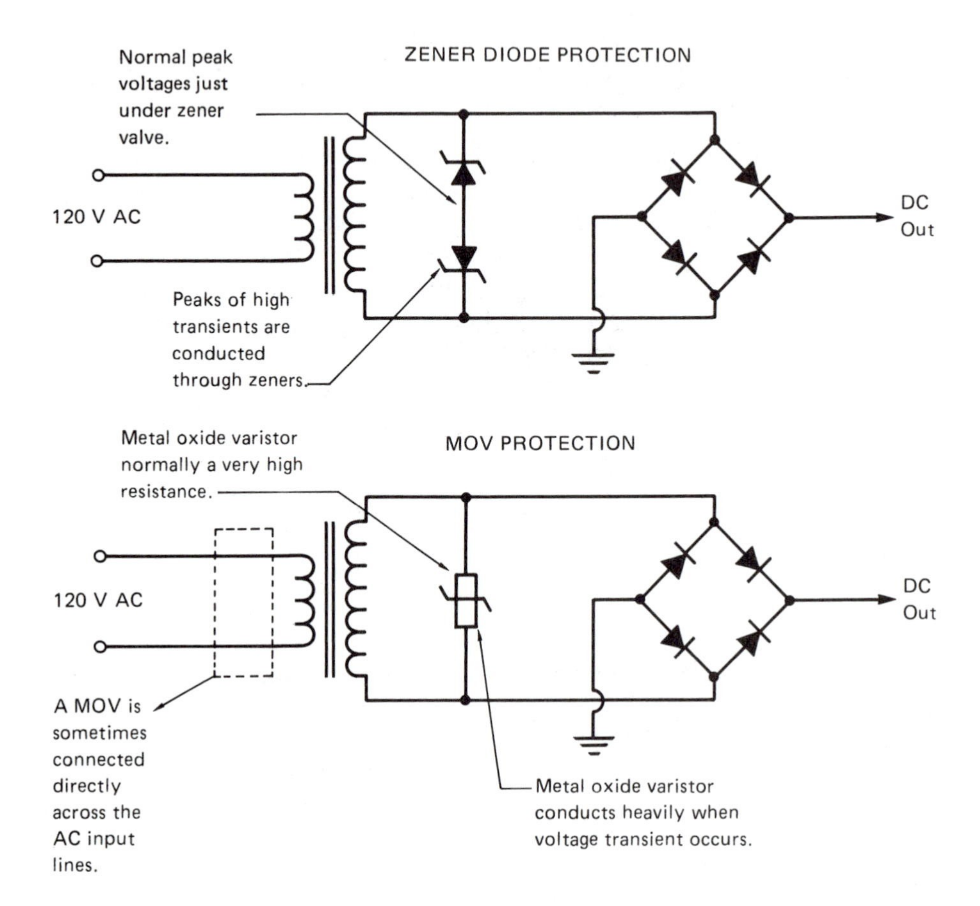

FIGURE 2-4 Common transient suppression circuits.

Typical equipment failures that may point to possible transient damage as the cause include shorted SCRs, rectifier diodes, and bridges. Most well-designed equipment operated from the 115-V AC or 220-V AC lines will include some form of transient suppression. These suppression components should be tested for proper operation if other semiconductors may have been damaged by transients, thereby indicating that the suppressors may be open or otherwise not performing their intended function (Figure 2-4). Most of these failures may be found by following the techniques in Chapter 4. Figure 2-5 shows some of the more common transient suppressor components.

OVERHEATED PART FAILURES

Resistors are made to throw off heat. The bigger they are, the more heat they can safely get rid of without damage to themselves. If hot resistors are mounted directly on a circuit board, they may discolor the board. In severe cases, they can burn or char the board.

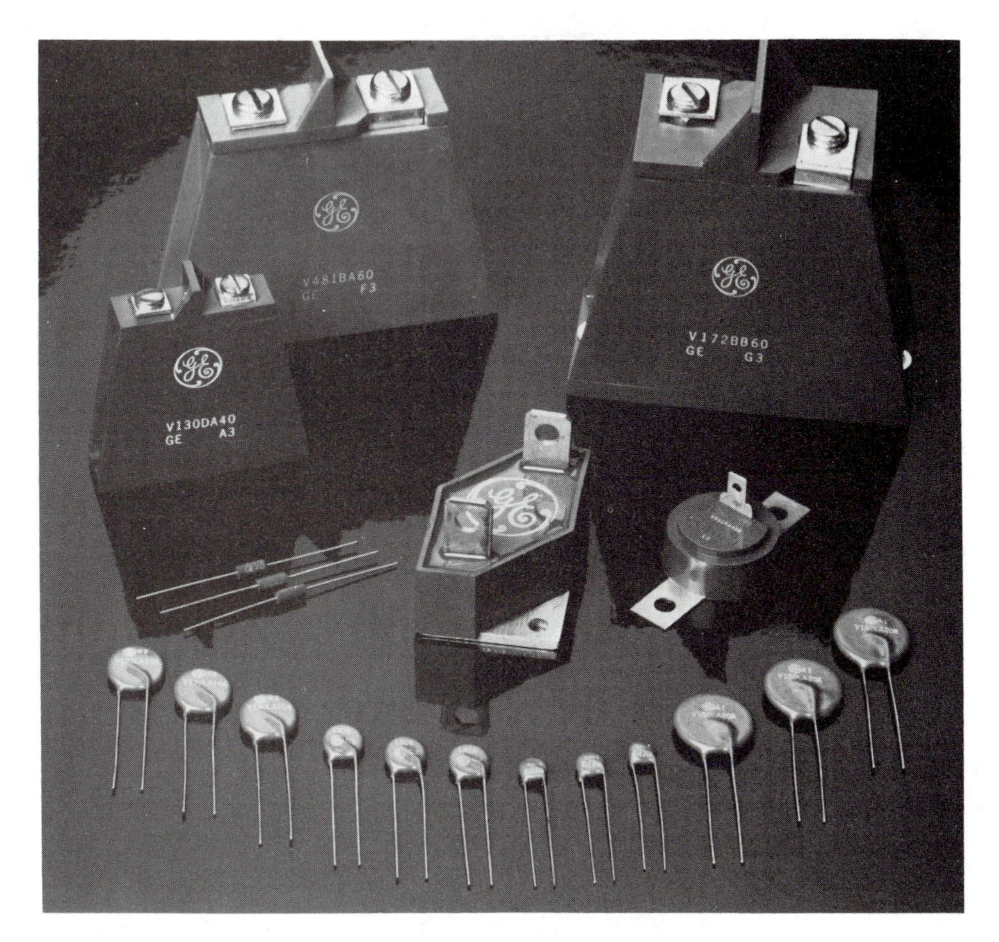

FIGURE 2-5 Examples of MOVs used in modern equipment. (Photo reprinted with permission of GE Semiconductor, General Electric Co.)

Small resistors should not get hot. They are used when the wattage rating is less than a watt, not enough heat to cause any problems with a printed circuit board or nearby wiring. Any small resistor that is burned is almost always caused by the failure of another component. The other component is usually a transistor or an electrolytic capacitor (Figure 2-6).

A large carbon-composition resistor may produce its own problems. A carbon-composition resistor operated at or near its maximum power rating can change its resistance value. In some circuits, this effect can accumulate, causing more heat and so on. It is not unusual to find a 10,000-ohm resistor that has decreased to 1000 ohms or less because of this. Resistors may also fail by increasing resistance or by opening.

Transistors can get hot in normal operation and fail. The cause is not always just the transistor, but is often defective biasing of that transistor.

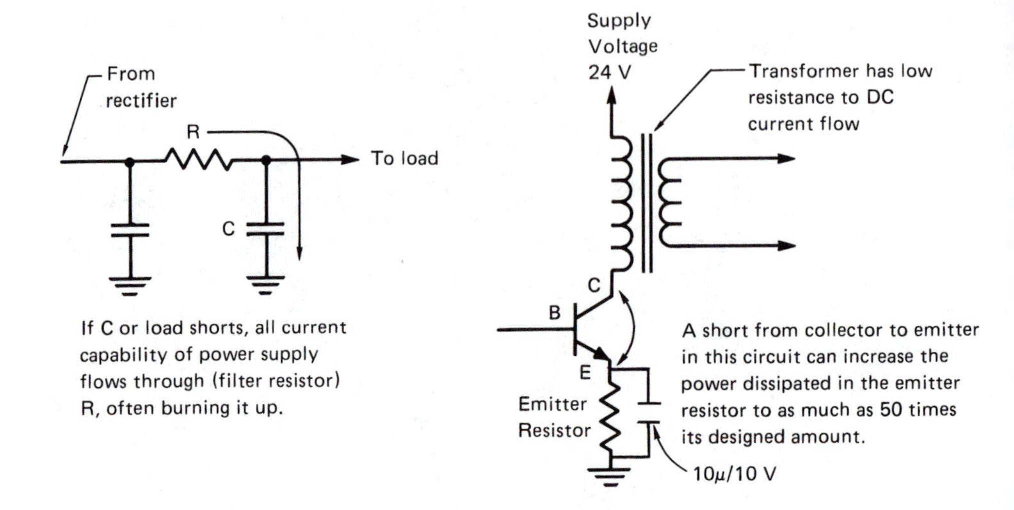

FIGURE 2-6 Failure of other components can cause resistor failures.

Capacitors should not feel hot. If hot capacitors are not being overheated by nearby components, they are probably very leaky. When replacing a capacitor, be sure that there is no associated failure causing too much voltage to be applied to the capacitor. This would be the case if the transistor in Figure 2-6 were to develop an emitter-to-collector short. The normal 5 V of bias might become as much as 24 V, severely stressing the capacitor if it has a voltage rating of only 10 V.

A burned printed circuit board can often be repaired. Charred portions of the board can accumulate moisture and dirt more easily than clean portions. The charred portions could even be conductive. Repair should consist of the removal of all the burned portions by cutting or filing. The remaining edges of the board should be sealed by lacquer, clear acrylic, or other suitable moisture-sealing product.

Burned wiring should be replaced. Even though the insulation may look intact, it is probably brittle and may flake off later and cause problems by shorting to other circuits.

If you suspect a component is overheating, you can get a good idea of component heat by holding your finger on the component for a few moments. If your finger is not sensitive enough, immediately put that finger to your upper lip, just below your nose. This area of the body is extremely sensitive to minute temperature changes.

HUM PROBLEMS

Hum problems have five general causes: (1) insufficient filtering in an AC-operated power supply; (2) an overload condition of that power supply;

(3) absence of a ground connection where there should be one; (4) presence of two common grounds where there should be only one; and (5) an open circuit in a high-gain, high-impedance circuit.

The electrolytic capacitors commonly used in the filters of power supplies can dry out with age and lose their capacity to store a charge. This effectively takes them out of a circuit and passes on any AC component to following stages. The cure is simple. Replace them. Testing them is also simple. Since the capacitor is suspected as being open, it is effectively out of the circuit already. Take a known good capacitor and temporarily parallel the bad one with the good one. If the hum disappears, you have found the problem.

An overloaded power supply will have similar hum problems to one having an open capacitor. Paralleling the filter capacitor with a good one will not cure the problem, although it may diminish the amount of hum to some extent. Along with this hum problem, an overloaded power supply would be expected to draw substantially more current than normal. This may be evident by the occasional blowing of fuses, the overheating of the power transformer, or both. Troubleshooting from this point should begin with Chapter 4.

Ground loops may also inject hum signals into a circuit due to lack of a single grounding point (see Figure 2-7). The current flowing from point A to point B will modulate the signal sharing the same path. Ground loops are caused by a resistance (which may be extremely small, perhaps 0.01 ohm) across which a hum signal can be developed by unrelated heavy currents. Ground loops are particularly troublesome when using high-gain amplifiers, such as microphone amplifiers, around high-current circuits, such as vacuum-tube heater wiring. It is for this reason that most microphones use a separate ground return (usually the shield of the tiny coax), which is grounded at the equipment end. Any other circuitry such as a push-to-talk switch uses a completely separate ground lead in the microphone cord.

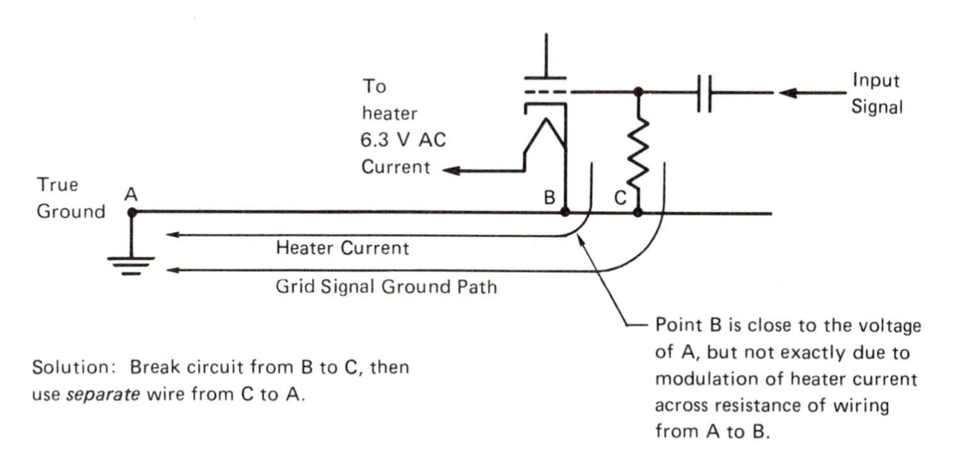

FIGURE 2-7 Example of a ground loop

DISTORTION PROBLEMS

Distortion refers to unwanted changes in a waveform from its original shape. Audio waveforms are among those that we wish to have amplified from a source (such as a tape player) to a load (such as a speaker) without changing the waveform along the way. A common distortion easily demonstrated is heard when a small transistor radio is turned up too loud. The waveform as seen on an oscilloscope is no longer rounded but is flattened on the top, bottom, or both (see Figure 2-8).

A distortion analyzer is an instrument that analyzes sine waves for distortion and provides a reading in percentage. Principally used in the design of new circuits, it is seldom used for routine troubleshooting. To use this instrument, measure the sine wave source (usually an audio oscillator) and record the percentage of distortion. Then apply the sine wave to the circuit under test and measure the distortion coming out of the circuit. If the circuit is producing distortion, the percentage indcated on the analyzer will have increased. This procedure would be used only in critical applications.

Distortion will usually be evident on an oscilloscope to the trained eye. Using a pure sinewave tone with minimal distortion on the input, trace the signal through the circuitry with the oscilloscope and watch for the stage that seems to introduce distortion. Since the trouble is often one of biasing, the technician would look for improper bias voltages. One of a pair of balanced or push-pull components such as transistors could also be bad, producing very pronounced distortion.

Distortion in RF circuits will produce harmonics of the fundamental frequency in use. In straight-through RF amplifier circuits the amplifying component is often operated Class C. Any frequency other than the fundamental is filtered out by the inductance and capacitance tuning of the stage. Production of harmonics is encouraged by frequency-multiplier circuits. Distortion is also deliberately used in mixer circuits to produce sum-and-difference frequencies. Unintentioned distortion of RF waveforms in transmitting equipment can cause

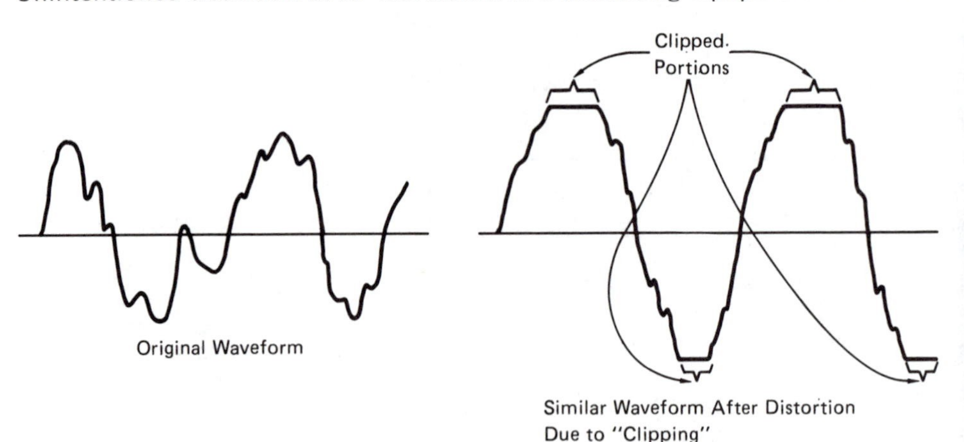

FIGURE 2-8 Amplitude distortion due to improper circuit operation.

interference on other radio frequencies, some of which can be far removed from the fundamental frequency. This problem is best detected by using a spectrum analyzer. This instrument allows the technician to see the unwanted frequencies along with the desired one. Changes are then made to the circuitry to reduce the unwanted frequency strengths to acceptable levels. It is common practice to include a filter on the output of transmitters to reduce the unwanted transmissions to low levels. Replacement of this filter will often cure the problem.

MULTIPLE PROBLEMS

Once a problem is identified and repaired, it is reasonable to assume that the equipment will work properly when assembled into its cabinet. This is not always the case. Once in a while in the course of finding and curing one problem, the technician may introduce another. The most common second problem is probably the broken wire. During the opening of equipment and dismantling of circuit boards from the chassis, the wiring takes some punishment. Watch for this to happen when dismantling and reassembling equipment. It is a sure indication of a technician-induced problem when the equipment was working fine after the repair but ceased operation when inserted into the cabinet. Watch also for pinched wiring that could cause a short to ground.

Multiple problems would be evident if a definitely defective component is discovered and replaced and yet the equipment does not work properly. Perhaps the technician has replaced a symptom component and neglected to find the principal cause of the problem. Troubleshooting must continue to find the second problem. If the second problem found bears no relationship to the first failure and there is little if any chance that the technician somehow caused the second failure during the troubleshooting procedure, the technician should suspect a possible massive trauma and additional damage. See "Massive Traumas" earlier in this chapter for more information on this subject.

MICROPHONICS

Microphonics were quite common in the days of the vacuum tube. The elements of a tube were subject to vibration, which would modulate the stream of electrons. Special tubes were manufactured to reduce this effect.

Semiconductors are not prone to microphonics themselves, but some other circuit elements can cause them. One case is the inductor used in some LC oscillators, particularly in the radio frequency range. This effect can sometimes be minimized by firmly anchoring each turn of the coil with special adhesives such as RTV.

NOISY CONTROLS

Noisy controls are probably most noticeable in audio volume control circuits; for example, moving the shaft or slider of the potentiometer makes a scratchy sound in the speakers. The temporary cure is quite easy. Apply an aerosol

product manufactured for this purpose through a long tiny tube provided with the can. Spray the liquid into the control and strike the target, the inside carbon track against which the slider operates. One such product is Tuner-Lube.®
However, the best and most permanent cure is replacement of the control.

OPERATOR-INDUCED PROBLEMS

Equipment operators do not always understand the controls as well as they should. Wrong assumptions or ignorance may play a part in a reported failure. When equipment works for the technician but the operator is still dissatisfied, it may be time for some tactful schooling in the operation of the controls. Be careful, however, that the problem is not an erratic intermittent, which may have the same appearance as an operator's improper adjustment of controls. The more complicated the equipment is, the more likely the operator will adjust it improperly.

THE TOUGH PROBLEMS

Every technician eventually encounters a problem that defies all efforts to correct. Nothing seems to go right, and all of the usual procedures don't help at all. What to do then? Consult this checklist carefully, and you may find the problem:

1. Have the batteries of either the equipment under test or of the test equipment gone dead? Are the battery connectors clean and making positive contact to the batteries?
2. Check the test equipment test leads and/or RF cables for intermittents or opens that may have occurred since starting troubleshooting.
3. Did you change ranges on some of the equipment under test or the test equipment and forget to change them back?
4. Try substituting known good test equipment for the test equipment in use at the moment. The test equipment may have failed. Try the same with the equipment under test. Remove the suspected unit and substitute a known good piece of equipment to verify the symptoms you are getting.
5. If you have access to known good equipment of the same type as the defective one, try comparing readings from good to bad unit to help in isolating the problem in the bad. Apply power to both units and take readings back and forth between them.
6. The problem you are looking for in a DC circuit may not be evident as a DC problem at all—it may be a totally unexpected RF problem that will not be noticeable from DC readings. This principle might apply to low frequency/audio problems when unintentional RF is present.
7. Get away from the problem for a while. Sometimes you get too involved and begin making dumb mistakes.

8. When you go back to the job, consider making notes on paper as you go, to assist you in keeping things straight in your head. This is a particular help when you are taking many voltage readings and have been trying to remember them and they've all gotten confused.
9. Consider asking the help of another technician. None of us has all the answers no matter how experienced. You could be overlooking something very simple that another person could point out.
10. Start your troubleshooting all over again, and take the voltage readings again. Something may have changed as you were working and that could scramble any logical approach you are trying to maintain.

SUMMARY

This chapter has covered some of the types of circuit problems that the technician might see in the course of a day's work. Familiarity with these problems and an ability to identify them instinctively will increase the technician's working speed.

REVIEW QUESTIONS

1. What is the easiest electronics problem to repair?
2. What would a broken paint seal on an internal adjustment indicate?
3. What is the key to repairing intermittent problems?
4. Name the three kinds of intermittents.
5. Name two components in common use to suppress voltage transients.
6. What does a hot capacitor mean?
7. Name three causes of unwanted hum.
8. How many suggestions are given to help solve a particularly tough problem?

chapter three

System Troubleshooting

CHAPTER OVERVIEW

This chapter points the way to identifying the defective equipment (often called a black box) or a defective printed circuit card operating within a much larger system. The technician may just be handed a defective unit. In that case this chapter is not applicable, and the procedures in Chapter 4 are the next step.

INVESTIGATE THE REPORT

Whether the initial trouble report is verbal or written, it is a very good idea to confer with the person making the report. Additional facts may come to light that can save considerable time.

To illustrate this point, consider the case of an aircraft that uses the microphone both as the communications transmitter and for intercommunication with the copilot. The initial report may have indicated that the transmitter was not modulating. Without further questioning, it would be reasonable to assume that the transmitter modulator was defective. Upon closer questioning, it could be revealed that the microphone doesn't work on the intercom circuit either, and therefore the technician would direct efforts to the likelihood that the microphone or its cord is the problem, rather than something in the transmitter. Note that the initial report in this case was honest but completely misleading.

Operator Problems

Unless the operator is thoroughly familiar with the equipment in use, the possibility of operator error should be considered. As the complexity of the system increases, so does the possibility of operator error. If practical, the technician should ask the operator to demonstrate the problem. The technician should then allow the operator to manipulate all of the controls while the technician watches for errors or controls missed by the operator. To do this effectively, the technician must be very familiar with the system and how it operates.

A related problem that could confuse the technician occurs when the operator reports a failure that is a matter of the operator's judgment rather than a definite problem. An example would be a report that the communications receiver output is weak. What is weak? This is a matter of judgment. Perhaps the operator is comparing the present level of audio output with previous levels or with the operation of similar equipment. The technician should attempt to verify the reported problem before dismantling the equipment or making any other rash moves.

Let's consider another example, a report of an inoperative transmitter. It is a very good idea to quiz the operator as to the frequency that was used, the transmitter location at the time of the supposed failure, and to whom the transmissions were directed. It may be entirely possible that the receiving operator's equipment was defective, the receiving operator stepped out for a minute for coffee, the transmitter and the receiver were not on matching frequencies, or that the propagation characteristics and locations of the transmitter and the receiver were such that communication was simply not possible or was at least very unlikely.

Look for the Obvious, Easy Cures First

Before concluding that the equipment will require detailed troubleshooting and must therefore be removed from the system for repair, it is a good idea to check for the obvious. Are all of the required switches in their proper positions? Is there an obscure selector switch, which is normally not used, in the wrong position? Is there a possibility that microphone and headphone cords are the problem? If so, a substitution of a known good set will quickly confirm it either way.

FINDING THE BAD UNIT

In this day of the computer, some systems are becoming completely computer driven, as are the isolation, testing, and rejection of circuit boards within the system. These systems are sometimes called integrated test systems.

The cards within these racks can be individually tested by commands entered at a maintenance and administrative position, a computer terminal. An operator trained in the equipment, with a knowledge of how telephone circuits work, can make many tests and inquiries about the system status, using only the keyboard. Audio levels can be checked, either in the switching station (Figure 3-1) or to the "outside world." Direct readout in decibels is available for any circuit the operator chooses. Defective cards are simply removed from the system, a substitute inserted, and the system is back in normal operation. The defective card is sent to the factory where automated troubleshooting equipment quickly finds the bad component. Computer-automated test systems are also often used in manufacturing electronic equipment and components where large numbers of tests must be run at high speed to keep up with production.

Without such sophisticated equipment, finding the bad equipment in a large system or the bad card in a rack of cards requires a thorough knowledge of how the whole system works. Another system example is a complete radar installation. As a general illustration, consider the block diagram of a small radar

FIGURE 3-1 Audio switching equipment at the telephone company. (Photo courtesy of GTE)

system in Figure 3-2. Before anyone can begin to find a problem in such a maze of electronics, it is necessary to know the functions of each of the main blocks. For instance, the technician would have to know what it is that flows from the antenna into the radar receiver. The technician would have to know that the flow from the receiver to the indicator is of video signals (signals from perhaps 10 MHz all the way down to nearly direct current). Similar knowledge would be required of the rest of the system.

The instruction book for the system may have procedures for isolating the defective card or equipment, using test points brought out at each card, etc. In this case, the instruction book will provide quick, easy ways to isolate the bad card. Another possible solution to finding the defective equipment is to use troubleshooting charts sometimes provided in the instruction book. Once the system is understood, troubleshooting charts are, as a matter of practicality, hardly ever used. It is too difficult to cover all of the possibilities necessary to make a troubleshooting chart sufficiently detailed much beyond the "black box" level.

With knowledge of the system, the technician is ready to begin elimination of the various blocks as probable causes of a system malfunction. In this example, let us assume the radar operator has reported that there are no targets on the indicator. The operator is an experienced operator with many hours on the equipment. This pretty well rules out the possibility of an operator error, and since there are no targets at all, it is not a matter of judgment on the operator's part. It would be prudent at this point for the technician to apply primary power to the entire system and verify the symptoms himself. Having verified that the indicator unit is apparently working properly (according to front panel checks of the indicator), he might then check the transmitter-receiver unit where he may find that a blown fuse indicator is lighted in the receiver power-supply circuit. Troubleshooting would then logically be directed to the receiver power-supply unit rather than the antenna, indicator unit, or the transmitter circuits.

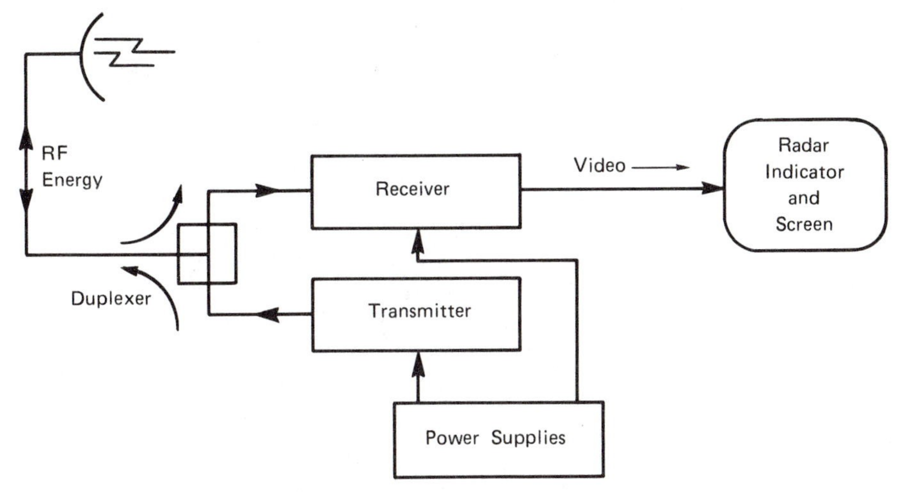

FIGURE 3-2 Sample block diagram of radar system.

Some applications have special test equipment dedicated to the specific system at hand. Aircraft electronics technicians use such instrumentation to provide the very special RF inputs and other signals necessary to test an entire system. An example is the aircraft's instrument landing system (ILS). This system is tested with a special instrument that contains several signal generators within it.

WIRE-BUNDLE AND COAX-CABLE PROBLEMS

If replacement of "black boxes" or circuit cards does not cure the problem in the system, it is quite possible that the interunit wiring or coaxial cables may be the cause.

Wire-bundle problems will have to be traced using system blueprints and continuity checks. The DMM with a continuity "beeper" is a valuable instrument to use when checking wire continuity. There is no need to watch the meter to see if there is continuity of the wiring.

Wiring extending long distances can be continuity checked in one of two ways. With some assistance, the far end of the wire in question can be grounded to the vehicle, aircraft, or equipment framework if all of the framework is bonded together throughout the installation. Then the near end can be tested for continuity to ground. An alternative way to handle the situation is to use a single conductor of the cable as a return lead, and make all continuity checks using that one lead for the return (Figure 3-3).

Besides testing for continuity of the wires in a cable, it is also wise to test for possible shorts from wire to wire and wire to ground. Testing for shorts to ground is important when the wire bundle might have a chafed portion and one or more of the wires might be contacting the metal frame of the aircraft or vehicle.

Testing for all possible combinations of shorting or grounding paths can be done efficiently by a series of tests. See Figure 3-4 for the method. In words,

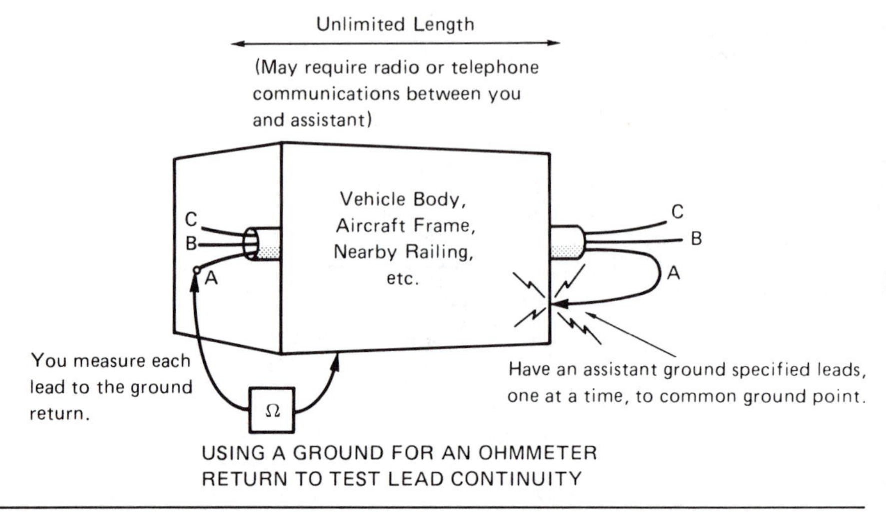

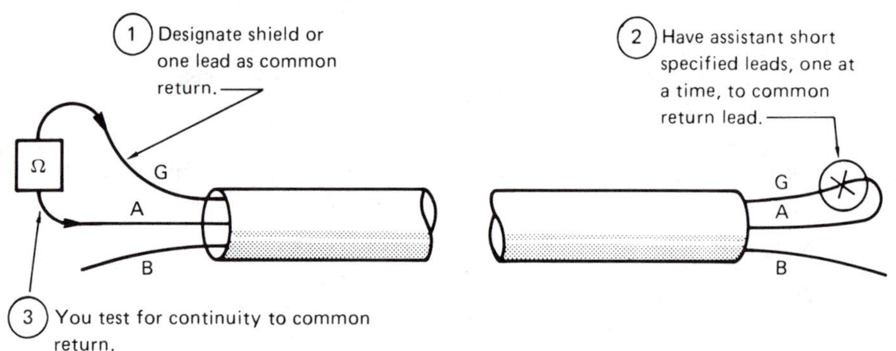

FIGURE 3-3 Cable continuity testing.

testing for shorts between all of the wires in a bundle is done by putting one ohmmeter test lead on the first wire. Taking the second lead, touch it to each of the following wires in the bundle in some logical order, one by one. After all of the wires have been considered, move the first ohmmeter lead down through the sequence one lead and again touch each of the following leads one at a time with the remaining ohmmeter lead. The third round, again come down one lead with the first ohmmeter lead, and continue looking for continuity with the second lead on all following leads. Note that the second ohmmeter lead never has to go to a wire *prior* to the first ohmmeter lead. This method checks for all possible combinations of wires with no wasted motion or checking for reversed combinations like wire B to wire A. That combination is tested once from A to B and never again.

Coaxial cables may be checked for opens, shorts, and also for sophisticated

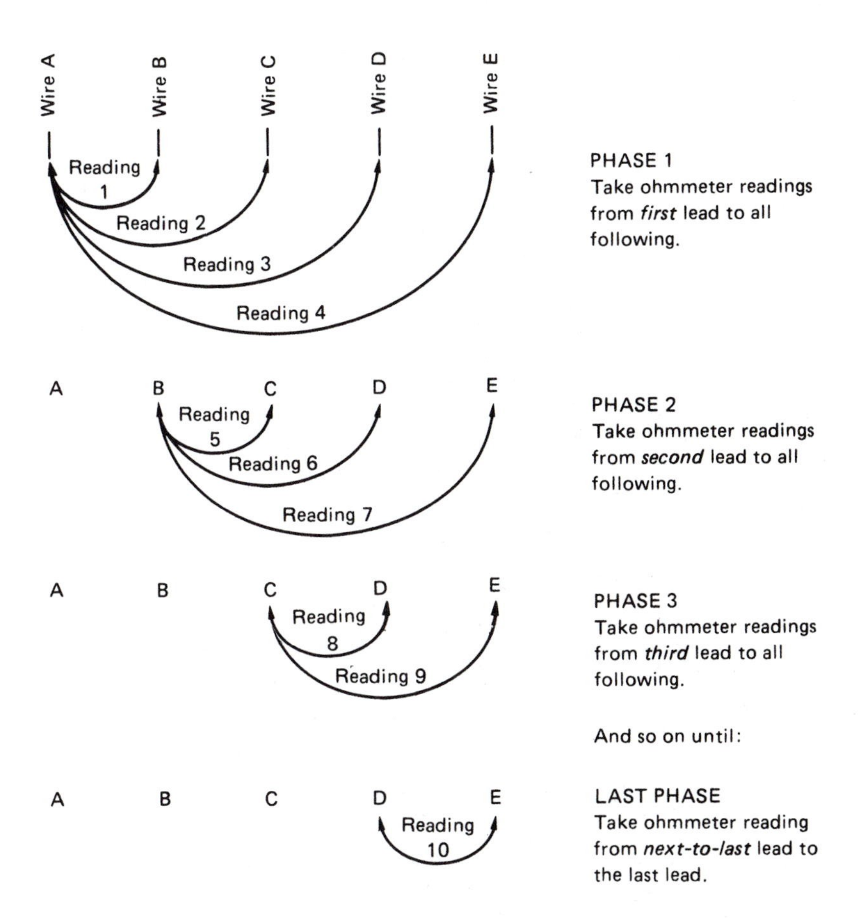

FIGURE 3-4 Detecting shorts between any two points.

problems, such as discontinuities caused by crushing or broken or frayed shielding, by using TDR (time-domain reflectometry) equipment. This is covered in greater detail in Chapter 19.

Antennas and feedlines can also be tested by applying a transmitter signal and noting a proper flow of power to the antenna using a directional wattmeter. A minimum of reflected power should be indicated. An antenna system that is causing problems may be divided for troubleshooting by disconnecting the coax where it connects to the antenna. Connect a dummy load of the proper impedance to the far (antenna) end of the coax and repeat the power test above. If the test now shows minimal reflected power with the dummy load on the coax, the problem is localized to the antenna (Figure 3-5).

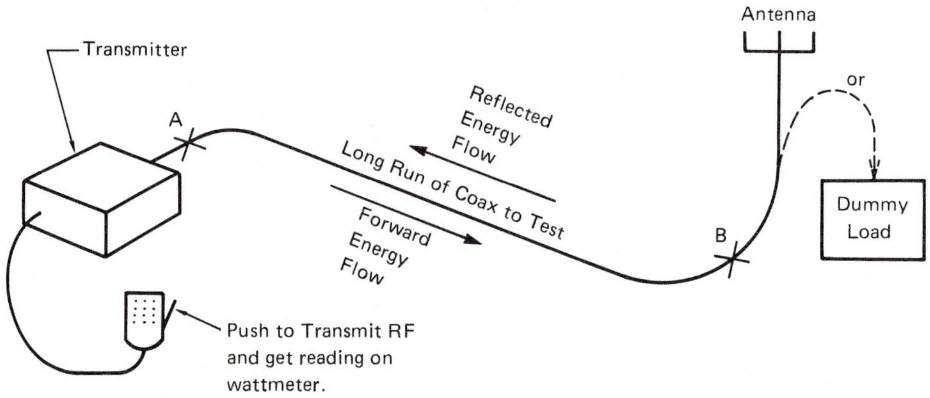

SAMPLE WATTMETER READINGS (APPROXIMATE) AND ANALYSIS

INTO ANTENNA:

	Point A	Point B	
Forward	11 W	10 W	Normal Coax and Antenna
Reflected	1 W	1 W	
Forward	14 W	13 W	Antenna Not Matched/ Out of Tune
Reflected	9 W	9 W	

INTO DUMMY LOAD:

Forward	10 W	10 W	Good Coax
Reflected	0 W	0 W	
Forward	14 W	0 W	Open or Shorted Coax
Reflected	14 W	0 W	

FIGURE 3-5 Using the wattmeter to test coax cables.

SIGNAL LEVELS AS SYSTEM OPERATION INDICATORS

Some large audio systems, such as telephone switching systems, are tested by comparing audio *levels* to a standard. Rather than stating the amplitude of signals in average or rms voltages, it is customary to refer to levels. A level is a reading of power expressed in decibels (dBs) up or down from an understood reference level. The common level in telephone audio systems is the 600-ohm, one-milliwatt standard. All levels of power are expressed in dBs from here.

The standard VOM is calibrated in dBs using this standard. When measuring the voltage across a normal 600-ohm load, the scale may be read directly in dBs. Larger voltages may require switching to a higher AC voltage range, where a specific amount of dBs is added to the indicated dBs. This scale is usually printed on the face of the VOM. If there is any DC level present at all, the VOM must be used on the "Output" function to block that component.

Fluke Instruments (Figure 3-6) offers a digital multimeter, the 8060A, that will read dBs directly on its digital readout. It has a range from −50 dB to +59

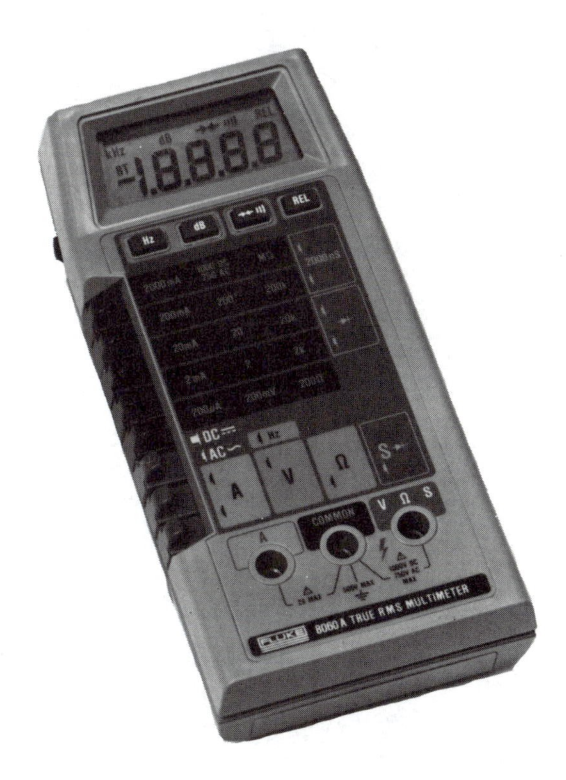

FIGURE 3-6 This portable digital multimeter has a special function to read out directly in audio decibels. (Reproduced with permission of the John Fluke Mfg. Co., Inc.)

dB. This instrument can also be programmed for different impedances and can be used to perform comparisons between power levels. These levels may be either "bridging," where the load (600 ohms) is already on the circuit, or "terminating," where the technician must terminate the circuit with an actual resistor before taking the level reading. Whether or not to use this terminating resistor is a matter of great importance in obtaining the proper level readings (Figure 3-7).

REPLACEMENT OF SUSPECTED UNITS OF THE SYSTEM

In a system consisting of small pieces of equipment, this procedure is called replacing the "black box." Aircraft systems use the black box replacement method so that expensive aircraft is not grounded waiting for repair of electronic items. A technician simply replaces the black box with a good spare one and the aircraft is ready to go again. The defective unit is tagged and sent to the avionics shop for repair. The same is often done in printed circuit card installations by replacing a defective card with a good spare card from stock. The defective card

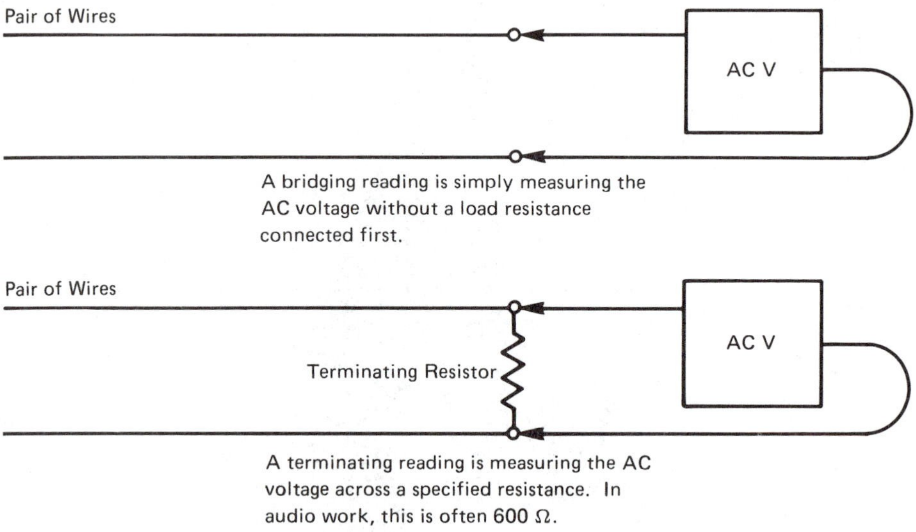

FIGURE 3-7 Bridging and terminating audio voltage readings.

is tagged for repair. This is the primary reason that printed circuit cards are provided with connectors. It is far easier to replace a card by pulling it from a connector than to unsolder the many wires that would be necessary otherwise.

VERIFY PROPER SYSTEM OPERATION AFTER THE REPLACEMENT

Once the suspected unit or card is replaced with a good one, it is important to verify that the system is back to normal operation; it is possible to have gotten a bad spare card or unit from stock. Or by accident, a good unit may be marked for repair and a bad one kept for a spare. This can happen very easily when several cards are brought from stock, unwrapped, and left lying about while a problem is being traced. When the technician finds a bad card, it could accidentally be shuffled in among the good ones. This is something the technician should be careful to prevent.

SUMMARY

Finding the bad unit or card in a large system is the first step in correcting the defect. Once the bad unit or card is found, the next step depends on several factors. If the repair facility has proper schematics, a mockup to energize the unit or card, and the required intercabling to other circuits, the troubleshooting progresses as covered next in Chapter 4. If schematics and a mockup are not available, and for digital ICs on printed circuit boards, the next step may be the passive tests of Chapter 16.

REVIEW QUESTIONS

1. Why should a technician not remove equipment immediately, based upon an operator's report, without further investigation?
2. What kind of diagram would be most helpful in finding a defective piece of equipment out of many operating together?
3. If a cable has five wires within it, how many readings would be necessary to find all possible combinations for wire-to-wire shorts?
4. Assuming a high reflected RF power reading is obtained on an RF wattmeter installed in a transmitter coax line at the rear of the transmitter, what two components could cause the reading?
5. What is the standard audio reference level, and at what impedance?

chapter four

Live-Circuit Testing

CHAPTER OVERVIEW

This chapter covers preparatory subjects necessary before applying power to a defective piece of equipment or circuit card. The schematic and circuit card extenders are discussed.

SAFETY AND CAUTIONARY NOTES

Electrical shock is an occupational hazard for electronics technicians just as getting burned is a hazard for the welder. With proper precautions, there is no reason for a technician to receive a shock. Receiving an electrical shock is a clear warning that the proper safety measures have not been followed.

Some of the most important rules are:

- Don't reach into energized equipment while it is operating. This is particularly important in the high-voltage, high-power stages of equipment such as transmitters. Make your tests on such equipment either while it is dead, or by turning off the power, discharging the filter capacitors, attaching test leads as needed, then reapplying power. Don't handle VOMs or DMMs when connected to high voltages.
- Discharging capacitors is done by shorting them. A special lead can be made for this purpose with a clip at one end for connecting to ground and a probe or other suitable terminal at the other. Most power-supply capacitors, the ones to be particularly careful with, are grounded at one end also. Capacitors that are not grounded ("floating") are best shorted using a screwdriver across their terminals. Sometimes discharging a capacitor produces a loud bang, but the action is harmless to the component.
- When working on live equipment containing voltages over approximately 24 volts, work with only one hand. Keep the other in a back pocket out of the way (Figure 4-1). A current of only 0.04 ampere passing through the chest can be fatal. Keeping one hand out of the way greatly reduces the possibility of passing a current through the chest. Sweating reduces the resistance of the skin to current flow and reduces the voltages that can be safely touched.

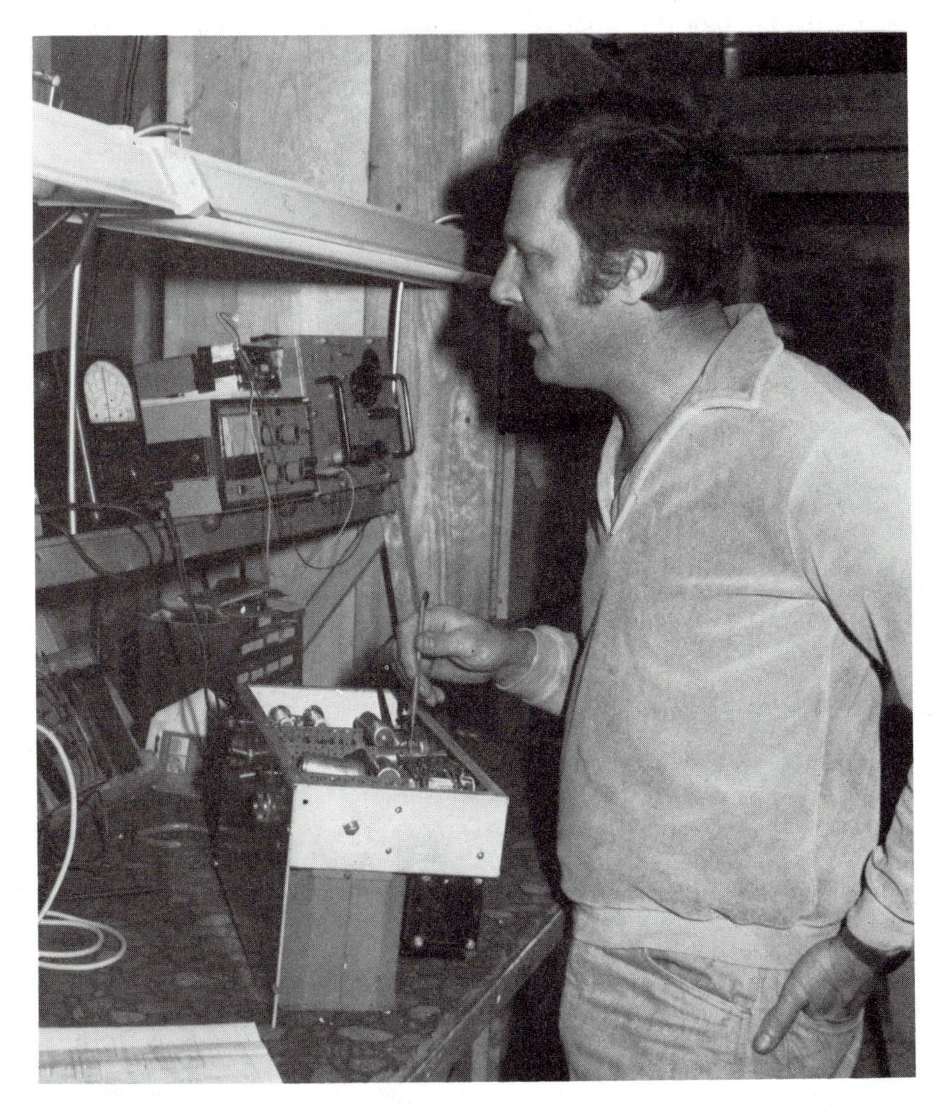

FIGURE 4-1 Troubleshoot live equipment containing voltages in excess of 24 volts with only one hand. Keep the other hand in a back pocket to avoid the possibility of passing a lethal current through the chest area.

- Remove metal jewelry before working on energized equipment. It is a severe shock hazard when working on high-voltage circuits; however, it is also important to remove jewelry when working on low-voltage, high-current circuits. These circuits can produce enough current flow through metal jewelry to cause it to become very hot.
- Discharge filter capacitors before working on dead circuits. The larger the physical size of the capacitor, the more energy it can store.

- If servicing tube-type receivers that do not use transformers, it is a good idea to use an isolation transformer (1:1 ratio) to reduce the hazards of working on a chassis that may be electrically hot. Incidently, the use of an autotransformer (e.g., Variac) does not provide power-line isolation.
- Cathode-ray tubes (CRTs) use very high voltages for acceleration of the electron beam. While one might assume that the anodes are highly positive and the cathode is harmless, quite the opposite is often the case. The anode may be grounded and the cathode may be very highly negative. Don't assume the cathode is safe to touch! Some CRTs are operated between these extremes with the anode fairly high positive and the cathode fairly high negative, all with respect to ground. These are also dangerous. Look at and understand the schematic, and don't assume anything.

Additional vacuum-tube safety notes are given in Chapter 20.

The following cautionary notes will help protect your test instruments and equipment under test:

- There are two very common ways for a VOM or DMM to be damaged: Using the meter on a live circuit while the meter is switched to either the resistance or the milliameter functions. Don't use the current scales without breaking the circuit under test and inserting the meter into the path of current flow.
- The same precautions apply to the use of a solid-state tester. It should *never* be used in an energized circuit. To do so is to invite disaster for the input circuits of the instrument.

USE OF CIRCUIT CARD EXTENDERS

Circuit card extenders allow access to the components of a circuit card under test. Without an extender it is difficult to make even routine voltage checks because other cards or the chassis are in the way (Figure 4-2). When using extenders, keep in mind that some do not provide for foolproof, proper insertion of the extender card into the frame connector or of the circuit card into the extender card. Be sure that the card is not reversed from its normal position. It is also very helpful to number and/or letter the traces right on the extender cards to assist in finding signals on them during troubleshooting. Use the identical lettering and numbering scheme that is used on the connectors into which the extender plugs.

USING THE SCHEMATIC DIAGRAM

The proper schematic diagram gives the technician the information necessary to troubleshoot and repair equipment. There is little point in trying to repair equipment without a schematic, particularly if the circuitry is extensive in size or complexity. This is much like trying to find a specific address in a strange city without the benefit of a roadmap. Street signs help and eventually, by checking

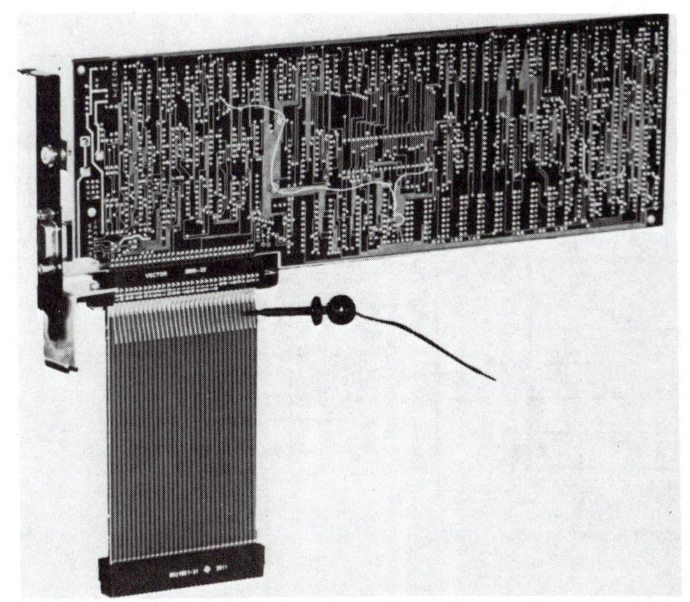

FIGURE 4-2 Using a circuit-board extender card to get the circuit card up where it can be reached with test instruments. (Photo courtesy of Vector Electronic Company, Inc.)

every address in the city, the address could be found, but is it worth it? It is better to have a map. A technician without a schematic can read part values and transistor numbers, but a great deal more information is needed to troubleshoot efficiently. It may take more than ten times the man-hours to repair equipment without a proper schematic.

The interconnection diagram is a special type of schematic used in servicing digital circuits. It is discussed in Chapter 13. Two other kinds of diagrams are sometimes provided but are seldom used by the technician. These are the practical wiring diagram and the functional block diagram (Figure 4-3). The standard block diagram of equipment is an important tool in determining which stage is malfunctioning.

A Schematic Display Suggestion

Technicians sometimes try to work on a piece of equipment and read a schematic on the same working surface. The circuit of interest invariably ends up underneath the equipment. The schematic should be fully visible while the job is under way. It is a simple matter to construct a small easel that holds a large flat backing at a slanted angle, upon which the schematic is placed. A piece of quarter-inch plywood about 2′ × 3′ makes a good backing. It is also helpful if the top of the plywood is fitted with a couple of large spring clamps used to hold piles of paper together. See Figure 4-4.

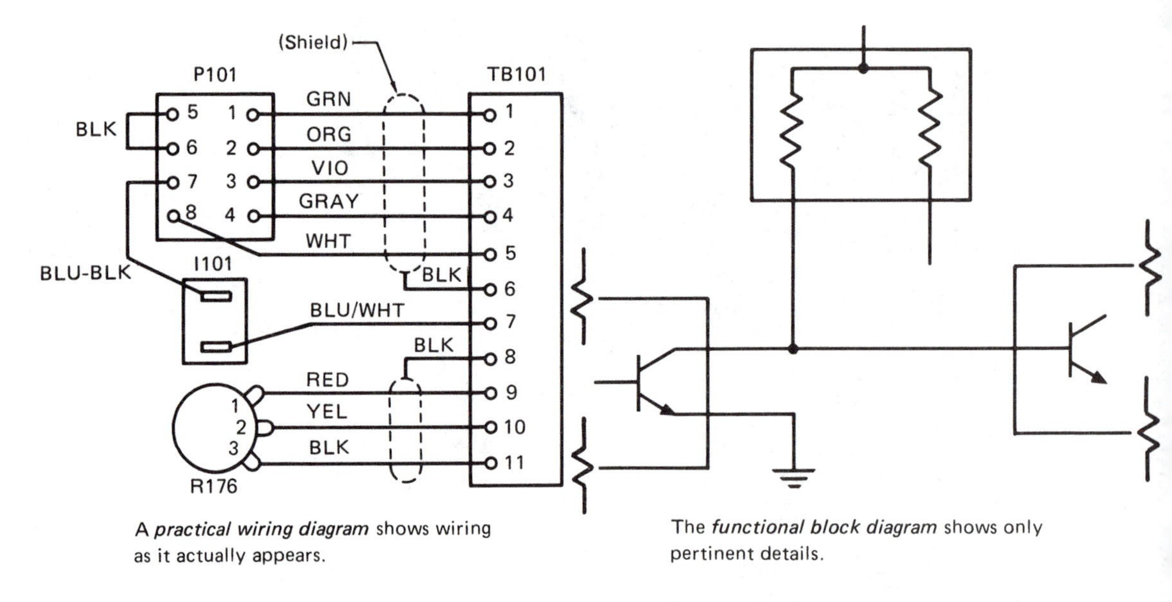

FIGURE 4-3 The practical wiring diagram and the functional block diagram.

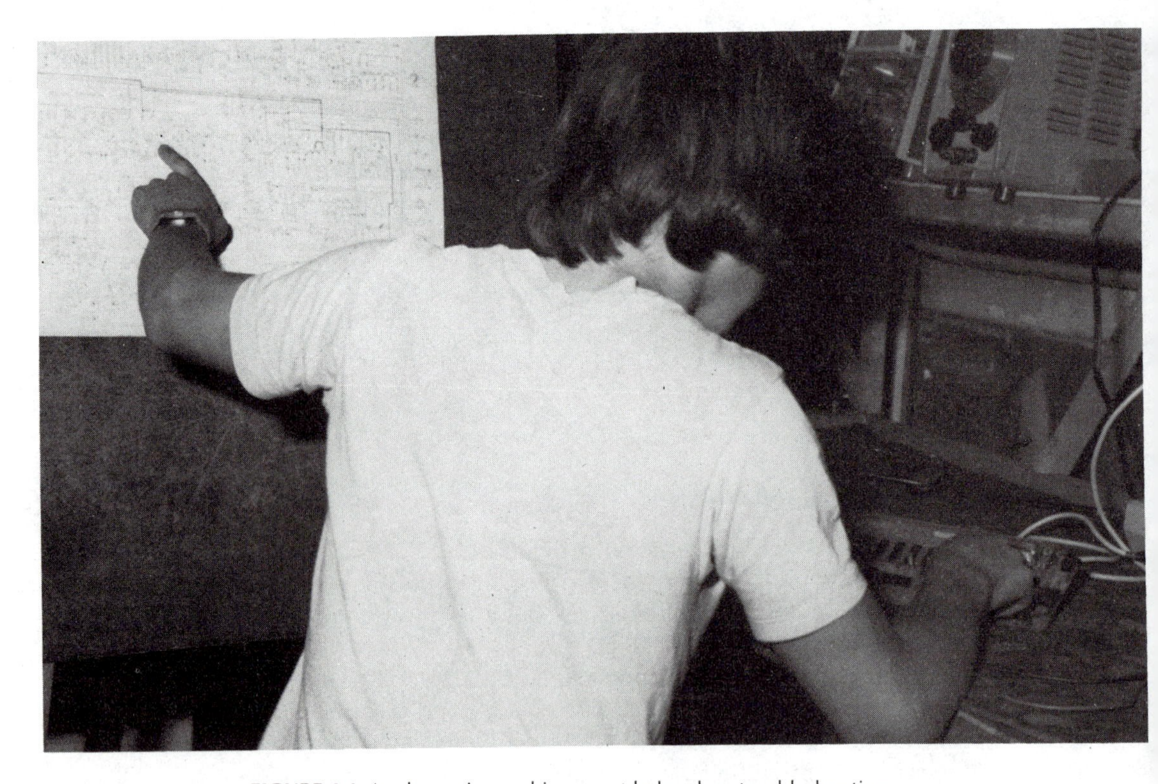

FIGURE 4-4 A schematic easel is a great help when troubleshooting.

Customize Your Schematics

It is a big help to color-code schematics. At first glance a schematic is confusing to anyone. A few common color codes help reduce this confusion a great deal. Tracing over the schematic lines with transparent alcohol markers with the following color codes is recommended.

- Ground or chassis connections in black.
- Voltage supply lines in red. If more than one supply is used, shades near red, orange, or pink may be used for the second and third supplies.
- Primary signal flows may be in other colors such as blue or green.

Add a touch of professionalism by putting the key to your color code somewhere on the diagram.

It is very useful to record *normal* voltage readings throughout the circuitry as appropriate. The power supplies should be well labeled this way. Circuits with signals other than sine waves, such as pulse generators, should be marked to include a small sketch of the waveform as seen on an oscilloscope. Be sure to include the vertical sensitivity setting and the sweep frequency setting of the oscilloscope to enable easy duplication of the pattern.

If a solid-state tester is available and if there are a great many identical circuits coming through the repair facility, it may be a good idea to make sketches of the CRT patterns of a *normal* board. Be sure at least to log the patterns obtained at the edge connectors. This will enable a quick initial check of the boards as they enter the shop. See Chapter 16 for details.

How to Read Schematics

Following are some important points to know when reading schematics.

The schematic usually flows from left to right as the signal flows through the equipment in question. A radio receiver schematic, for instance, would show the antenna connection at the left side. The signal would be processed through the receiver from left to right, ending in the speaker output on the right side of the diagram.

Although the lines in the schematic represent the circuit, the physical layout of the circuit may appear very different. Consider each end of a line on the schematic as sharing exactly the same voltage at each end and everywhere in between. Figure 4-5 illustrates this point. Lines that connect and lines that cross on the schematic without connecting are shown in Figure 4-6.

Figure 4-7 shows the circuit symbols commonly used to indicate earth ground, chassis, and circuit common. A ground refers to a connection to the earth itself. Chassis common is the metal of the framework of the equipment. The chassis and earth grounds are usually connected. Circuit common is usually the negative side of the power supply in the equipment. Some telephone company equipment connects the positive to circuit common. (ECL-integrated [emitter-coupled logic] circuits also are an exception in that they are operated

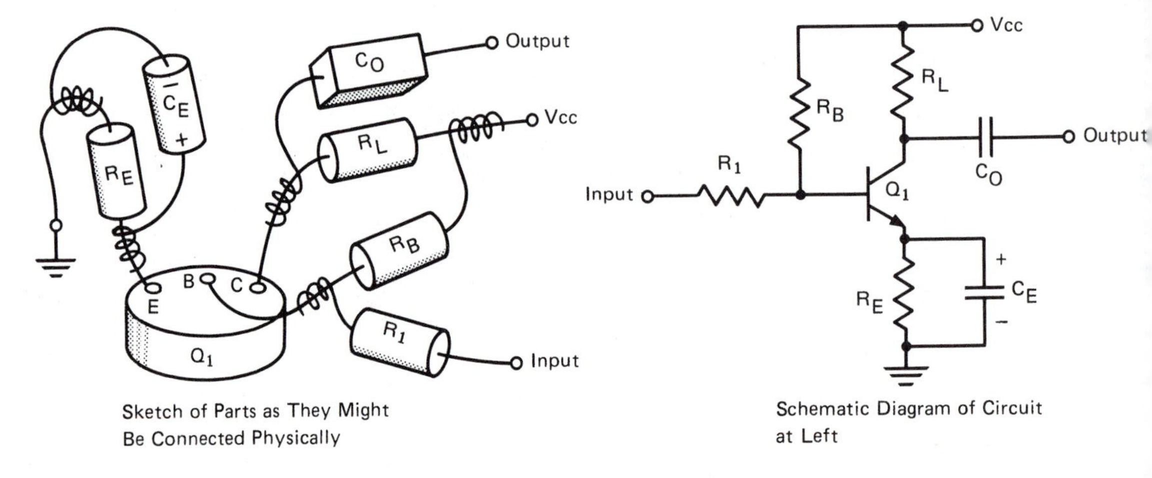

FIGURE 4-5 Comparison of physical and schematic appearances of a circuit.

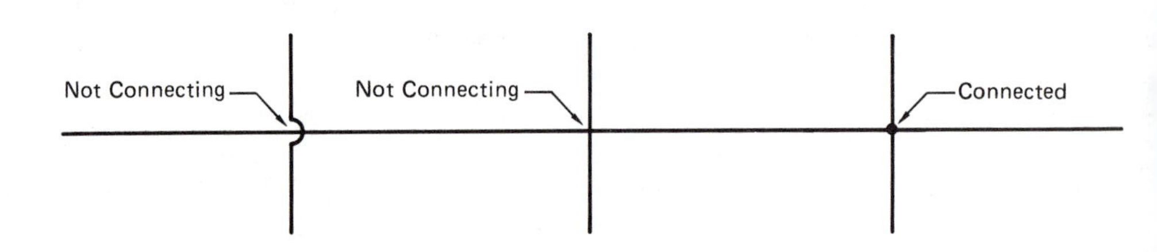

FIGURE 4-6 Schematic symbology for connecting and nonconnecting lines.

FIGURE 4-7 Ground, chassis, and common circuit symbols.

"below ground" at up to −5.2 V.) The circuit common may or may not be connected to the chassis of the equipment. If it is not, it is called a *floating common* or *floating ground*.

Components are often referred to by use of a letter and number combination to save space on the schematic. The parts may be found listed in a separate parts listing where values, tolerances, and other details regarding the part may be given. The letter portion of the reference designation is coded as follows:

B	Blower	P	Plug
C	Capacitor	Q	Transistor
CR or D	Diode	R	Resistor
F or X	Fuse	S	Switch
I	Indicator	T	Transformer
J	Jack	U	Integrated circuit
K	Relay	V	Vacuum tube
L	Inductor	TB	Terminal board
M	Meter		

Sometimes only a minimum of information is given right on the schematic in lieu of using a letter/number key. In this case only the most important information is given. Resistors are given with their resistance values and, in the case of larger resistors, also with their power ratings. Capacitors are given with their capacity values and possibly the voltage ratings. Transistors and ICs are given only their identification numbers. Inductors are given their inductance values only. Figure 4-8 is an example of such a listing.

IF NO SCHEMATIC IS AVAILABLE

There are several things that may be done if a repair is needed on equipment for which no schematic is available. The defective component might be identified quickly by passive testing with a solid-state tester without benefit of a schematic. Chapter 16 discusses the use of this instrument for this purpose.

It may be practical to use a schematic that is similar to the proper schematic. For instance, a CB transistorized 27-MHz transmitter is pretty universal in principle, and the schematic for another transistorized CB transmitter might be close enough to use in finding some problems. Of course, circuit details would be different, and this would have to be taken into account by the technician.

If another, identical unit in working order is available, cross-checking of voltages at critical points during operation can often reveal differences that can lead to solving the problem. In this case, the better the technician knows the general layout of the equipment, the faster the troubleshooting will be. In other words, if the problem is in the modulator section of a transceiver, it makes life much easier if the technician knows which transistor stages are used for the modulator. Time is saved by not checking stages that are not involved in the problem.

Another possibility is that a schematic might be available from the manufacturer of the equipment. In such a case it may be best to put the equipment aside until the diagram is received.

How to Draw Your Own Schematic

One last possibility, and the one that is the most time-consuming, is to hand-draw a schematic of the equipment. To draw an entire schematic would be

MAIN		
IC1	IC	upc577H
IC2	IC	upc575CZ
IC3	IC	TA7045M
Q1	Transistor	2SA639
Q2	FET	3SK40M
Q3	FET	3SK40M
Q4	FET	2SK49H2
Q5	Transistor	2SC945P
Q6	Transistor	2SC945P
Q7	Transistor	2SC945P
Q8	Transistor	2SC1571G
Q9	Transistor	2SC945P
Q10	Transistor	2SC1571G
Q11	Transistor	2SC945P
Q12	Transistor	2SC945P
Q13	Transistor	2SC945P
Q14	Transistor	2SC945P
Q15	FET	2SK44D
Q16	Transistor	JA1050
Q17	Transistor	2SC1528
Q18	Transistor	2SC1947
Q19	Transistor	2SC2053
Q20	Transistor	2SC945P
Q21	Transistor	JA1050
Q22	FET	3SK40M
Q23	–	–
Q24	Transistor	2SC945P
Q25	Transistor	2SC945P
Q26	Transistor	2SC945R
Q27	Transistor	2SC945R
Q28	Transistor	2SC1571G
Q29	Transistor	JA1050
Q30	Transistor	2SC1571G
Q31	Transistor	JA1600G
Q32	Transistor	JA1600G
Q33	Transistor	2SC945P
Q34	Transistor	JA1600G

D1	Diode	1SS55
D2	Diode	1SS55
D3	Diode	1S2473
D4	Diode	IN60
D5	Diode	IN60
D6	Diode	IN60
D7	Diode	IN60
D8	Diode	IN60
D9	Diode	1S1555
D10	Diode	IN60
D11	Diode	IN60
D12	Diode	1S2473
D13	Diode	1SS53
D14	Diode	1S1555
D15	Diode	1SS53
D16	Diode	1SS53
D17	Vari Cap	1S2688C
D18	Diode	IN60
D19	Diode	IN60
D20	Diode	1SS53
D21	Diode	XZ096
D22	Diode	1SS53
D23	Diode	1SS53
D24	Diode	1SS53
D25	Diode	1SS53
D26	Diode	1SS53
D27	Diode	1SS53
D28	Diode	SR 10N-2R
D29	Diode	1S2473
D30	Diode	1SS53
R23	Trimmer	3K FR-10
R71	Thermistor	33D28
R89	Trimmer	100K FR-10
R106	Thermistor	23D29
R113	Thermistor	33D28
R132	Trimmer	5K FR-10

C1	Ceramic	0.01 50V
C2	Ceramic	0.001 50V
C3	Ceramic	0.001 50V
C4	DIP	10P 50V
C5	Ceramic	0.01 50V
C6	Ceramic	0.001 50V
C7	Ceramic	0.01 50V
C8	Ceramic	0.01 50V
C9	Ceramic	0.01 50V
C10	Ceramic	0.01 50V
C11	Ceramic	0.01 50V
C12	Ceramic	0.01 50V
C13	Milar	0.039 50V
C14	Ceramic	2P 50V
C15	Milar	0.039 50V
C16	Milar	0.01 50V
C17	Milar	0.039 50V
C18	Milar	0.039 50V
C19	Milar	0.039 50V
C20	Milar	0.01 50V
C21	Chemicon	10u 16V
C22	Milar	0.01 50V
C23	Milar	0.01 50V
C24	Milar	0.056 50V
C25	Milar	0.056 50V
C26	Milar	0.01 50V
C27	Milar	0.01 50V
C28	Milar	0.01 50V
C29	Milar	0.001 50V
C30	Chemicon	10u 16V
C31	Milar	0.056 50V
C32	Milar	0.056 50V
C33	Milar	0.056 50V
C34	Milar	0.056 50V
C35	Milar	0.002 50V
C36	Milar	0.056 50V
C37	Chemicon	4.7u 25V

FIGURE 4-8 Sample parts list. (Courtesy of ICOM America, Inc., Bellevue, Washington)

impractical for complex equipment, but for simple units and for small areas of direct concern it may be considered. This is a particularly attractive alternative if the equipment is a unit upon which the technician will often be called to work. Knowing color code is a big help in identifying components. Semiconductor reference books giving transistor and IC pinouts are necessary for this job. If the circuits are on double-sided printed circuit boards, the job becomes about four times more difficult than on a single-sided board.

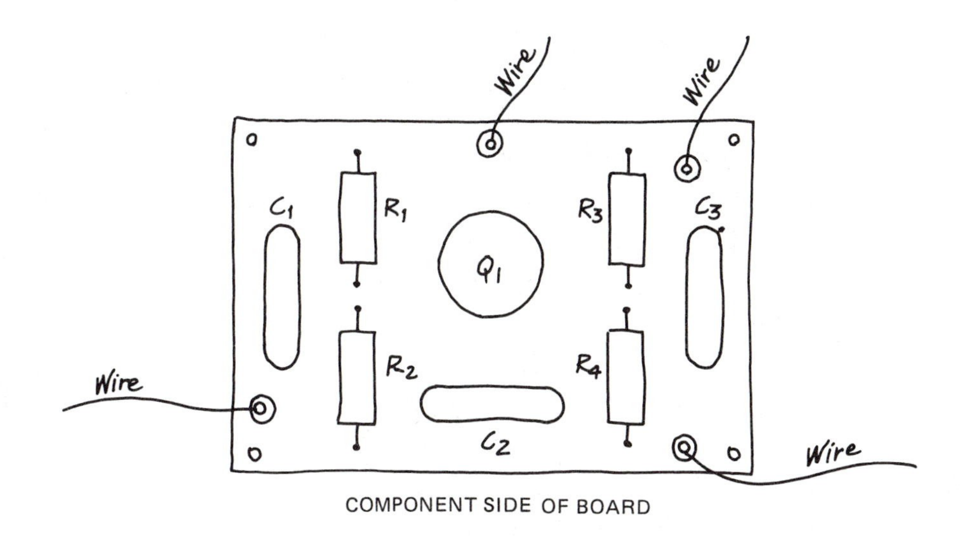

FIGURE 4-9 The first step in drawing a schematic: Identify each part with a unique part number.

Drawing a schematic of a physical circuit is a matter of practice. Start by drawing schematics, using standard symbols, of simple circuits of perhaps only a half-dozen components. Progress to more complicated circuits as you gain experience and confidence. The first step in drawing a schematic is to draw a somewhat scaled-up picture of the components you will be identifying. Label the resistors R_1, R_2, etc., and the capacitors as C_1, C_2, etc. This drawing will help you cross-identify the schematic symbols you will be drawing with the physical components (see Figure 4-9).

The second step is to identify the power-input Vcc or B+ lines, the ground bus, and if possible, the input to the circuit in question. For power supplies this would be the power input leads, and for amplifiers the signal input. Using this physical point as a starting point, trace the circuit through and mark each component you meet that connects *directly* to this point. Don't try to draw several circuits at once, but just identify all of the components connected to this first trace or wire. Remember that if a wire goes between two points, those two points connect to each other on the schematic in any way you choose, especially for this first try at a schematic. At this point you may have two or more components that each have a single lead connected to a common point. Label the schematic with each of these components using the arbitrary identifying code of R_1, C_1, or whatever you have assigned on the physical layout drawing (see Figure 4-10).

The next step is to select *one* of the components and "go through" it, identifying the lead on the other end. Take this point and again draw all of the components that connect to this lead. You may find that this end of the component connects to an unused end of one of the components you drew in the first set (see Figure 4-11).

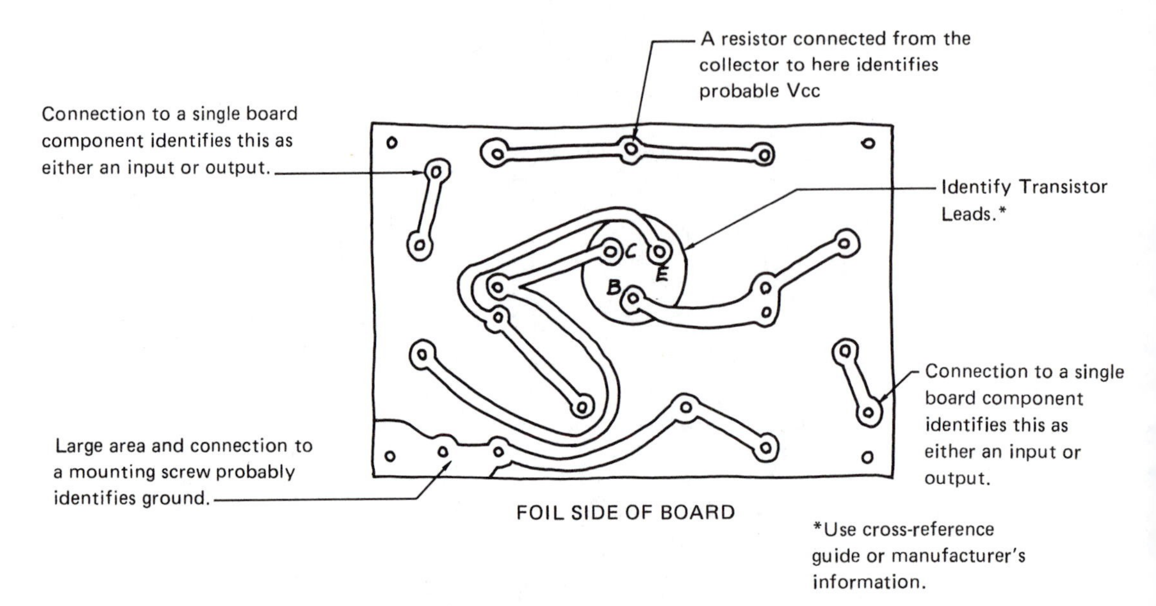

FIGURE 4-10 The second step: Identify input, output, ground, and Vcc leads.

So you don't forget branching wiring, make a large black dot at intersections of wiring. Come back to them as soon as you have finished drawing all of the components on a branch. The circuit components make convenient "waiting points" for you to make sure that you have not forgotten one of the branches or components along the way. Be sure to label the components in each case with the arbitrary identifying labels (see Figure 4-12).

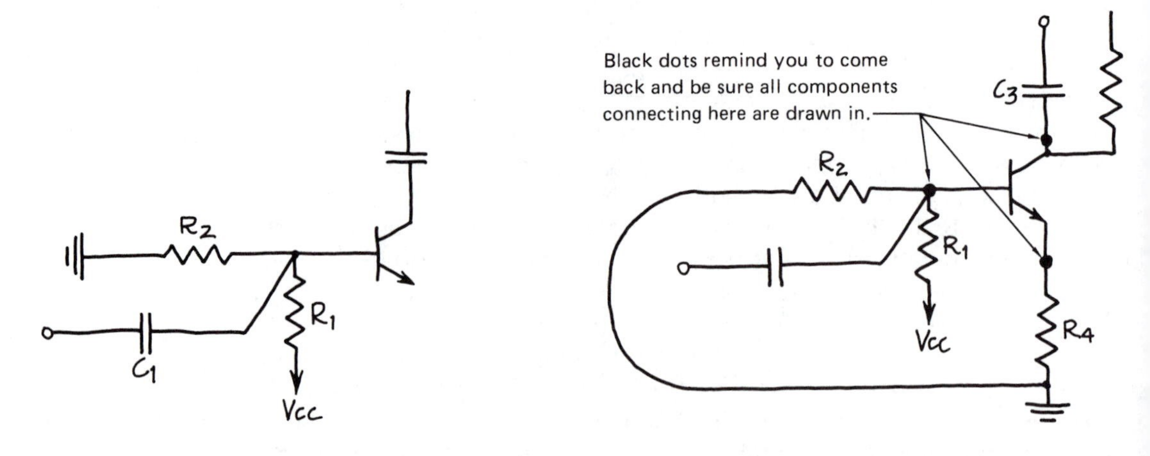

FIGURE 4-11 The third step.

FIGURE 4-12 Fourth step: Make sure you include all connections.

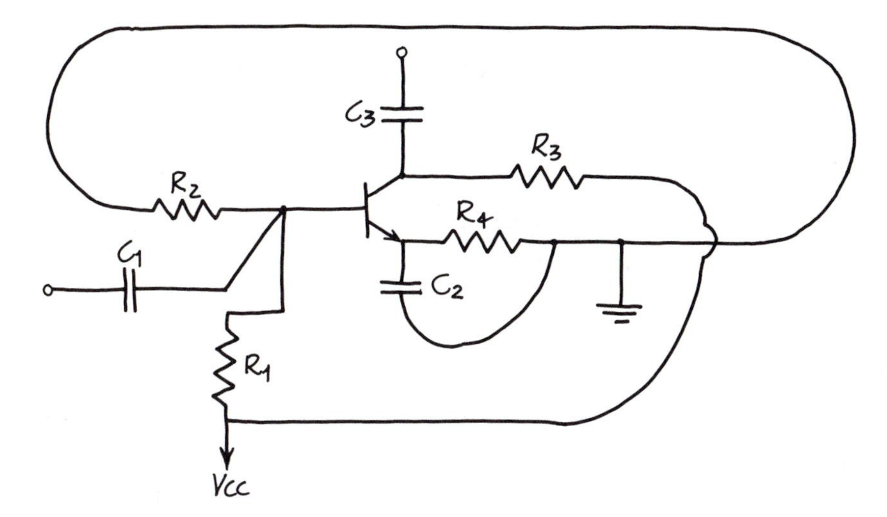

FIGURE 4-13 Fifth step: The complete mess.

Once all of the components and the interconnecting leads or wires have been drawn, you will likely have an atrocious-looking mess. This is normal. You now have an intermediate drawing. This drawing can be cleaned up considerably by redrawing it at least once, perhaps many times, depending upon the priority and permanence of the job desired. See Figure 4-13.

The last step is to straighten out the messy drawing. This is done by inspection and logic. Identify the Vcc or power-input lines and draw them generally across the top, from left to right. Starting at the left, draw in the circuitry as identified for the general flow of signals. Grounds may be drawn with a simple symbol wherever they are encountered, without the necessity of connecting all such points with a common line. Once the circuit begins to look like something familiar, you may be able to draw the circuit more easily if you have handy a

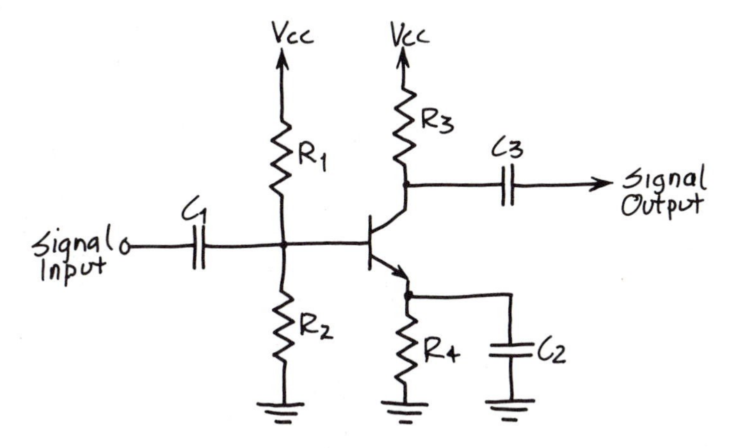

FIGURE 4-14 Sixth and last step: Cleaning up the mess in Figure 4-13.

similar schematic, or you may draw the schematic according to standard circuits, such as amplifier stages. Differences between the standard circuits and the circuits in question are then easier to identify. Sample circuits are available in the reference books you may already have. The Motorola series of reference books has many sample circuits used specifically with Motorola products. The TTL and CMOS cookbooks have circuits involving those IC families, and many of them are complete, do-it-yourself projects. Consulting these references will often uncover a circuit very similar, if not identical, to the one at hand. Figure 4-14 shows the end result of the example used thus far.

SUMMARY

This chapter paves the way for the technician to begin troubleshooting inside the equipment. Chapter 5 begins the subject of internal troubleshooting.

REVIEW QUESTIONS

1. What is the most common way to damage a multimeter?
2. Can some circuit card extenders be inserted backwards?
3. Name the two kinds of diagrams most used by technicians.
4. Name three ways to make your schematic diagrams work for you by modifying them.
5. What component would be marked with a "U" on a schematic diagram?
6. Name three alternatives when a schematic diagram is not available.

chapter five

DC Troubleshooting

CHAPTER OVERVIEW

This is one of the most important chapters in this book. The concepts of voltage troubleshooting presented here are the very foundation of troubleshooting. Most electronic failures are detectable by an improper voltage. There may be no voltage where there should be, voltage where it doesn't belong, or the voltage measured may not be what it should. The procedure for voltage troubleshooting assumes that there are no signals coming into the equipment from test equipment and that the equipment is turned on for voltage testing.

GENERAL DC TROUBLESHOOTING

Voltmeter Test Probes

The probes used with any voltmeter should be sharp and should either be of very small diameter or have a long, narrow taper with insulation down to the very tip. The sharp tip will pierce any contaminants on a PC board, including residue of soldering flux and circuit coatings of various types. The sharp tip is also very good at preventing the probe from slipping from one trace to another, or worse, shorting two traces together. Standard test probes should be fitted with insulation such as heat-shrinking plastic (Figure 5-1). Huntron Inc. manufactures test probes that meet these requirements (Figure 5-2). Note that the probes can be retracted or extended to reach far into circuitry where other probes may not be able to reach.

Test leads are subject to considerable punishment in normal use. An intermittent test lead can produce strange symptoms that can very easily mislead the technician into thinking the equipment under test is causing the problem. Test leads should be checked frequently for continuity using the lowest ohm scales. Short the leads together and gently pull on the wires, especially where they enter the terminations on each end. A reading of more than one tenth of an ohm indicates bad leads.

It may be convenient to provide the black (negative) voltmeter test lead with a small alligator clip. It is convenient to attach this lead to the circuit common. This leaves one hand free to leaf through schematics, for example,

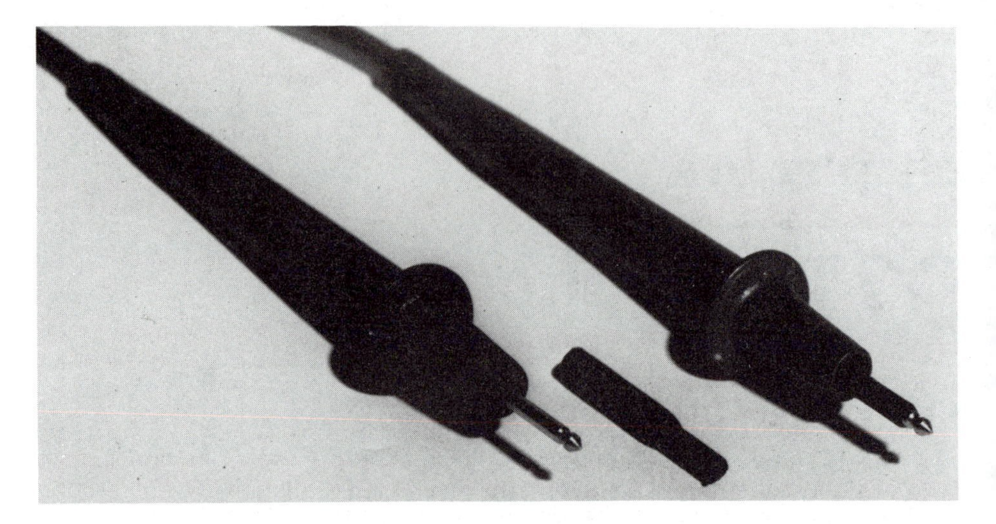

FIGURE 5-1 Test probes with heat-shrinking plastic insulation to help avoid accidental shorting of the circuits under test.

FIGURE 5-2 These test leads have sharp, long tips to avoid circuit shorting on printed circuit boards. (Photo courtesy of Huntron Instruments Inc.)

while the other hand is occupied in taking voltage readings with the red (positive) lead. In the tests that follow it will be assumed that the voltmeter in use has the negative lead connected to the circuit common.

For safety in taking voltage measurements and to prevent errors when using the ohmmeter functions, make it a habit never to touch the exposed tips of the test probes.

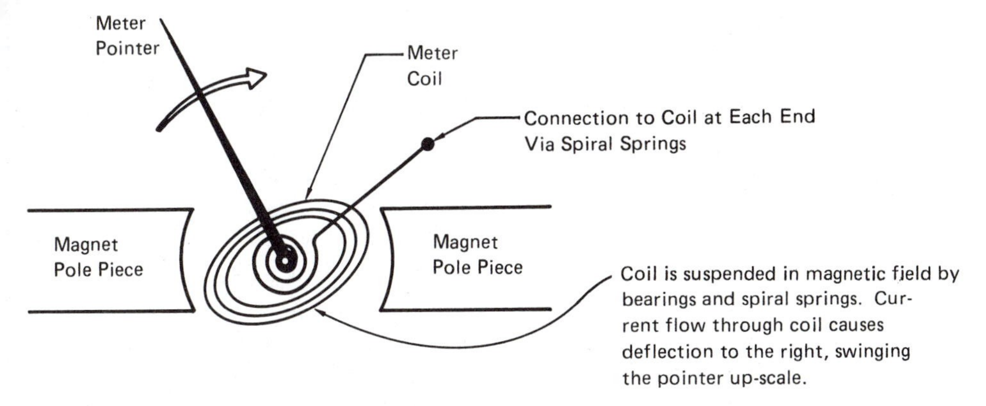

FIGURE 5-3 D'Arsonval movement used in most analog meters.

Using the VOM for Voltage Readings

The volt-ohm-milliammeter (hereafter called VOM) is an instrument in common use. The simple, inexpensive VOM can detect a great many failures, especially in DC circuits. The VOM uses a d'Arsonval movement. This is basically a coil that is free to move within a constant magnetic field. A pointer is attached to the moving coil, and as the current directed through the coil is increased, the coil and the pointer are deflected by the distortion of the combined magnetic fields involved (Figure 5-3). The meter movement is current-operated and draws its operating current from the circuit under test. Since the coil has a relatively low resistance, resistors are used in series to limit the current at higher voltage inputs. Changing these resistors, called multiplier resistors, results in giving the meter different voltages for full-scale readings. A series of multipliers with a switching arrangement is shown in Figure 5-4.

Since the VOM uses current from the circuit under test, this instrument can load a high-impedance circuit and give a voltage reading that is not correct, indicating voltage lower than is actually present when the voltmeter is not connected. There is a quick way to tell if the voltmeter is loading the circuit: Take a voltage reading on two different voltage ranges. If the indicated voltage changes appreciably downward on the lower of the two ranges, it is loading the circuit under test.

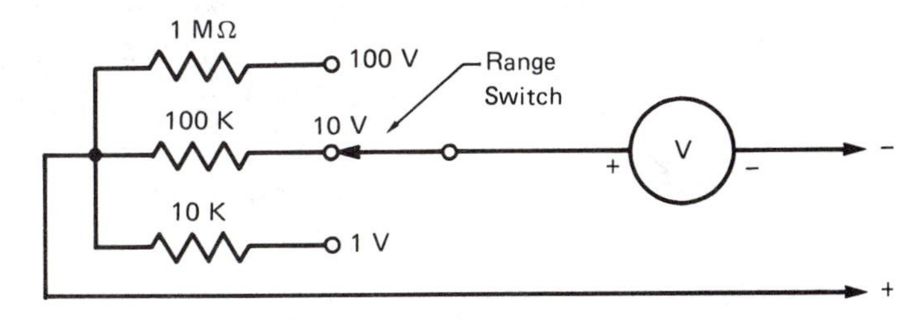

FIGURE 5-4 Simplified voltmeter with range switching.

This loading effect can be a serious disadvantage in low-voltage, high-impedance circuits. Several instruments were designed to minimize this loading effect and still use the d'Arsonval meter movement. The first of these was the vacuum-tube voltmeter (VTVM). By using the high-impedance input of the vacuum tube to sample a circuit under test, it was possible to drive the meter movement with the plate circuit of the tube. Later the transistorized voltmeter (TVM) and the field-effect transistorized voltmeter (FETVM) were developed using more modern components. These components made these instruments truly portable, a big advantage over the VTVM (Figure 5-5).

The accuracy of a VOM for troubleshooting seems to be an overemphasized specification. A VOM that is within five percent will be quite sufficient for most routine voltage troubleshooting. In critical applications, however, the digital multimeter is better to use. When using a VOM for precision readings, select a range that puts the needle somewhere in the upper two thirds of the scale. Look at the meter straight on. If there is a mirror behind the needle, aligning the needle with its reflection reduces parallax errors caused by looking at the needle from the side.

FIGURE 5-5 Examples of analog transistorized VOMs that minimize the loading effects of a conventional VOM. (Courtesy of Simpson Electric Company, Elgin, Illinois)

Whenever the VOM is handled a great deal, such as taking it to a new job site or into the field for servicing, it should be switched to a specified range setting that will afford the best protection to the meter movement. The meter movement coil can be shorted out, which will put a magnetic brake on the meter needle. This helps prevent the needle from banging on the end pegs. Some meters use a "Transit" position and others an "Off" position for this purpose.

The VOM and FETVM with their analog meter movements do have one advantage over most digital multimeters. Slow changes in voltages are more easily interpreted when using an analog indicator. The tuning of some circuits requires monitoring of voltages that change with adjustments made during alignment procedures. These slow changes are easier to see on an analog display than they are to interpret in the changing digits of a digital meter.

Using a Digital Multimeter for Voltage Readings

Today the digital multimeter is the latest and best all-around meter to use. On the voltage ranges, it has very little loading effect, typically up to as much as 10 million ohms (Figure 5-6). A good quality 20,000-ohm/volt VOM, by comparison, has an input resistance of only 60,000 ohms on its 3-volt range.

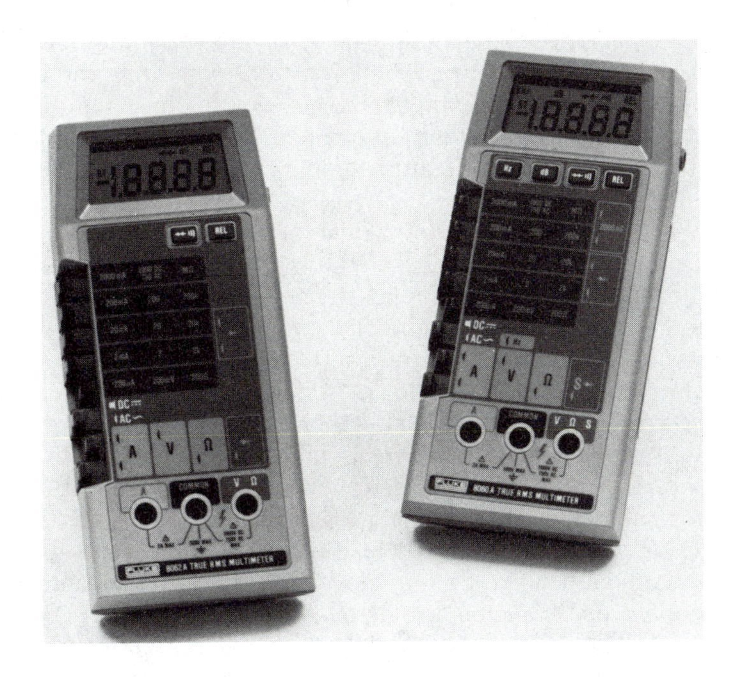

FIGURE 5-6 Two digital multimeters meant for portable work. (Reproduced with permission of the John Fluke Mfg. Co., Inc.)

The digital multimeter (hereafter called a DMM) has many advantages over the VOM:

1. The DMM has a higher input resistance (this loads the circuit under test less). Typical DMM input resistance is 10 million ohms.
2. The DMM is better for measuring precision voltage sources since the accuracy of a DMM is commonly one percent and sometimes as good as 0.03 percent.
3. Another advantage is the repeatability of DMM readings. A constant source of voltage will yield the same reading no matter who may be operating the DMM. Operator interpretation is not a factor in the accuracy of a reading as it is with the meter movement.
4. Some DMMs will give readings regardless of the polarity of the input DC voltage. The polarity is shown as (+) when the input voltage is positive on the red lead, (−) when positive is applied to the black lead (reversed from normal polarity). This avoids the frequent reversal of the test leads necessary when using a VOM with circuits using both positive and negative power supplies.
5. The DMM will measure much smaller voltages with accuracy. Reading a voltage of 155 millivolts is very difficult and inaccurate with a VOM, but simple with a DMM.
6. Some DMMs are available with an autoranging feature. This is an attractive feature when troubleshooting. The technician does not have to reach over and change the voltage range switch while holding a probe with the other hand. Manual range changing can be a problem sometimes. Because of the light weight of most meters, when you turn the switch, the whole meter twists around, and the range change switch doesn't move after all. Autoranging eliminates this annoyance. It is desirable to provide for bypassing the autoranging feature, holding the DMM on a selected range if desired. More on this later.
7. Another feature, available on the Fluke Model 77 digital multimeter (Figure 5-7) for instance, is the ability of the meter to hold a reading once it is taken, even with the probes taken off the circuit. This allows the technician to give full attention to the placement of the test probes without the need to watch for a reading on the DMM. When the DMM has seen a constant voltage for ½ second, it "locks" the display on the DMM and sounds a beep. The test leads may then be disconnected and the reading on the DMM observed at leisure. To take another reading, the technician just applies the probes to a new voltage. The DMM will sense a new reading and lock it into the display as before.
8. The whole Model 70 series of Fluke meters also has a bar graph display, which is used much like a meter movement. There are advantages of using this model with its analog display over using a VOM or FETVM. First, the analog display reacts faster to changes. Second, the analog display will automatically switch ranges, a feature a meter instrument does not duplicate. Third, the analog display automatically displays a (+) or a (−) without the necessity of test lead reversal.

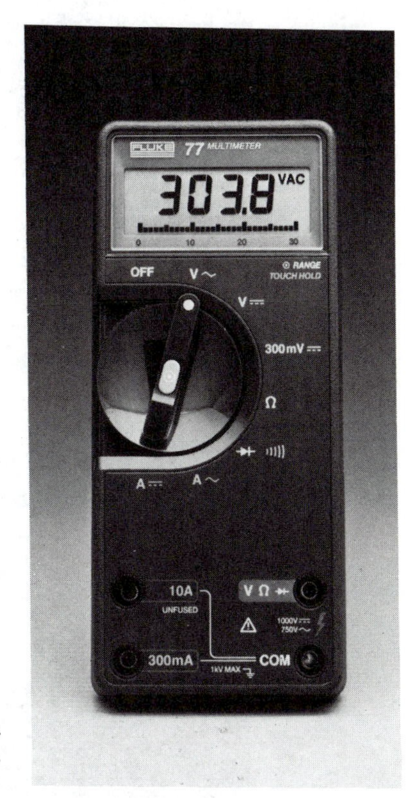

FIGURE 5-7 The Fluke Model 77 digital multimeter. (Reproduced with permission of the John Fluke Mfg. Co., Inc.)

There are two disadvantages to a DMM: First, it requires a battery to operate, whereas the VOM simply draws power from the circuit under test. A dead battery can make an expensive DMM useless until another battery or a 115-V AC battery eliminator can be obtained. Battery life is a major consideration in selecting a DMM. One DMM model has a stated battery life of 100 hours, or four days of continuous operation. The Fluke Model 77 instruction book claims "in excess of 2000 hours"—almost three months of continuous operation. Second, most DMMs have a voltage limitation that could be called a "hard limit" of 1000 volts, because a substantial overvoltage could severely damage the instrument.

Using an Oscilloscope to Measure DC Voltages

The oscilloscope can be used to measure DC voltages if its vertical amplifiers are DC coupled. Inexpensive oscilloscopes cannot be used in this way because their vertical amplifiers are capacitor coupled and are thus only usable on AC signals. A look at the front panel of an oscilloscope will tell if it is suitable for DC measurements. It will have a switch labeled "DC/AC". This switch will be located near the vertical input connector. The same switch may also have "GND" as a third position (Figure 5-8). The absence of a switch like this indicates an instrument good only for signals of perhaps 60 Hz or more. The application of DC

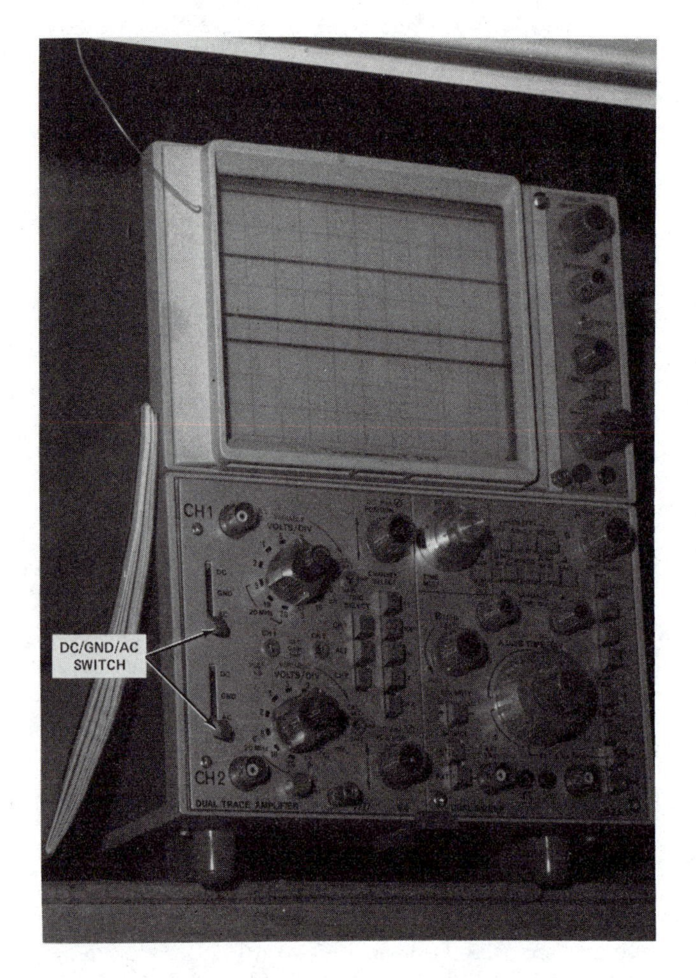

FIGURE 5-8 The DC/GND/AC switch on a DC-coupled oscilloscope. Look for the DC/AC switch to identify an oscilloscope that may be used to measure DC voltages.

to this kind of scope will result in the upward or downward deflection of the sweep for only a moment, after which the sweep will return to the original position.

Whenever an oscilloscope is operated, the spot on the screen should not be allowed to remain stationary, particularly at high intensities. A bright, stationary spot can burn the phosphor on the screen, resulting in permanent damage. Keep the spot moving or turn down the intensity to a very low level.

Using the oscilloscope to measure DC voltage is a bit of overkill. It is usually more convenient to use a DMM or VOM for DC troubleshooting. If the oscilloscope is the only instrument at hand, it can be used. The vertical positioning control should usually be adjusted so that the center of the screen represents ground. If the circuit under test has no negative voltages to measure, it will give pictures twice the size to set the ground level at the bottom of the graticule (gridwork) of the screen rather than at the middle.

A test is necessary to see if the trace position is stable when changing ranges to avoid having to reset the vertical positioning each time the range is changed. Set the input switch to "GND". This prevents outside signals from affecting the amplifiers of the scope. Put the trace in the center of the screen with the vertical positioning control. Change the range switch through all its settings. The center position should not change appreciably. If it does (and sometimes the trace can even leave the screen because of gross misadjustment) find the adjustment called "DC BAL" (often a screwdriver adjustment). Proper adjustment of this control will minimize the shifting of the trace as the range switch is changed. This adjustment should be made whenever a shift of the DC level with range changes is noticed.

When using the oscilloscope to measure DC voltages, it is probably best to set the sweep speed range switch once, then leave it alone. Since there is no vertical variation of the trace due to an AC signal, changing the sweep speed switch affects only the flicker of the horizontal line. Any sweep speed fast enough to avoid flicker is suitable, perhaps 1 ms/cm or 100 Hz, depending on how the horizontal range switch is labeled.

When measuring DC voltages, it is seldom necessary to change the vertical position of the trace, the zero reference line. This should be set at the center or the bottom of the screen; to change it often invites confusion. Changing the vertical sensitivity of the oscilloscope during testing can also be confusing. Once the sensitivity is set up so that the maximum circuit voltage is at the top of the screen, the range switch setting should be noted. This is the "basic setting." If lower settings are used when looking at small voltages, the switch should be immediately returned to the basic setting. This will help keep the relative amplitude of the signals more clearly in mind.

The input circuits of oscilloscopes usually connect the grounded side of the probe to the power-line ground. In some circuits this could create a problem and could even damage the circuit. Before connecting the ground side of a probe to a circuit common, be sure that the circuit will not be damaged by connection to the power-line ground. In some isolated cases it may be necessary to use an isolation transformer on either the circuit under test or the oscilloscope itself. Generally, however, this is not a problem since most circuits use a power transformer, which provides any isolation needed.

The oscilloscope should be used with an appropriate probe for measuring DC voltages. With a good probe, the full gain of the vertical channel can be used without the pickup of hum and noise that would occur if the oscilloscope were used with two wires for the input and ground connections. If a probe is not available, two wires could be used, but noise and hum pickup will be excessive at high gain settings. Under such conditions, the technician may not be able to tell which signals are coming from the circuit and which from stray signal pickup.

The oscilloscope has an advantage over a DC voltmeter for measuring voltages during troubleshooting: It will show AC signals riding on DC levels. In some cases the unexpected AC signal shown on an oscilloscope can lead the technician directly to the cause of a problem. This situation otherwise might escape notice for some time if using only a voltmeter. The oscilloscope will also amplify very small DC voltages that a VOM could not show.

Using the VOM/DMM for Current Measurements

Current measurements are not often used in troubleshooting. Two possible exceptions to this are the use of the current tracer instrument, covered in Chapter 13, and the clamp-on ammeter discussed later in this chapter. **CAUTION: Do not apply voltage to an ammeter.** The internal resistance of an ammeter is near zero. Besides shorting the voltage source, the full current capability of the source is pumped through the instrument.

Measuring current usually means breaking the circuit and putting the meter into the circuit so that all of the current flows through the meter (Figure 5-9). Breaking a circuit for troubleshooting purposes is very seldom done. Where current measurements are routinely made, there will be provisions made for easy insertion of the meter, most often a normal-through jack. A normal-through jack will allow the normal circuit current to flow through itself via contacts built into it. When an ammeter is plugged into the jack, the switch contacts open and the current flows through the ammeter (Figure 5-10).

When you insert an ammeter into a circuit, you are also inserting a certain amount of resistance. Although this resistance is very low, the voltage drop across it can cause disturbance of low-voltage circuits and thereby cause inaccurate current readings. This is called *burden voltage* and may be as much as 2 volts. If the circuit behaves differently after the ammeter is inserted, the meter is causing a change in the circuit by this means.

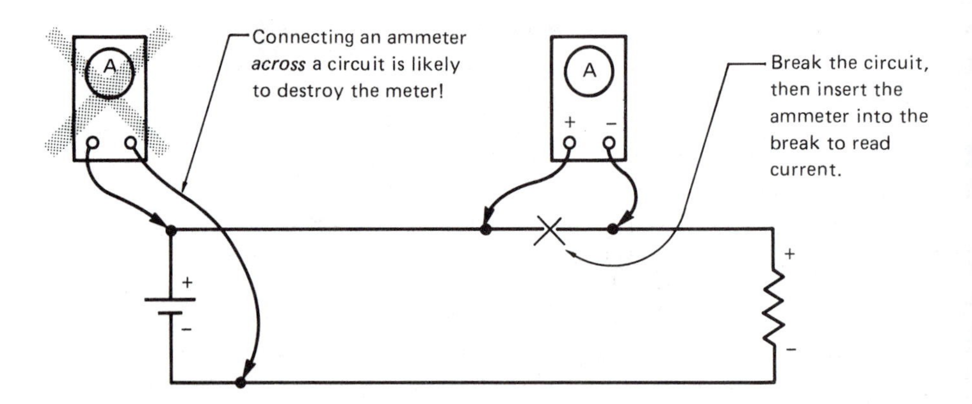

FIGURE 5-9 How to connect an ammeter properly and also how to quickly ruin one.

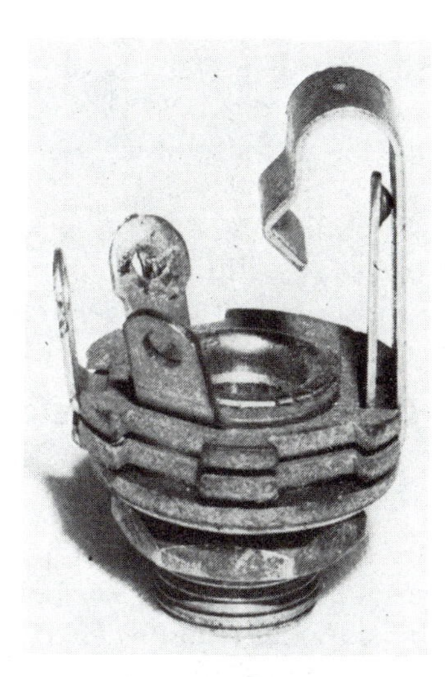

FIGURE 5-10 A normal-through phone jack commonly used for current measurement.

An alternative way of measuring current that does not involve opening the circuit is to use a resistor of known value installed in the circuit. This resistor is called an ammeter *shunt* and may be installed in the circuit by the manufacturer or inserted by a technician for temporary use. In many applications it may be left in the circuit after the tests are made, for future use. This resistor must be of a low enough value to prevent disturbing the circuit operation. Its only purpose is to provide a small output voltage for a measurement. The DMM is preferred for this measurement because it is able to accurately measure small voltages. Using a DMM allows the use of a much smaller resistor in the circuit, thus disturbing the original circuit less. See Figure 5-11.

The amount of current measured by a shunt is calculated by Ohm's Law ($I = E/R$): Divide the voltage read by the value of the resistance. As an example,

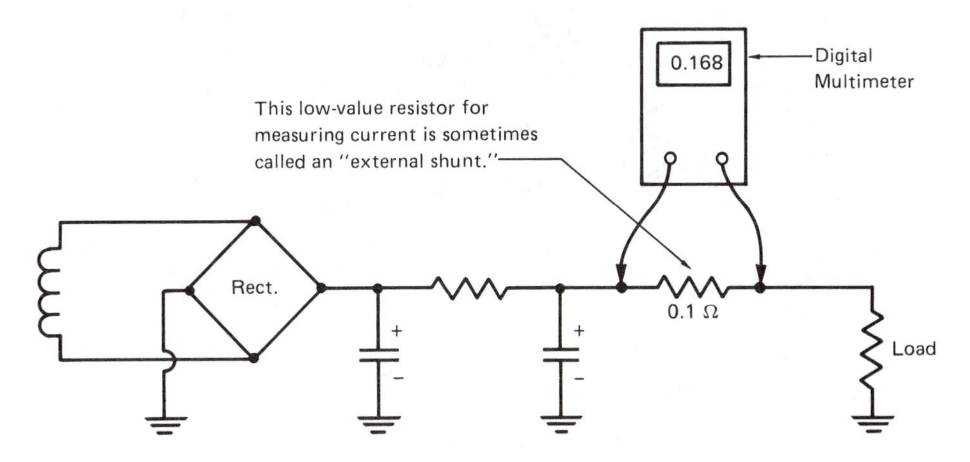

FIGURE 5-11 Using a low-value resistor to monitor current without having to break the circuit.

suppose that a resistor of 0.1 ohm had a measured voltage of 168 millivolts. This would yield 0.168/0.1 = 1.68 amperes of current flow in the circuit. Note that the circuit did not have to be opened and an ammeter inserted. Note also that we measured *voltage* rather than current. The DC current flow through any resistance can be calculated using this method. It is a very practical way of measuring current in circuits quickly and without damage to the circuitry that would occur if they were cut open to insert an ammeter.

The Clamp-on Ammeter

The DC current flow through an individual wire may be measured, if it is above about one ampere, by the use of a DC clamp-on ammeter or a similar accessory for a DMM (Figure 5-12). These instruments are not often used in troubleshooting, partly because they require a single wire to monitor, while most modern circuits are printed on an insulated backing. Also, their lower limit of one ampere is a restriction when circuit currents are well below this level. A similar clamp-on accessory is available for oscilloscopes. This small clamp-on will measure currents as low as 2 milliamperes, but it, too, is seldom used in troubleshooting.

Resistance Measurements

Using the VOM or DMM resistance functions on active circuits is a sure way to damage the test instrument, the circuit under test, or both. **Caution: Never apply voltage from any outside source to the probes of a resistance meter.** Testing circuits with the ohmmeter is restricted to *dead circuits only* and is covered in Chapter 16.

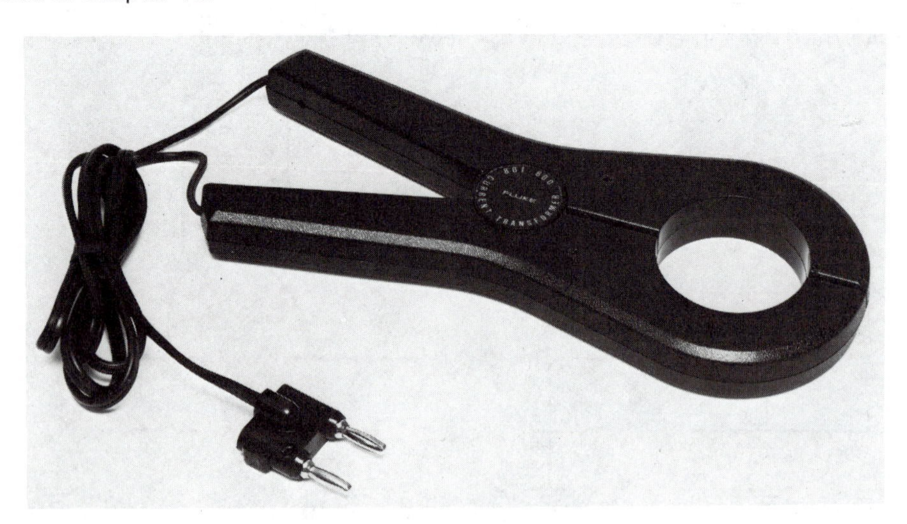

FIGURE 5-12 Clamp-on AC-measuring accessory for a Fluke DMM. (Reproduced with permission of the John Fluke Mfg. Co., Inc.)

DC VOLTAGE MEASUREMENTS TO FIND THE PROBLEM

DC voltage measurements are the major means of locating electronics problems. In the great majority of cases, a defect in circuitry will also shift the circuit voltages out of the ordinary, thus making it possible to find the defects by analyzing "normal" versus "abnormal" voltages.

In the course of troubleshooting with a voltmeter, it is necessary to *visualize a schematic diagram as DC would see it.* Figure 5-13 illustrates this concept. Note that the capacitors in the figure are not drawn in the equivalent circuit. This is because a good capacitor is an open circuit to DC—it just isn't there. The inductors are shown as low resistances because DC sees them only as their equivalent low values of DC resistance, the resistance of the wire within them.

By visualizing circuits as their DC equivalents, the technician can estimate what voltage can be expected at any point within the diagram. This is necessary because very few voltages are given on the schematic and many voltage values must be estimated. When a voltage far from that expected is encountered, the technician can then recheck the estimate of the proper voltage to be confident of it. If there is still a large discrepancy, the cause must then be determined.

Understanding the division of a voltage across series and parallel loads is essential to the ability to "guesstimate" expected voltages. In the words of the mathematician, the voltage will divide in proportion to the ratio of the resistances. Put more simply, if the resistors are equal, there should be an equal split of voltage between them. If they are not equal, the higher resistance will have more voltage across it than the other. In any parallel situation, all of the parallel components must have an identical voltage across them. It is very important to grasp the concept involved here. This feel for proper voltage division must become intuitive, "second nature," when using voltages to troubleshoot circuitry.

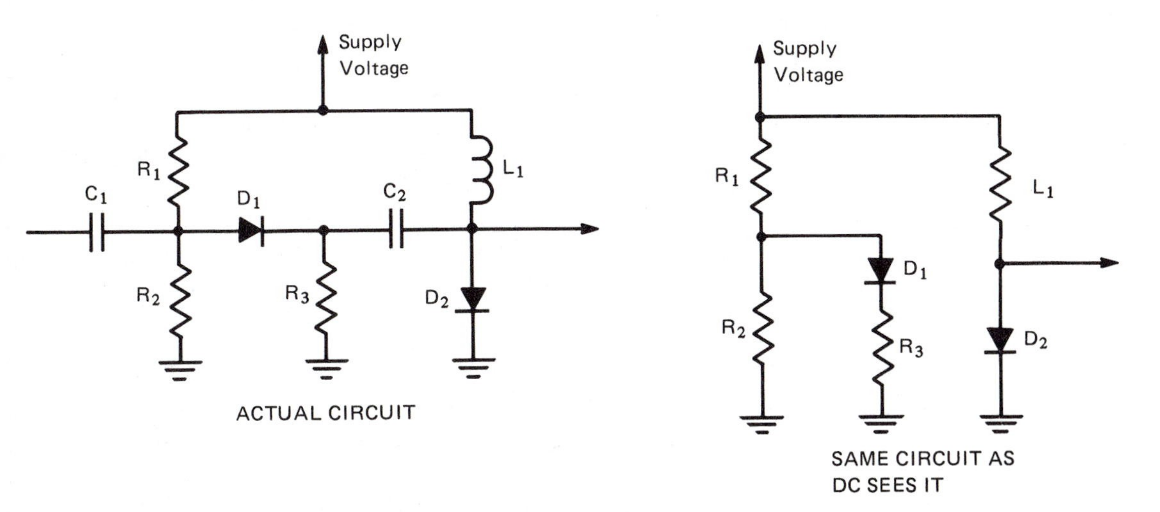

FIGURE 5-13 Actual circuit and same circuit as DC sees it.

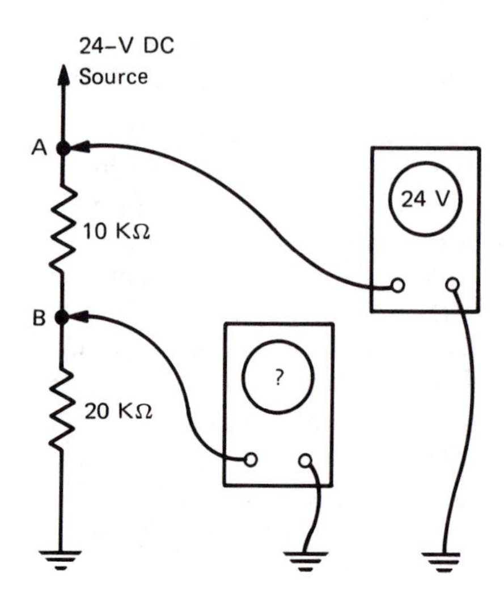

FIGURE 5-14 Estimating voltages not listed on the schematic.

How can you estimate what voltages to expect at a particular point in a circuit? This involves the preceding concepts of series and parallel components and how the voltages behave, and some judgment. Consider the circuit of Figure 5-14. Note point A in the figure. This voltage has already been provided. It may have been on the schematic or obtained from a voltage table in the instruction manual. If only a table of voltages is provided, it is a good practice to put the table values directly on the schematic for quicker reference. But how about the voltage at point B? This voltage will have to be estimated by the technician.

The ability to estimate unlisted voltages in a normal circuit speeds the troubleshooting process for the technician. The voltage at point B of Figure 5-14 should be approximately 16 volts above the circuit common. This value is arrived at by applying the principle of ratio and proportion. If the ratio of the resistance values is two to one (in our example, a reasonable approximation of 22 K to 47 K), then the voltage ratio will also be two to one, with the total being 24 (the value of the supply voltage). If 47K:22K is about 2:1, then E1:E2 will be about 2:1, and their total must be 24. Therefore, an estimate of the voltages would be 16:8, since 16 + 8 = 24, the circuit total voltage. The larger voltage will be dropped across the larger resistor. If the 24-V DC supply was a few percent higher, the voltage at point B would also be that percentage higher than the estimated 16 volts.

Remember that the actual resistors do not have the precise resistance that is marked on them. Carbon-composition resistors have a tolerance rating that allows them to be off as much as 5 percent, 10 percent, or 20 percent from their marked value. Metal-film resistors will be closer to their marked value, commonly within 1 percent. Considering resistor values might be off by as much as 10 percent as an example, a technician troubleshooting according to the estimated voltage at point B (Figure 5-14) would probably accept voltages from 14 to 18 volts before suspecting something was really wrong. As a matter of practicality, technicians never take the time to actually calculate the possible voltage swings in view of maximum and minimum resistances possible and the percentages of source voltage variation. The ability to estimate voltages can be developed by experience, a bit of intuition, and a good "feel" for Ohm's Law.

However, if the circuit under repair is a critical one, the technician should accept less deviation in all readings. For example, if the circuit of Figure 5-14 were in the input circuit of a precision circuit such as an analog meter, the voltage at point B would be very critical. A percent either way from the calculated value could determine whether the instrument were in calibration or not.

Another clue that an experienced technician would take into account is the type of failure for which the equipment is on the bench versus the departure of voltage from normal. For instance, if the equipment is defective because it simply quit working, there is no reason to expect that a voltage only 10 percent from normal is the cause of the failure. The same low voltage reading, however, could have a direct bearing on a case of marginally acceptable performance.

A MOST IMPORTANT CONCEPT

The principles that follow are the very heart of this chapter. The reader should understand these principles very thoroughly, for they are to be applied whenever tests are made on equipment using a voltmeter, and often when using other test instruments that record voltage levels, such as oscilloscopes.

When the Measured Voltage Is Too Low

See Figure 5-15 for the basic circuit referred to in the following discussion.

If the voltmeter in Figure 5-15 reads a voltage that is too low, there are five possible reasons why this might be so:

1. The source itself is low in voltage.
2. The resistance across (parallel with) the voltmeter probes is too low.
3. The resistance back toward the source is too high.
4. The voltmeter is loading the circuit down.
5. The voltmeter is defective.

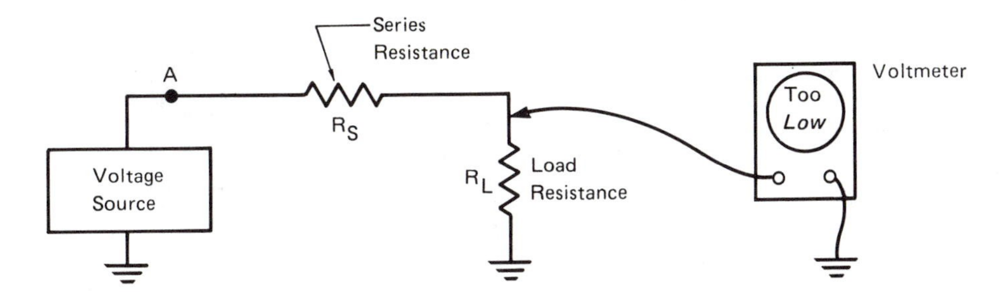

FIGURE 5-15 Analyzing cause of a voltage that is too *low*.

These are fundamental rules and should be memorized!

Let's examine each of them. Possibilities 1, 2, and 3 warrant particularly careful attention since they are by far the most common causes of low voltage readings in electronic circuits. Possibility 1 is that the source itself is too low in voltage. Check the source at point A for abnormally low voltage. This may be low line voltage or weak batteries. The internal resistance of the source itself may have increased and be causing the problem. (This is what happens with worn-out batteries.) A good voltage reading at the source eliminates possibilities 1 and 5.

Possibility 2 occurs when the circuit resistance *parallel* with the meter, R_L (across the meter probes), has *decreased*. This is typically the case where a circuit has an overload of the circuitry, or in the extreme case, a dead short across it.

Possibility 3 is that a resistance in *series* with the meter (looking back toward the voltage source) has *increased* in resistance. R_S of the circuit represents any resistance in series external to the source. In extreme cases, this could be a completely open circuit.

Possibility 4 should be considered when using a VOM, with its inherent low resistance, across high-impedance, "weak" circuits, especially on the lower voltage ranges. Some circuits, such as vacuum tube grid circuits, FET, and CMOS circuits, have very high impedance and are susceptible to this type of error. Refer to the definition of high impedance in Chapter 1.

Possibility 5 is that the meter has been overloaded (had its needle slammed into either end-stop), dropped, or is otherwise giving readings that are too low because of problems within itself. It might also be a case of the meter needle dragging on the meter glass. The best check of this possibility is to substitute a known good meter for comparison of readings.

When the Measured Voltage Is Too High

If the voltmeter in Figure 5-16 reads a voltage that is too high, there are five possible reasons why this might be so (mostly common-sense reversals of the reasons for low voltage):

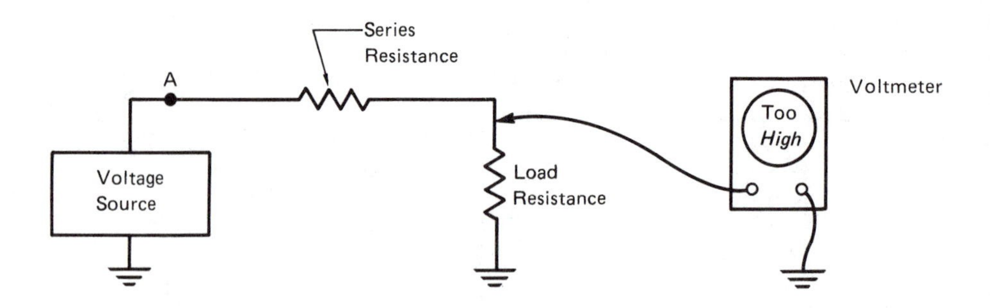

FIGURE 5-16 Analyzing cause of a voltage that is too *high*.

1. The source itself is too high in voltage.
2. The resistance across (parallel with) the voltmeter probes is too high.
3. The resistance back toward the source is too low.
4. The voltmeter is *not* loading the circuit down enough.
5. The voltmeter is defective.

Let's examine each of these. Possibilities 1, 2, and 3 again warrant careful attention for the same reasons as previously mentioned. Possibility 1 is that the source itself is too high in voltage. Check the source for an abnormally high voltage. A good voltage reading at the source eliminates Possibilities 1 and 5.

Possibility 2 is relevant when the circuit resistance *parallel* with the meter, R_1 (across the meter probes), has *increased*. This is typically the case where a circuit has an open or reduced load on the circuitry.

Possibility 3 is that a resistance in *series* with the meter (looking back toward the voltage source) has *decreased* in resistance. R_S of the circuit represents any resistance in series external to the source. Practically speaking, this is often a voltage source filtering resistor or a voltage regulator's pass transistor shorting collector to emitter.

Possibility 4 should be considered when using a digital voltmeter or FET amplifying meter, with its inherent high-impedance, across high-impedance circuits. There are still a few schematics around that specify the use of a lower impedance meter for taking voltage readings in fairly high-impedance circuits. In this case, the readings given depend upon a certain amount of loading by the meter. Use of a better meter will read a truer voltage, but higher than specified on the schematic.

Possibility 5 is that the meter is giving readings that are too high because of problems within itself. Again, the best check of this possibility is to substitute a known good meter for comparison of readings.

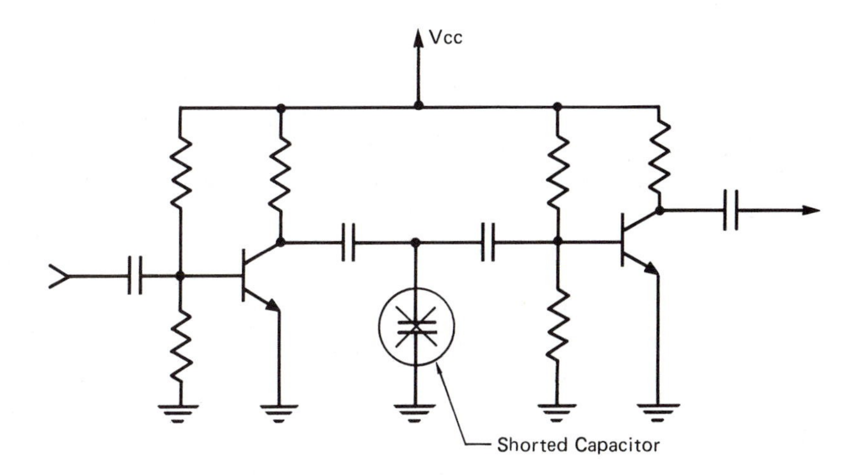

FIGURE 5-17 Example of circuit defect that DC voltage readings can't detect. No DC voltages would be affected, but the circuit would not amplify signals.

These principles of voltage troubleshooting apply to AC and other signal tracing methods as well. Missing signals, for instance, may be considered as voltages that are too low—possibly indicating a short to ground where the probe is sampling or an open component back toward the source.

In troubleshooting with the voltmeter most problems may be located, but not all of them. A circuit problem such as shown in Figure 5-17 will not respond to troubleshooting with a voltmeter. It can be found easily by using an oscilloscope, however. This is covered in the next chapter.

OPENING THE EQUIPMENT

Always unplug equipment before removing the cabinet or case. It is very easy for the case to contact dangerous voltages when being removed. Taking equipment out of its case is no challenge so long as it is done slowly, watching for wiring that can get caught and damaged by pulling. Speakers mounted to the case are an example of wiring that can be torn loose in this process. Good equipment comes with small plugs on such items to allow them to be disconnected during servicing.

Keep in mind when disconnecting wiring that if the parts needed for repair are not available, it may be some time before the equipment is reassembled. If there is any chance at all that wires might not be placed on the same terminals from which they came, write down on a piece of paper the layout of the wiring and where it connects. Keep these notes with the equipment.

When removing screws from equipment, be careful to note the length of each as it is removed. Some screws may need to be shorter than others to prevent them from going too far into the equipment and possibly damaging or shorting something. The best way to save the bolts and to be sure they are returned to the proper holes is to take off the covers and immediately put the screws back into the holes from which they came. This is a very good idea if the equipment may not be reassembled for some time. A second-best alternative is to put the bolts into a small container that can be kept with the equipment. The worst thing that can be done with loose screws is to lay them somewhere on the workbench. At least one will find its way onto the floor and will be lost.

TIMELY SHORTCUTS

Once the cover is off the equipment, it is quite possible that the problem can be found quickly by a very close visual inspection. Look for burned or broken components or anything else that is out of the ordinary, such as loose or broken wiring and poor soldering joints. If there are plastic ICs on the boards, look closely at the center of each for raised dimples indicating overheated and probably shorted ICs.

Besides a close look at the board, the smell of burned components may lead to the problem. Resistors and transformers have their own characteristic smell when burned, and the technician will come to recognize them.

SUMMARY

This chapter covers in detail the basic principles of troubleshooting using only the voltmeter. A thorough knowledge of the principles of this chapter will enable a technician to troubleshoot almost every power-supply problem.

REVIEW QUESTIONS

1. Where is the black voltmeter lead usually connected?
2. Name a quick way to tell if a VOM is loading a circuit.
3. What kind of voltmeter is easiest to use when accuracy is important?
4. What is the difference between the AC and DC switch settings of the vertical amplifier input to an oscilloscope?
5. What is the rule when connecting an ammeter or milliammeter to a circuit under test?
6. What is the name for a small-value resistor inserted into a circuit for the purpose of measuring current?
7. How much current is flowing through a resistor of 0.1 ohm if there is a 0.4-volt drop across it?
8. Generally speaking, how should a capacitor "look" to DC voltages?
9. If two resistors of 100 ohms and 1000 ohms are in series across 100 volts, what *approximate* voltage would be across the 100-ohm resistor? Do not use a calculator, only a mental estimate.
10. Name five reasons why a voltage might be too *low*.
11. What should be done before removing equipment covers?
12. What is the first step in troubleshooting once equipment covers are removed?

chapter six

Fixing the Power Supply

CHAPTER OVERVIEW

Most failures in electronics involve the power supply. The power supply produces heat, and heat contributes to failures. When a load fails by shorting, it is the power supply that takes the brunt of the punishment.

It is important when working with power supplies to have on hand power diode specifications. A good reference is Motorola's *Rectifiers and Zener Diodes,* #DL125. Another is Motorola's *Linear and Interface Integrated Circuits,* #DL128. This book covers regulator ICs and other chips used in power supplies. See Appendix III for ordering information.

FINDING THE OPEN FUSE

If there are only a few fuses in the equipment being repaired, checking them visually or with an ohmmeter may be the quickest way to test them. However, an open (blown) fuse may or may not "look" bad. If there are many fuses involved, it may be quite a chore to find the open one. Depending upon whether power is available, there are two methods that can be used.

First, if the main power can be applied, an open fuse might be revealed by a built-in indicator, usually a neon lamp across the fuse. A lit lamp means voltage across the fuse. Since the fuse should be a short, a lit lamp means the fuse is open. If there are no indicators provided, an open fuse may be detected quite readily by using a voltmeter (*not* an ohmmeter) across the fuses while power is applied. The meter must be set to the proper scale for the voltages involved with the fuses. The load normally fed by the fuse *must* be connected and turned ON (see Figure 6-1). This method of finding an open fuse is particularly useful for vehicular installations where the power is not easily turned off without disconnecting the battery. To find the bad fuse in an automobile that has a 12-volt system, for instance, just turn on the device that has no power and measure the voltage across each fuse until you see an indication of about 12-V DC. That fuse is open, or the holder is not making contact with the fuse.

Second, if power cannot be applied for one reason or another, the open fuse may be located by using an ohmmeter or a solid-state tester. However, *be sure* the power is completely off! Use the lowest scales, "RX1" or "Low," of either instrument.

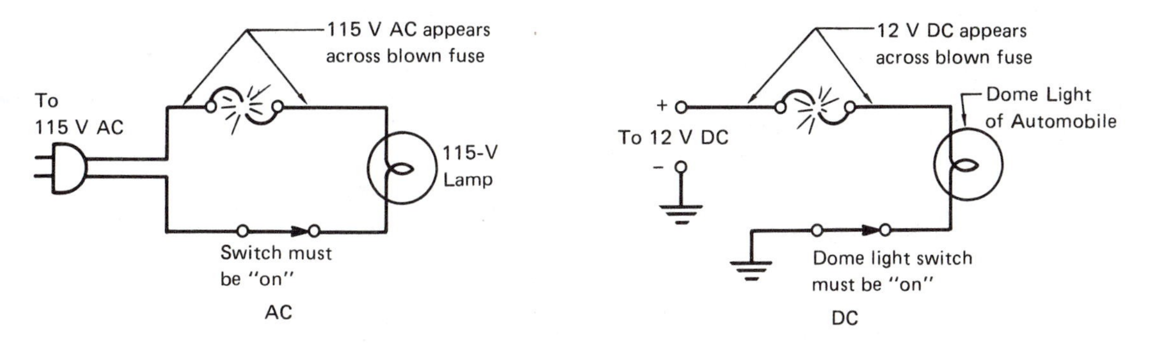

FIGURE 6-1 Finding a blown fuse with power "on" using appropriate voltmeter.

ANALYZING FUSE FAILURES

Fuses can open in normal operation. Current surges can progressively weaken them, particularly when turning on equipment. The inrush of current may weaken a fuse to the point that it eventually fails, usually when the equipment is turned on that one last time. This kind of failure is more or less normal and is corrected by putting in another fuse of the *same* rating. Look at the fuse that is removed. Assuming it is clear glass, you can get a clue as to why it failed by close examination. If the fuse element (the tiny wire inside) is simply broken or has been melted into tiny globules inside, then the chances are good that the fuse simply got tired as explained above. If, on the other hand, the fuse has vaporized and coated the inside of the glass with a silvery mirrorlike deposit, then there is a gross overload and replacing the fuse without finding the cause is wasteful.

If the fuse has blown violently, there is a good chance that the problem can be found with the use of an ohmmeter without the application of power. Using an ohmmeter or solid-state tester if you have access to one, check the power-supply filters for short circuits. Test also for shorts across the load terminals, and test each of the rectifiers for shorts.

Don't install a fuse of higher rating unless you are willing to gamble on the probability of burning up something for the lack of proper fuse protection. Also, use a fuse of the same type. The slow-blowing fuses are made to absorb current surges. A fast-blow fuse will likely burn open immediately in a slow-blow fuse circuit. A slow-blow fuse installed in a fast-blow circuit will allow very high, short-term current surges and may allow circuit damage. *General rule:* Replace with the identical type and current rating as recommended by the equipment manufacturer. The voltage rating of a fuse is an indication of the voltage under which it can operate without exploding if there is a short-circuiting current flow. It is also an indicator of the insulation available end to end when the fuse blows.

A small possibility exists that the fuse holder itself may be causing the fuse to open. This can happen in high-current applications if the fuse holder does not make solid contact with the fuse. The passage of sufficient current can cause the holder-to-fuse resistance to heat. This heat can be conducted into the fuse and, in combination with the small amount of normal heat within the fuse, cause the fuse to open. Be sure the holder makes good contact, especially in high-current applications. Keep the fuse clips clean and tight on the fuse ends. If the holder is loose it can often be resprung for a good fit by removing the fuse and squeezing the clips together a bit.

THE POWER-SUPPLY BLOCK DIAGRAM

Power supplies have reasonably standardized block diagrams. Since the technician will spend a good deal of time repairing power supplies, the block diagrams must be learned thoroughly (Figure 6-2).

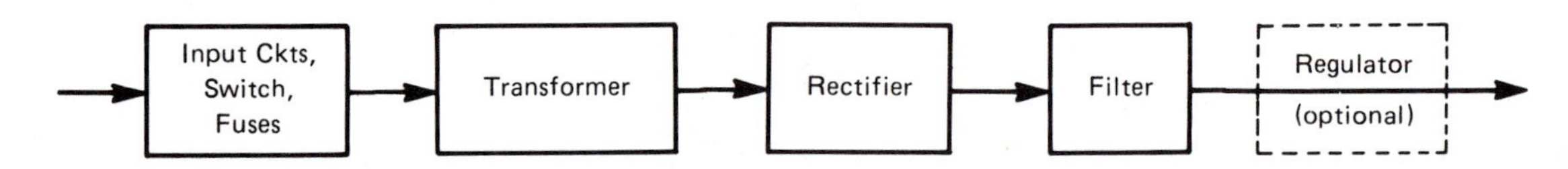

FIGURE 6-2 Block diagram of typical power supply.

Input Circuits

Power-supply input circuits accomplish one basic purpose: They provide either a sine wave or a square wave to a transformer. They consist mainly of a fuse, switch, and a transformer in the cases of the AC-operated power supply and the DC/DC inverters. In the inverter circuit, the input power is a DC source that must be chopped up into a square wave before application to a transformer (Figure 6-3).

Transformer

The power-supply transformer performs up to three functions: It provides any needed increase or decrease in voltage over that applied to its primary; it provides DC isolation between the primary voltage source and the secondaries, and it may additionally provide several output voltages simultaneously.

During any dead-circuit troubleshooting, the technician should remember that the windings of power transformers have a very low resistance and will appear to be shorted so far as the ohmmeter indicates. This is a good reason to prefer the live-circuit, output-voltage testing of power transformers. Transformers rarely fail by themselves. They more often fail because of some other circuit failure that overloads them, such as a shorted rectifier or power-supply filter capacitor.

Rectifiers

The power-supply rectifiers conduct at the proper times to convert the pulsations of current from the transformer into one-direction output current. In voltage multipliers they conduct at such times as to add the voltage across

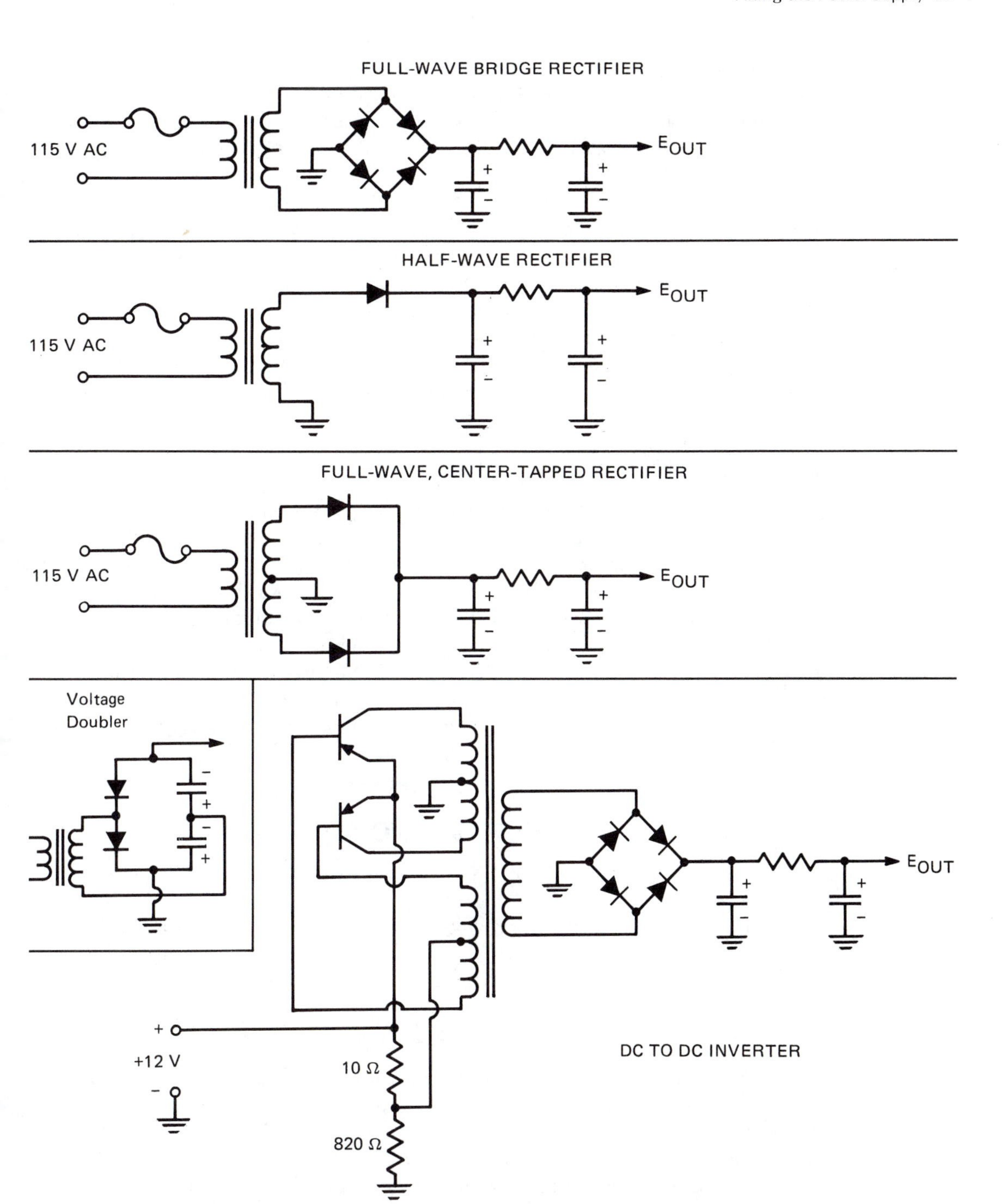

FIGURE 6-3 Power-supply circuits.

multiple capacitors connected in series. The shorted rectifier diode will place a short across the transformer secondary at least half the time, depending upon the circuit used. This will usually cause the power transformer to overheat and may cause the transformer's eventual failure if a fuse does not open to protect the circuit from damage.

A voltage doubler, tripler, or other multiplier power supply will contain many diodes. Troubleshooting them is done by applying power and testing for the proper output voltage. If it is too low, and the source voltage is correct, then each diode may be checked by using an AC voltmeter across the diode. An open diode will indicate too high a voltage, and a shorted one no voltage at all. In a similar manner, the capacitors may be tested for shorts. A capacitor without a voltage across it may be shorted. The circuit should be turned off and the capacitors checked for shorts or leakage using the procedures given in Chapter 16.

Filter Capacitors

Power-supply filters smooth out the DC pulsations that come from the rectifiers. These capacitors are easily identified as they are almost always the largest capacitors within the equipment. An electrolytic capacitor should show no signs of overheating. If it does, immediately suspect a short circuit within it. Electrolytics often fail by shorting. Occasionally they can fail by drying out and becoming an open circuit. In this case there will be a large AC voltage across the capacitor, which would be very low if the capacitor were operating properly. Incidently, electrolytic capacitors can blow up quite violently if internally shorted and power is relentlessly applied to them. Keep your eyes away from possibly shorted capacitors or wear good eye protection.

APPLYING INPUT POWER TO THE POWER SUPPLY

When performing the following tests, be ready to quickly disconnect the input power to the equipment upon the first indications of smoke, heating of any components (be careful not to get burned—thermally or electrically—by an innocent-looking component), or any unusual sounds such as crackling or frying.

Power Supplies Operating from DC Power Sources

These power supplies are operated from low-voltage DC sources such as the 12 volts used in vehicle applications. They do not require transformers and rectifiers as do the AC power supplies. DC power supplies are simpler and easier to troubleshoot. Figure 6-4 shows an example of such a power supply. Before applying power to this kind of power supply, be absolutely certain that you will be applying power with the proper polarity! A great deal of damage can be done, particularly if there is no reverse polarity protection diode (Figure 6-4). Be sure to use a proper fuse to protect the equipment in the event of a short.

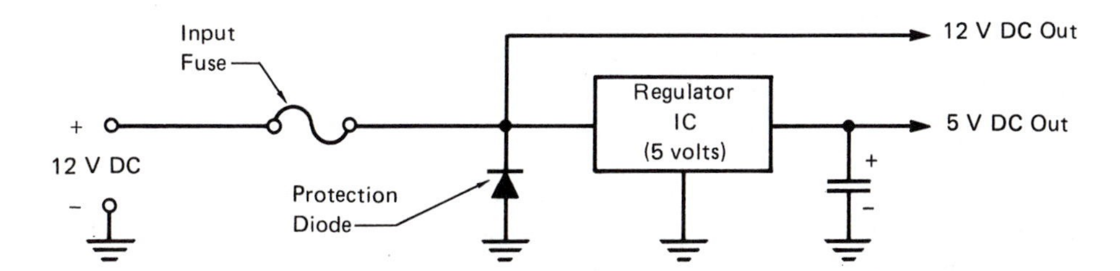

FIGURE 6-4 Typical 12-V DC power input circuitry with one regulator and input protection diode to blow fuse if input polarity is applied in reverse of normal.

If one is available, a variable DC power supply should be used to "ease up" to the proper input voltage. The technician should watch closely the current being supplied during this procedure and should stop immediately if there are any problems. Before using this alternative method of applying power, be aware of the *crowbar circuit.*

The crowbar circuit is designed to deliberately short and blow the input fuse if the supply voltage exceeds a preset maximum or falls to a preset minimum voltage. This circuitry is used on some special equipment that could be damaged if the input voltage were to go too high or fall too low. If the equipment has a similar circuit set to blow the input fuse when the input falls too low, the fuse will blow every time you bring the input voltage up slowly from zero. You will have to begin with the full voltage for which the circuit was designed. See Figure 6-5.

Troubleshooting the DC to DC Inverter Power Supply

This kind of power supply was in very common use when 12-volt vehicle battery voltage had to be increased to high voltages to operate vacuum tubes. It is now used in some digital applications to produce voltages other than those

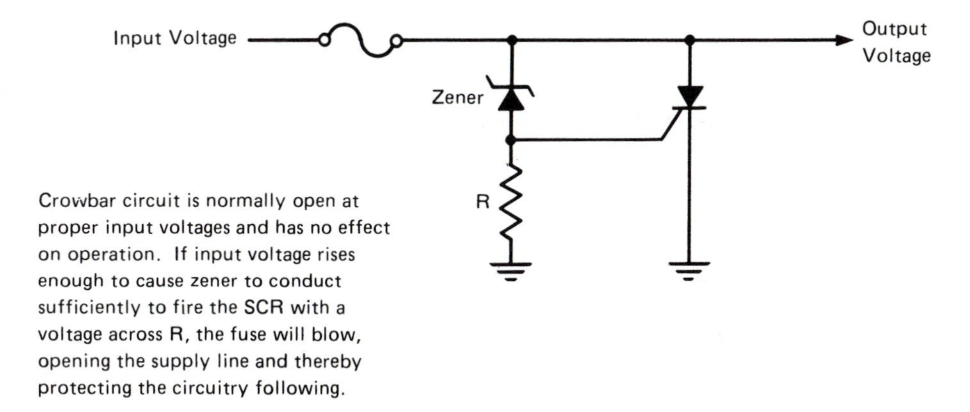

FIGURE 6-5 Sample crowbar circuit.

provided directly from a main power supply. For instance, a computer application could have a main power supply of +5 volts, and from this source other voltages and polarities such as −5, +12, or −12 volts could be generated.

In a discrete transistor power DC to DC inverter, the input DC is applied to a pair of transistors, where it is chopped into a square wave that is then applied to a heavy winding on the power transformer. The two transistors are a free-running push-pull oscillator. Another winding provides the positive feedback for oscillation.

Troubleshooting the power DC/DC inverter begins with eliminating the transistor side of the inverter as a possible problem. If the power supply is severely loaded, there may not be sufficient feedback for the transistors to oscillate. This gives the power supply an inherent current-limiting feature. Overloading must be eliminated as a possibility by disconnecting the load on the transformer output and substituting a suitable load to absorb the output of the transformer for a few moments, long enough to verify whether or not the transistors and associated circuitry are working. For instance, if the DC to DC inverter is rated at an output of 150 V DC and 100 watts, a suitable load would be a 220-V 100-W light bulb. If the output is 24 V DC at 300 watts, a resistor of about 3 ohms and 250 watts would be a good load. The load does not necessarily have to operate the inverter at 100 percent load, but should load to at least 50 percent of full load to ensure that the inverter is capable of reasonable output voltage. Figure 6-6 shows this principle.

If the transistor side of the inverter is defective, the transistors should be checked first as the most likely problem. (Chapter 16 describes how to test these

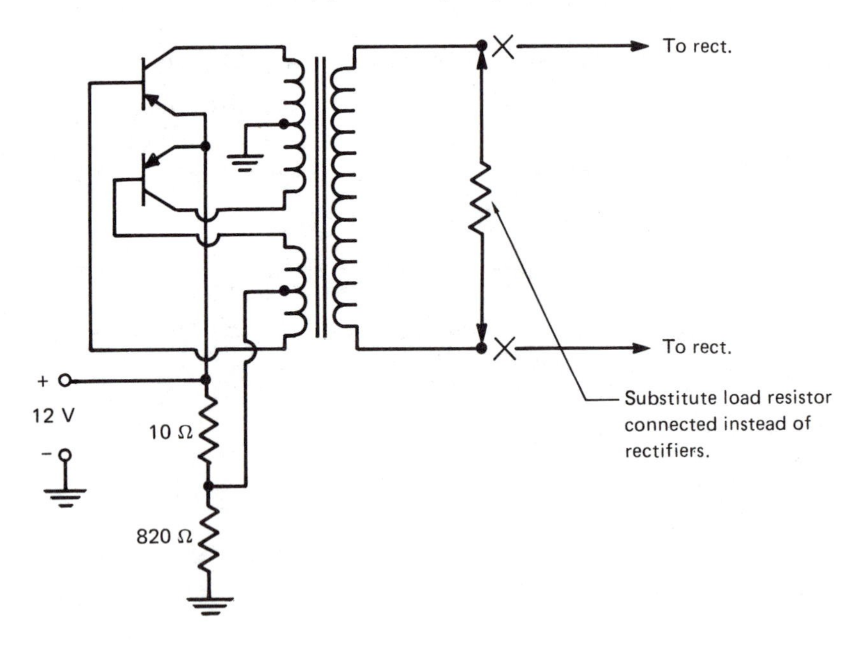

FIGURE 6-6 Isolating problems in the DC-to-DC inverter to the transistor oscillator section.

components in a dead circuit.) If, however, the transistors apparently drive the transformer well, troubleshoot the power-supply output circuits as explained in the sections on troubleshooting an overloaded (shorted) power supply (page 81) or troubleshooting an inoperative power supply (page 82), whichever is appropriate.

AC Power Supplies

Applying power to a defective AC-operated power supply can be done in one of several ways.

First, plug in the power supply and supply it with fuses while troubleshooting. The shorted power supply will blow the fuses as long as the overload exists, or it may simply "smoke out" the problem. This is the least desirable method and one that can be very stressful on the equipment, but one that can work under conditions where the technician has no variable transformer or test lamp to use. When using the fuse-blowing technique, it is best to disconnect circuits one at a time beginning from the left, or input end, of the circuit and proceeding to the right. As each stage is eliminated as the possible cause, it is reconnected, the next stage disconnected, and power reapplied. This procedure continues through the power supply until the fuse blows. The circuit just connected into the circuit is the one that is shorted. See Figure 6-7 for an explanation of this procedure.

Using a variable transformer is another and preferable method of applying power to equipment having a shorted AC power supply. After each stage is disconnected for elimination as the possible overload, the input voltage is brought up slowly and the current monitored. There will be much less current drawn when the defective component is disconnected. See Figure 6-8 for an illustration of this method.

Instead of using an AC ammeter to monitor the current being drawn by the power supply, a series 115-V AC lamp (on 115-V AC input lines) can be used as a

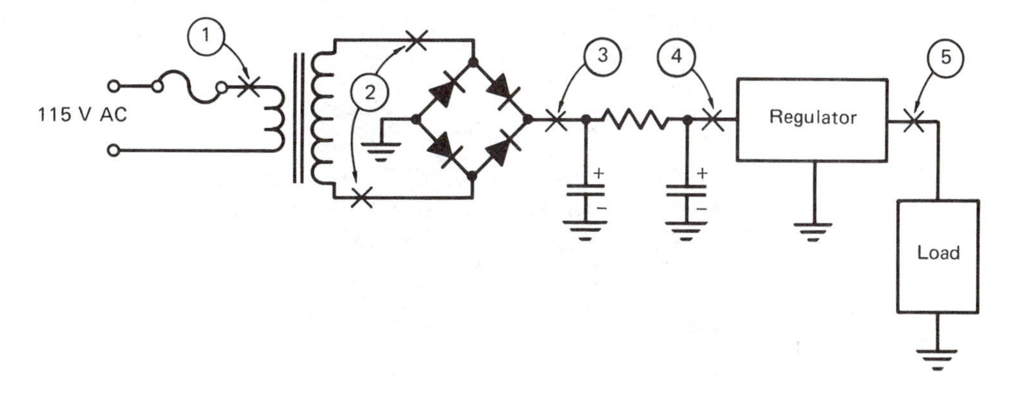

FIGURE 6-7 Finding short in power supply without a voltmeter.

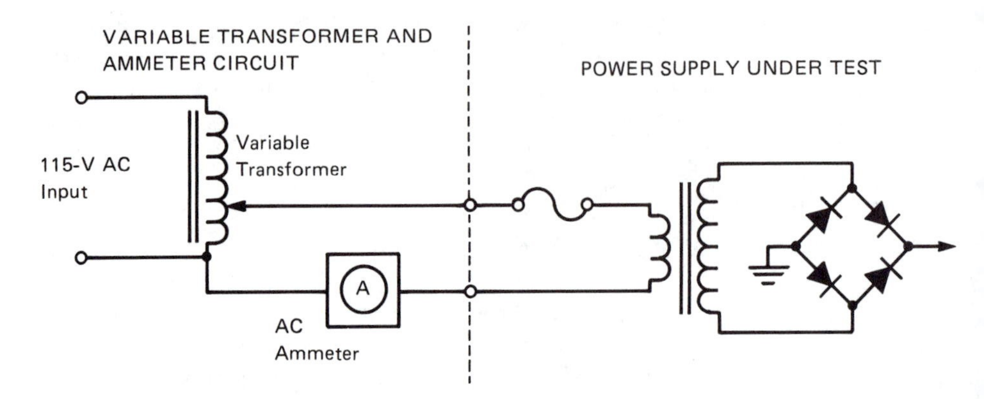

FIGURE 6-8 "Easing up" the input voltage to a shorted power supply to avoid further damage and enable further troubleshooting.

'soft fuse." If there is a short in the power supply, the current simply lights the lamp rather than damaging the equipment under test. The lamp should have a rating of approximately the power input of the equipment. Thus, if you are troubleshooting an audio amplifier of 100 watts input operating from the 115-V AC line, use a 100-watt lamp in series. This is not a hard-and-fast rule, as there is considerable leeway. Using a lamp of twice this rating will allow more current to the equipment under test if you so desire, and a smaller lamp will reduce the maximum possible current flow to less than the larger lamp. The use of this "soft fuse" makes it possible to apply the full line voltage without the use of a variable transformer.

An alternative to using the soft fuse is to monitor the current as the voltage is brought up with the variable transformer, using a clamp-on ammeter (Figure 6-9).

FIGURE 6-9 Using a clamp-on ammeter for troubleshooting AC power distribution problems. (Courtesy of Simpson Electric Company, Elgin, Illinois)

TROUBLESHOOTING AN OVERLOADED (SHORTED) POWER SUPPLY

The following procedure in tracing a short in a power supply is basic to the very nature of troubleshooting. Split the circuit in the middle and see if the problem is still there. Depending on the outcome of this test, the decision is made as to which way to go on another half-split, narrowing the problem down to a fourth of the equipment, and then again to an eighth, and so on, until the defective component is found. This principle should be understood thoroughly.

A shorted or overloaded power supply is evident by one or more of the following identifying clues: blown fuses, a louder than normal hum coming from the power supply, any overheated power-supply components, smoke from anything in the equipment, and sometimes small crackling or frying sounds. If the technician is using a "soft fuse" as described above, bright lighting of the lamp indicates a shorted power supply.

The first step in localizing the problem is to remove the load on the power supply. This is conveniently done in some circuits by pulling out all of the printed circuit boards except those in the power supply itself. If the overload is still present, the field for consideration is considerably reduced. For the following discussion, see the schematic diagram of a typical power supply in Figure 6-10.

Apply increasing power to the power supply until the overload is evident. Turn off the supply. Break the circuit between the rectifier and the filter, at point A. This step breaks the circuit into two parts, the left and the right, and makes a break in about the center of the circuitry. This is called the half-splitting method of troubleshooting and is the fastest method to use.

Apply power again and see if the overload is still present. If it is still overloaded, either the rectifier or the transformer is causing the problem. The short cannot be in any of the components to the right of point A, because they have been eliminated by breaking the circuit. Let us for a moment assume that the overload is still there. Turn off the power. Reconnect point A, and break the

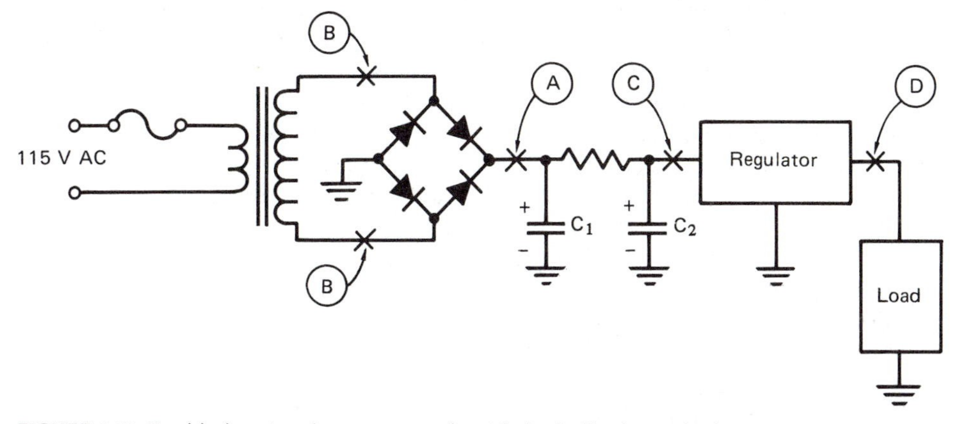

FIGURE 6-10 Troubleshooting the power supply with the half-split method. Note that the order of isolation points is different from Figure 6-7.

circuit between the rectifier and the transformer, at point B. Reapply power and see if the overload is still there. If it is, the transformer is shorted; if not, the rectifier is shorted. Be sure you understand this simple principle, as it is recommended throughout the rest of the troubleshooting in this book.

If the overload was gone back when point A was first broken, then the overload had to be caused by something that was disconnected, a component to the right of point A. This might be the filter, the regulator, or the load itself. Reconnect point A and break the circuit at point C. Reapply power and see if the circuit is overloaded. If it is, filter capacitor C_1 or C_2 is shorted, and if not, either the regulator or the load is shorted. If you have determined that the overload is in the regulator or the load, turn off the power and reconnect point C. Break the circuit at point D and make the overload test again. If the overload is gone, the circuits beyond the power supply are overloading it; if the overload is still there, the regulator is causing the problem.

If the equipment is constructed so that the load may be disconnected, such as with circuit board cards, these may be removed to verify that the overload can be cleared. If the overload is in the circuitry supplied by the power supply, and if the circuitry is a printed circuit board, Chapter 14 might be consulted now as a shortcut.

TROUBLESHOOTING AN INOPERATIVE POWER SUPPLY (NO OUTPUT VOLTAGE)

Some failures are evident by the fact that the power supply has little or no voltage output. Be aware of the current-limiting circuitry covered later in this chapter. The output of a current-limiting power supply will seem to indicate that the power supply is defective when it is really just idling along, preventing any damage by reducing its voltage output to a safe level.

If the power supply has no voltage output at all, check the fuse carefully. The best way to do this, with the power applied and the power switch of the equipment "on," is to use the appropriate voltage and function of the voltmeter and see if there is full supply voltage across the primary fuse. For a power supply operating from the 115-V AC line, use the 300-V AC scale of the voltmeter (or the next highest range from 115-V AC) and test for voltage directly across the fuse clip connectors. If the fuse is blown, the full line voltage will show across the fuse. In a similar manner for a 12-V DC system, test for the supply voltage across the fuse with an appropriate scale to measure the 12-V DC voltage. A good fuse will have negligible voltage drop across it.

The next step in troubleshooting an inoperative AC-operated power supply is to see if the power transformer has any AC output voltage from the secondary winding. If it does, next check the rectifiers for a proper DC output; if it does not, then the transformer may be open. This can be checked by verifying that 115-V AC is present at the input of the transformer, measured as physically close to the transformer primary winding as possible. If there is input voltage and no output, it is conclusive proof that either the primary or the secondary winding is open. Do not discount the possibility that the transformer was burned opened because of an overload. Before replacing a power transformer be sure to test the rectifiers and filters for shorts. If the rectifiers of an AC-operated supply have an AC input voltage but no output DC, they are probably open. Again, give thought to *why* they might have failed. They may have opened due to a short circuit farther down the circuit. This possibility can be checked with an ohmmeter as covered in Chapter 16.

TROUBLESHOOTING THE POWER-SUPPLY REGULATOR

Regulators are commonly used in electronic equipment. They come in two general types, series and shunt (Figure 6-11). The series regulator is, for all its apparent complexity, a variable series resistor between the power supply output and the load. This series "resistor" may be a power transistor (in this case it is called a *pass transistor*) or a tube. It may also be a power transistor contained in a regulator IC. The shunt type "pulls down" the power supply voltage to a specified voltage.

Some regulators have additional circuits that sense the output current and reduce the output voltage if the current demand exceeds a certain preset limit. These current limiters can cause confusing symptoms if the technician is not aware of their presence. The best way to troubleshoot such a power supply is to substitute an acceptable load on the regulator instead of the circuits normally operating from it. If the regulator output voltage comes up to normal across the substitute load resistor, then the circuits are loading the regulator to the point that the current-limiting circuit reduces the output voltage. If, however, the power supply still does not produce the proper output voltage, then the regulator or something previous to it is defective (Figure 6-12).

A short to ground through a regulator will produce the same symptoms as a shorted filter capacitor (page 76). An open regulator will have a good voltage at

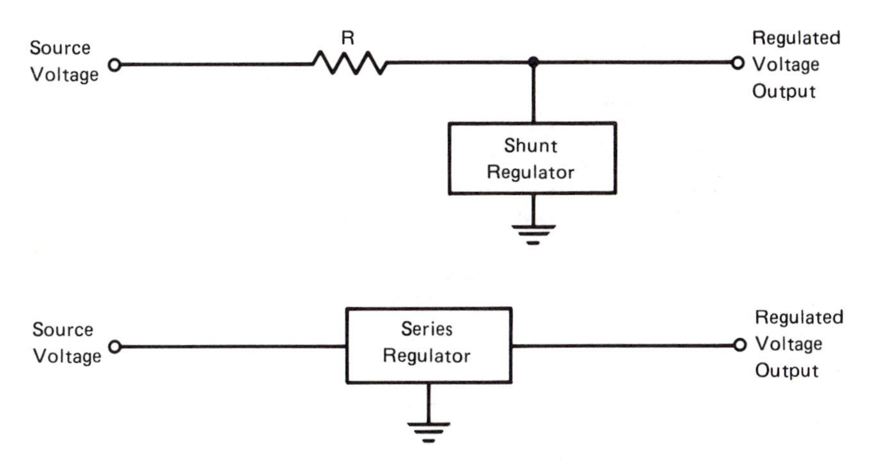

FIGURE 6-11 Block diagrams of shunt and series regulators.

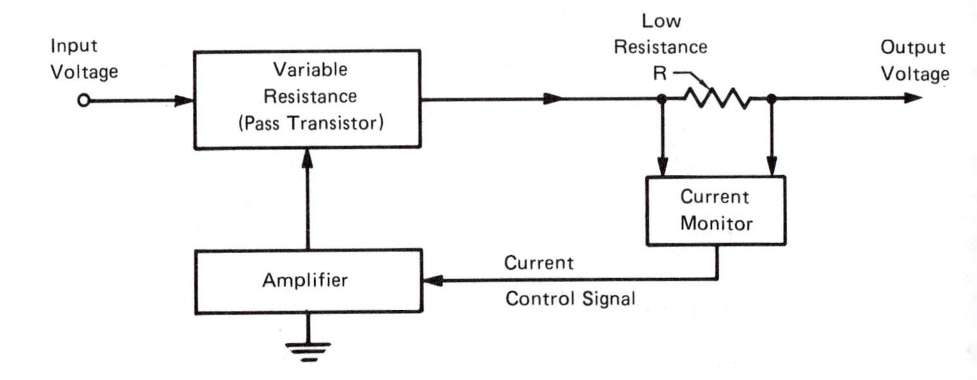

FIGURE 6-12 Simplified current-limit regulator.

its input and little or no output voltage. Some voltage regulators can be made to fail if they are overloaded. Those regulators that have no internal current-limiting features are prone to failing by this means. An open regulator should suggest to the technician to check the load for shorts or overloads.

The Series Regulator

The series regulator works on a simple principle. By sampling the output voltage and comparing it against a reference, the resistance of a series element is varied to maintain a constant voltage. Amplification makes the output very stable and accurate at a specified voltage (Figure 6-13). The first step in analyzing failures of series regulators is to measure the input and output voltages. If

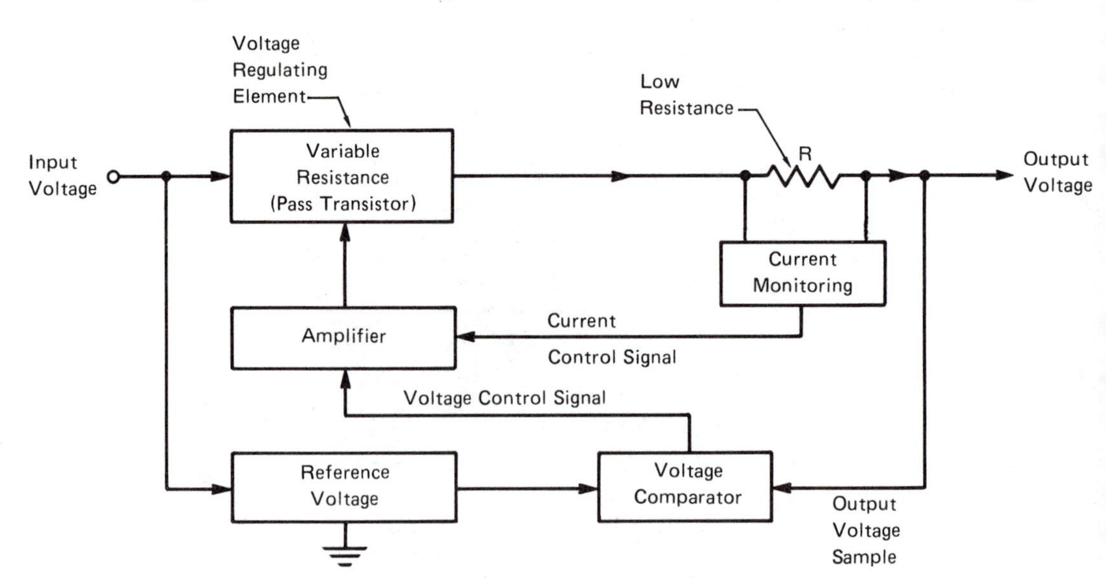

FIGURE 6-13 Block diagram of voltage regulator and current regulator circuits.

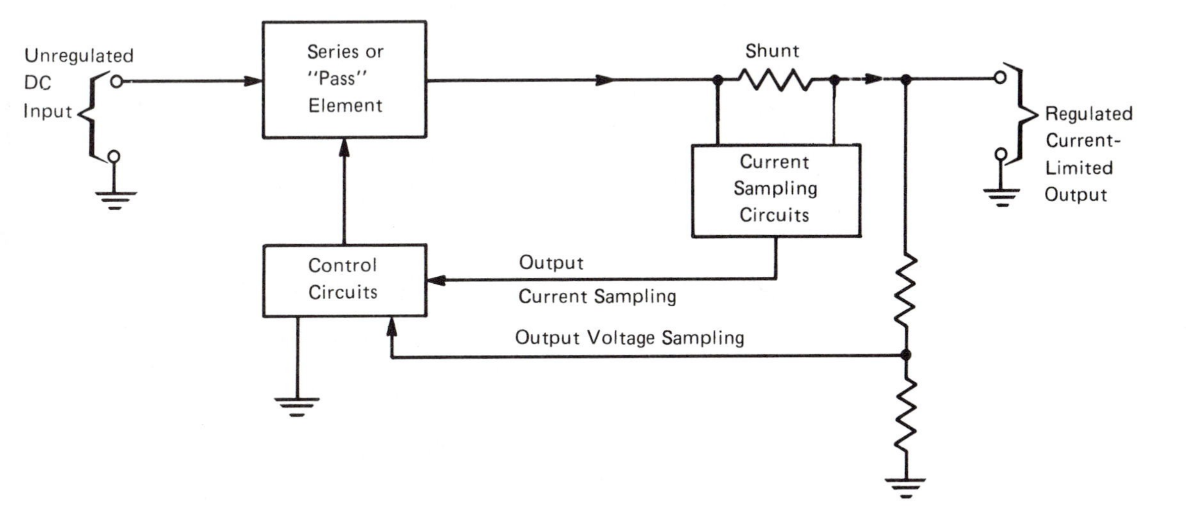

FIGURE 6-14 Block diagram of voltage-regulated supply with current-limiting feature.

the output voltage is too high, the regulator is defective. If the input voltage is correct but the output voltage is too low, the regulator may or may not be doing its job.

Before assuming that the regulator is defective, the possibility of current limiting should be considered. Some regulators sample the output current and reduce the regulator output voltage if the current drawn is above a threshold value (Figure 6-14). There is another variation of the current-limiting circuit called a foldback circuit. This circuit will reduce both the regulator voltage and the output current to very low values if the regulator has seen an overload. This essentially turns off the power supply so that no further damage will occur, yet leaves enough current flowing to sample the overload. If the overload is removed from a foldback power supply, the output voltage recovers (Figure 6-15).

Troubleshooting the Series Regulator

The presence of either the foldback or the current-limiting features may make a regulator seem defective when the real problem is an overload. These circuits will make it necessary to break the circuit between the regulator and the load to ascertain which is causing the problem. Open the circuit, apply power, and measure the output voltage of the regulator. It may be necessary to connect a reasonable load to prevent oscillations of the regulator. Be sure that any load attached does not overload the regulator, or the limiting circuits will keep the output voltage low.

Series regulators most often fail by opening or shorting the series element. The IC, transistor, or tube is often required to throw off quite a bit of heat, and heat will take its toll. A check of defective series regulator circuitry should include checking the reference voltage and checking for the proper bias on each

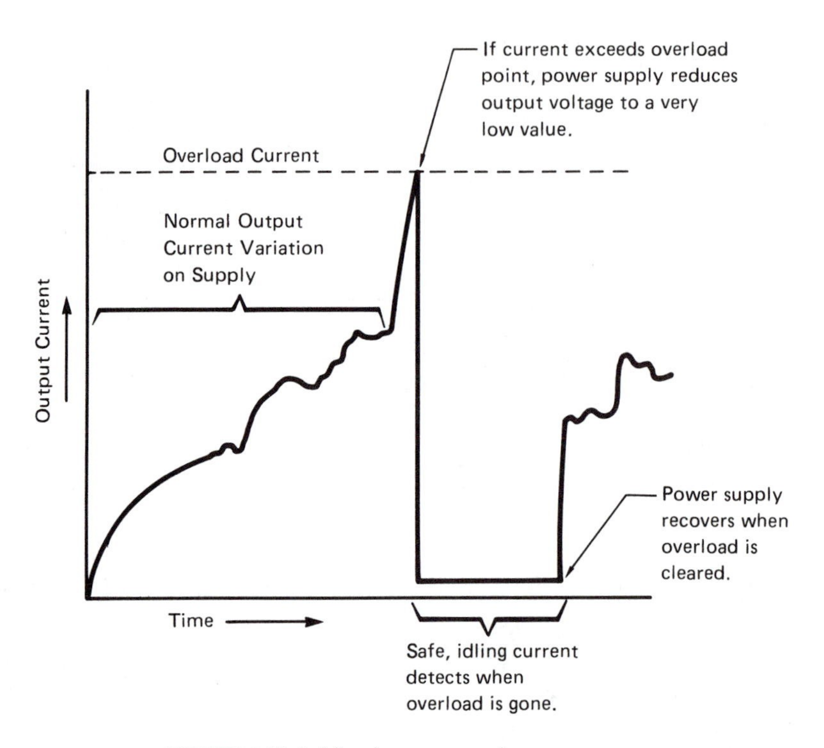

FIGURE 6-15 Foldback power-supply output curve.

of the discrete components. See Chapter 9 for discrete series regulators, Chapter 10 for regulator IC chips, and Chapter 20 for vacuum-tube series regulators.

A rare possibility that may cause power-supply problems is an oscillation either in the regulator or in other outside circuitry that is influencing the output of the power supply. Only an oscilloscope connected to the power supply will detect this rare defect. It is caused by a lack of sufficient filter capacity and probably indicates that the output filter capacitors connected to the regulator are open.

The Switching Regulator

Varying the resistance of a series transistor is one way to regulate the output of a power supply; turning the transistor full-on and full-off at a very rapid rate accomplishes the same thing—control of the output voltage. A large capacitor across the output of the transistor averages out the pulsations to a DC level. This method has an advantage over the variable-resistance method—the series transistor runs much cooler, since it is either on or off, never in-between. The regulator is therefore more efficient, wasting very little power as compared to the variable-resistance method.

The switching regulator must turn the series transistor on and off at a rate high enough to allow easy filtering. An oscilloscope will show the switching waveforms (Figure 6-16).

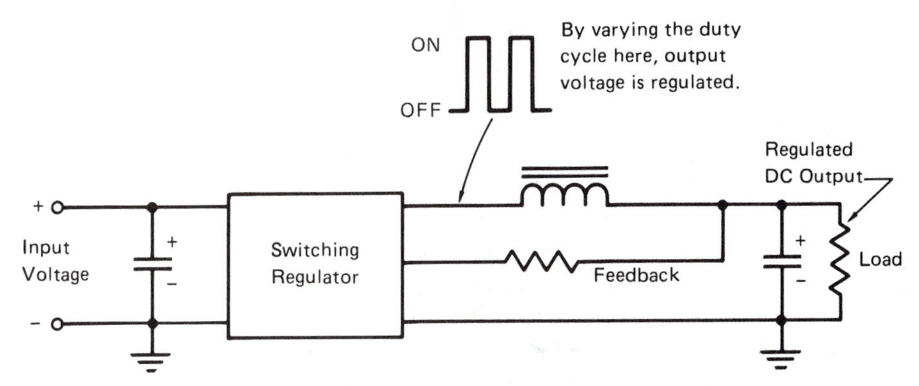

FIGURE 6-16 Switching regulator.

The Shunt Regulator

The shunt regulator maintains a constant output voltage by varying a load on a voltage source through a fixed series resistance (Figure 6-11).

Troubleshooting the Shunt Regulator

The first step in analyzing failures of this regulator is to measure the input and output voltages. If the output voltage is too high, the regulator is defective. If the input voltage is correct but the output voltage is too low, the regulator is probably all right, but the load is demanding too much current for the regulator to operate. This would effectively remove the shunt regulator element from the circuit. Shunt regulators most often fail by opening or shorting of the shunt element. This is covered in Chapter 9 for discrete shunt regulators and Chapter 20 for vacuum-tube shunt regulators.

BATTERIES AS A POWER SOURCE

Batteries have some real disadvantages. They don't last forever. They are heavy. But they are with us to stay, so the technician should be familiar with how they fail.

Dry Cells

Dry cells eventually develop a high internal series resistance so that they are no longer capable of supplying the necessary current to the load. They have to be discarded, and the sooner the better because they can physically deteriorate and leak corrosive liquids that can severely damage electronic equipment. Most readers have seen this kind of damage in flashlights or small portable radios.

Ni-Cads

Nickel-cadmium (ni-cad) batteries are a distinct improvement over the primary (nonrechargeable) batteries in many ways. Ni-cads can be recharged many times and with proper care can be used for years without problems. In most applications that use these batteries, there is a human element—the batteries have to be put on charge and later taken off charge after a reasonable length of time. Lack of charging of course means that the battery is not ready for full service, if at all. Overcharging, on the other hand, reduces their service life. The more severe the overcharging, the less the battery will give in service life. Too high a charging current and too long a period of time on charge are both factors in the severity of charging. Consult the manufacturer of the equipment or of the cells in use for the proper charging current and charge time for full charge. Ni-cads should be charged at the maximum rate of one-tenth the ampere-hour rating of the cells. For example, if a ni-cad of flashlight size is rated at 4 ampere-hours, the maximum charging rate would be 400 milliamperes for a period of 14 hours. This overcharges 40 percent to allow for the inefficiency of the charge/discharge cycle.

Ni-cad batteries will sometimes fail by shorting internally. This is usually caused by allowing a bank of cells in series to discharge to the point where a weak cell will begin to have current flow through it in the wrong direction. This is very hard on ni-cad cells and should be avoided. A shorted cell can sometimes be more or less reclaimed by "zapping" it with a strong current in the proper charging direction. The cell then apparently works properly and will accept a charge. However, keep in mind that for maximum reliability in critical applications such cells should be replaced rather than returned to service.

Ni-cads are available in small vented forms that appear similar to common flashlight-variety cells. These cells are usually vented with a small hole in the end to prevent explosion if the cell is charged at too high a rate, when the cell may internally produce gasses too rapidly for them to be reabsorbed within the cell.

Ni-cads are also available in a wet type of construction, much like the common lead-acid battery of automotive use. The electrolyte is a strong alkali, liquid caustic potash, which will react extremely violently if mixed with acid. For this reason, the maintenance facilities for ni-cads should be *completely isolated* from similar facilities for lead-acid cells. Wet ni-cad batteries should be maintained by keeping the connections tight (be extremely careful not to short any of the cells with a wrench!); the connectors and other fittings may be kept reasonably free of corrosion by applying a light coating of oil to them. Distilled water should be added only when the cells are fully charged, to a level predetermined by the manufacturer. The reason for filling only when charging is that the cells then produce gasses that force liquid out of the plates. Filling at any other time may result in overflow of the electrolyte when the cell reaches full charge.

Ni-cads, like lead-acid cells, produce hydrogen and oxygen gasses while charging. Since these two gasses are explosive in combination, keep all flames away from charging cells.

Lead-Acid Cells

Lead-acid cells (car-battery types) are not often used in electronics because of the corrosive fumes and liquids in them. As a matter of maintenance troubleshooting, the technician should remember that a reading of 1260 or above on a hydrometer indicates a good battery. This number is influenced by temperature, however. A temperature correction should be applied when the battery is very cold or unusually warm. These cells require a level of electrolyte that covers the plates. The full capacity of the cell is reduced as the electrolyte is allowed to drop, and permanent damage to the plates will result if they are allowed to dry.

Gel-cells are a variation of the lead-acid cells, but have the electrolyte suspended in a gelatinous state rather than as a free-flowing liquid. This helps prevent some of the spilling and corrosion problems. Specific maintenance instructions should be closely adhered to as to charging and discharging rate.

SUMMARY

This chapter has dealt with troubleshooting the power supply, the most common failure in electronic equipment. Once the power supply is again functioning normally, additional problems may be found by tracing signals through the equipment. If the power supply was the only problem, the reader is referred to Chapter 17 for parts replacement and verification.

REVIEW QUESTIONS

1. When testing for blown fuses with a voltmeter, how should the power switches be set to make the test valid?
2. Under what conditions may a fuse be replaced with one of higher rating?
3. Draw a schematic of a bridge rectifier operated with a transformer from 120-V AC, complete with filter. Be sure diodes are drawn with proper direction of the schematic arrow and indicate the output polarity.
4. Name two advantages of using a "soft fuse" in testing a shorted power supply.
5. Briefly, what does the crowbar circuit do if it is tripped?
6. What is the electrical mechanism whereby batteries fail to deliver enough current, yet their terminal voltage without load is okay?
7. How can an internally shorted ni-cad battery sometimes be saved, at least for noncritical use?
8. Which wet-cell type, the lead-acid or the ni-cad, produces explosive gases?

chapter seven

How to Trace Signals

CHAPTER OVERVIEW

This chapter deals with the systematic narrowing down of a problem within a piece of equipment or on a printed circuit board to the stage that is causing the problem. The completely random, uneducated approach to troubleshooting is called "shotgunning." Just stand back and make a wild guess. Replace parts until the failure symptoms disappear. As a troubleshooting method, this is useless. The experienced technician who is thoroughly familiar with the equipment to be repaired can often make a very accurate appraisal of a failure by simply noting all the symptoms and comparing them to problems and cures in the past. This is not shotgunning—it is the result of much experience and consists of matching the symptoms with a large catalog of failures within the technician's memory. It is this form of troubleshooting that is the fastest and that can make the most money for the technician and the employer.

TROUBLESHOOTING DIGITAL CIRCUITS

If the equipment under repair is a digital board, it is appropriate to proceed now to Chapter 13, where specific troubleshooting for this special kind of electronics is explained. The concepts and instruments used with digital circuits are quite different from those used in analog troubleshooting.

TROUBLESHOOTING ANALOG EQUIPMENT

There are five basic kinds of analog circuits. Recognizing them will help give you an overall picture of the circuitry (Figure 7-1). The first kind of circuit is the *originating* circuit. From the energy of the power supply, the originating circuit generates a signal. Oscillators of all sorts come under this classification. Signal tracing these circuits is relatively easy. All you have to do is see if the circuit is producing the signal it is supposed to, usually with an oscilloscope.

The second type of circuit is the *processing* circuit. This may be an amplifier of some sort, but it can also be a loss-producing (passive) stage such as a crystal filter. In a processing circuit a signal is applied to the input and the output is examined to see if the signal is coming out the other end as it

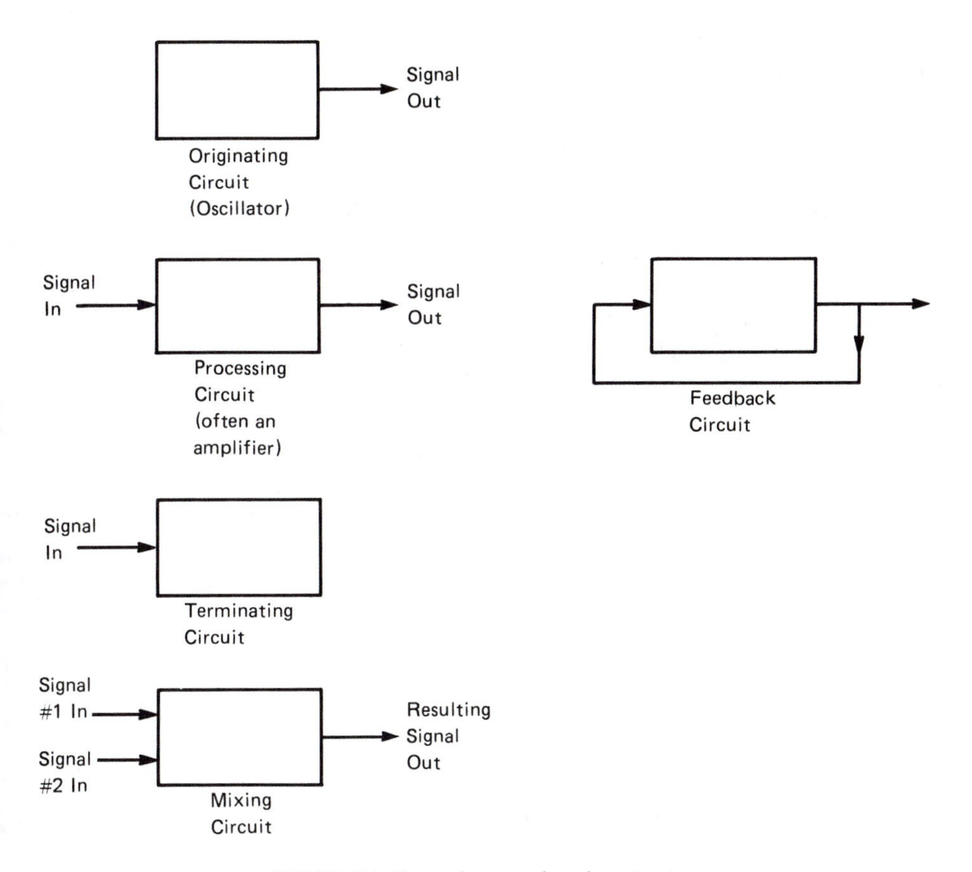

FIGURE 7-1 General types of analog circuits.

should. An example is an audio amplifier stage where a small signal goes into the input and comes out the other end larger in amplitude. In this case you would apply a signal to the input of an audio amplifier and look at the output with an oscilloscope.

The third circuit is the *terminating* circuit or the end-effect stage. A TV picture tube might be considered such a stage. The result of all of the previous processing is the terminating stage. With a proper input test signal, one should see or hear the proper output, often without connecting any special load instrument on the output.

The fourth type is the *mixing* or converging stage. Two or more signals are combined to produce a single output. A radio-frequency mixer stage is a good example of such a circuit. When tracing signals through these stages, the technician must remember that it takes both signals on the input to produce the proper output. The absence of one signal produces no valid output.

The last type of circuit is the *looped* or *feedback* circuit. The output of one or more stages is fed back into an input again. Circuits such as voltage regulators and the phase-locked loops of frequency synthesizers are good examples. This type of circuit can be difficult to repair because a change in one circuit will cause changes in the other circuits involved, some quite far from the failure. Often the best way to troubleshoot a looped circuit is to break the loop to prevent feedback, then troubleshoot the circuits as processing stages.

THE BLOCK DIAGRAM—FINDING THE BAD AREA

You should find a block diagram somewhere in the equipment instruction book. Occasionally the manufacturer will assume that everyone knows all about their product and will be able to troubleshoot it without one. For common equipment such as 27-MHz transceivers or AM radios, a block diagram for other, similar equipment may often be used. Special built-in features such as frequency synthesizers will mean departures from standard block diagrams.

The block diagram details the function of each stage in the overall design of the equipment. In the block diagram of a simple radio receiver, each stage has a specific function and accomplishes one necessary step toward the final output (Figure 7-2). The block diagram of the AM receiver is shown because of the reader's familiarity with it and its controls. The principles to follow will apply to any piece of equipment. By using the block diagram of the AM receiver and learning the *concept* of how to localize a problem to an area using nothing more than the front-panel controls, one can then apply the same concept to any other equipment. The intelligent application of this concept is basic and must be learned well.

USING A BENCH POWER SUPPLY

Some equipment comes with self-contained power supplies such as batteries or AC-line-operated power supplies. In these cases you wouldn't need a

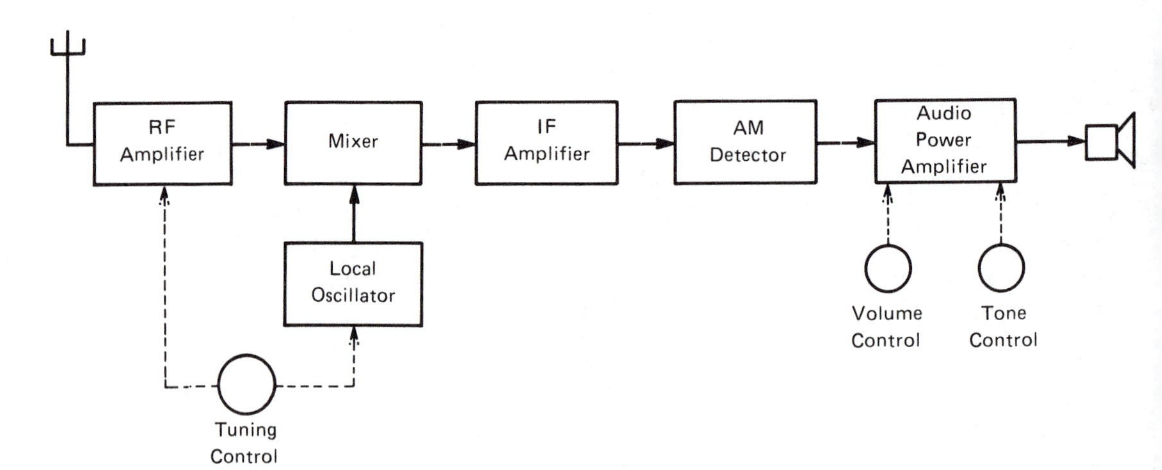

FIGURE 7-2 Block diagram of AM receiver.

bench power supply. But if you are asked to service equipment that does not contain a power supply, you will have to provide voltages at specified levels to particular points in the circuit to make it operate. Then you would require an appropriate bench power supply.

Some circuits require more than one source of voltage. Analog ICs may require both positive and negative supplies for operational amplifiers, for instance, and some digital circuits such as some computers require both 5 volts and 12 volts. In each case a separate external supply will be needed to meet each of the voltage requirements. Bench power supplies are of several different types:

1. unregulated fixed or variable voltage supply.
2. regulated fixed or variable voltage supply.
3. regulated fixed or variable voltage supply with fixed overcurrent protection.
4. regulated fixed or variable voltage supply with adjustable overcurrent protection.

Any of these supplies may be either fixed in voltage output or adjustable, depending on its design.

Unregulated fixed or variable voltage supplies can be used in testing cases where the current demand is constant and the power supply is able to supply the required current without overheating. A good example of this type of power supply is the small transformers designed to plug into the 115-V AC line right at the plug, and which have a small two-wire cord to a load such as a small tape recorder. These small units may put out either AC or DC, depending upon whether the manufacturer has placed the required rectifier at the transformer end or in the equipment itself. The output voltage of these units is not critical, and the load is relatively constant. Inexpensive battery chargers are also examples of unregulated power supplies. They usually don't even have a filter, letting the battery under charge receive pulses of DC direct from the rectifier.

The regulated fixed or variable voltage supply is an improvement over the unregulated supply because the output voltage is held relatively constant as the load increases—to a point. This point is the maximum current capability of the supply. Once the current demand exceeds the designed rating of the supply, the output voltage falls while the load may increase further. This overload current may or may not damage the power supply, depending upon other design features that monitor the amount of current demanded and take corrective action accordingly. Without extra circuitry to protect the power supply, it can be provided with a fuse to open the circuit to the load or to turn off the primary power to the supply.

The regulated fixed or variable voltage supply with fixed overcurrent protection incorporates circuits to monitor the amount of current drawn by the load. If the load current increases past a set amount, the voltage then decreases to whatever is necessary to hold the amount of current constant at the set point. This is an important point to understand. By reducing the voltage, the power supply is protected from current overload.

FIGURE 7-3 A high-quality bench power supply featuring digital readouts of both voltage and current. (Photo courtesy of Viz Manufacturing Co.)

An alternative way to protect the supply from current overloads is to turn off the supply output to the load if the load demand exceeds the design value. This type of power supply is called a *foldback* power supply. If the load is removed, the output voltage of the foldback power supply may recover by itself, or it may require resetting by a pushbutton or by turning off the primary power and turning the supply on again.

The most versatile power supply for test bench use is the regulated fixed or variable voltage supply with adjustable overcurrent protection. Figure 7-3 shows a power supply from Viz Manufacturing that incorporates these features to the best advantage for test bench use. Using this power supply as an example against which to compare other power supplies, it will provide an adjustable voltage up to 40 volts. Once the voltage is set with the "V" control on the left of the panel, the output voltage will remain at that value, regardless of the load current drawn, until the current reaches the value set by the "A" (amperage) control on the right side of the power supply. As the current demand increases beyond the set value, the voltage will decrease to hold the current at the level and no more.

This power supply will completely shut down the output if the overload is very large. Thus, a short circuit on the output causes the power supply to turn off the output voltage and wait for a "reset" button to be pressed.

A variable power-supply output voltage should be monitored, especially when setting up the initial values for operation. The better power supplies will provide output metering. This avoids tying up a test meter for such a routine chore. The Viz power supply has *two* digital readouts, one for the output *voltage* and one for monitoring the output *current*. Besides being very handy for routine power-supply work, both these readouts can be switched to electrically isolated front-panel connectors for use in monitoring voltages in external circuits. This is particularly handy if several voltages in the circuit under test need to be monitored at one time.

The details of setting up a voltage-regulated, adjustable-current power supply such as this may vary in detail, but generally the procedure involves the following steps:

Setting up the supply for resistive loads:

1. Determine the operating voltage and the maximum current demanded by the circuit under test.
2. With the output circuit of the supply *open,* turn the voltage control to a few volts. This step is necessary to have a voltage causing a current flow for the next step.
3. With the output circuit of the supply shorted, set the maximum current-limit value with the current control.
4. Remove the output short and adjust the voltage control to the proper value for the circuit under test.
5. Turn off the power supply or switch off the output circuit.
6. Connect the power supply to the circuit under test.
7. Turn on the supply. The circuit under test now has the proper amount of voltage applied and if there is a malfunction, the current-limiting feature of the power supply will protect the circuit from possible further damage.

A bench power supply can be put to use in many different applications including some component test procedures. They may also be used for special jobs that demand a bit more care than simply providing voltage sources for resistive loads. Charging batteries, for instance, requires particular care. *Even a momentarily reversed connection of a battery to a regulated power supply can severely damage it!* The electronic control circuits will not tolerate a forced reversal in polarity of the output terminals. Connect batteries for charging by connecting + to +, and − to −. The color codes do match, red to red, black to black.

There are two ways to charge batteries: with constant current or to a set voltage. Ni-cad batteries are best charged from a fully exhausted state by applying a charging current through them of one-tenth their ampere-hour rating for a period of 14 hours. Lead-acid cells, on the other hand, are often charged at heavy current until the final, fully charged voltage is reached, at which time the current is tapered off to maintain the set voltage at the battery terminals. The setup procedure when using a regulated fixed or variable voltage supply with adjustable overcurrent protection varies slightly.

Setting up the supply for constant-current *charging of batteries:*

1. Determine the fully charged battery voltage and the maximum charging current to be allowed.
2. With the output circuit of the supply open, turn the voltage control to a few volts. This step is necessary to have a voltage causing a current flow for the next step.
3. With the output circuit of the supply shorted, set the maximum charging current value with the current control.
4. Remove the output short and adjust the voltage control to the proper value for approximately 50 percent over the value of a fully charged battery.

5. Turn off the power supply or switch off the output circuit.
6. Connect the power supply to the battery to be charged. Be certain of polarity!
7. Turn on the supply. The battery will charge at the amount of current set by the current control. It is up to the operator to turn off the power supply in about 14 hours to prevent overcharging.

Setting up the supply for constant-voltage *charging of batteries:*

1. Determine the fully charged battery voltage and the maximum charging current to be allowed.
2. With the output circuit of the supply open, turn the voltage control to a few volts. This step is necessary to have a voltage causing a current flow for the next step.
3. With the output circuit of the supply shorted, set the maximum charging current value with the current control.
4. Remove the output short and adjust the voltage control for the exact final terminal voltage of a charged battery.
5. Turn off the power supply or switch off the output circuit.
6. Connect the power supply to the battery to be charged. Be certain of the proper polarity!
7. Turn on the supply. The maximum charging current will be supplied until the battery voltage reaches the final charged voltage, at which time the current will decrease to a trickle-charge amount, sustaining the battery at final full-charge voltage indefinitely without damage.

NARROWING THE PROBLEM TO AN AREA

Apply power to the equipment under test. If a fuse blows or for any other reason the equipment cannot be operated, the power supply should be examined and the problem located as covered in Chapter 6. Once the equipment can be turned on, proceed with the steps to follow.

Note on the block diagram exactly where the front-panel controls connect into the blocks. In our example of the AM radio, note that the volume control connects into the circuitry between the AM detector stage and the audio power output stage. Note too that the tuning control connects into two blocks at the same time (as shown by the dotted line). Varying the tuning control makes changes in both the RF amplifier stage and the local oscillator stage. With the equipment operating, or as nearly so as possible, vary each control and note the effect that they have on the receiver output, the speaker.

Sample Problems

For example, if in the example above the speaker was completely quiet and no amount of manipulation of the controls could produce any output at all, consider where the problem could be: Could it be in the antenna? Not likely, since there would be at least a little noise or hissing coming from the speaker when the radio was turned up to maximum volume. Could it be in the tuning sections? This is not likely, for the same reasons. Could the problem be in the detector stage? Again, not likely, but for a different reason. If the detector stage and everything past it were working well, there would be at least a tiny bit of scratching sound as the volume control was varied up and down.

The technician could at this point reasonably assume that either the final audio amplifier, the power supply, or the speaker was defective. By simple manipulation of two controls *and some logical thought,* the amount of circuitry to check in further detail has been reduced to perhaps one fourth of the total circuitry!

Let us consider different symptoms for the same radio: Assume that the speaker has hiss coming from it that varies up and down with the volume control, and that changing the tuning control has no effect on the hiss. Since there is a signal (noise) that responds to the setting of the volume control, it is reasonable to assume that the power supply and all of the circuits from the volume control to the right, including the speaker, are working acceptably well. Therefore, only the circuits to the left of the volume control are suspect. Now note the amount of hiss that is coming from the speaker when the volume is turned up high. Here is where a bit of technical judgment and familiarity with the equipment will come into play. If there is a "lot" of hiss, plenty of volume to it, it would be reasonable to assume that the IF amplifiers are doing their thing in providing plenty of gain. If there is very little hiss, this may not be the case. For our example, we will assume that the hiss is quite high in volume, and that the trouble must therefore lie in the RF or oscillator circuits. At this point this is all that we can do without going farther into the circuitry.

It should be evident from the above examples that it is important for the technician to use the block diagram and the front-panel controls to at least make an educated quess as to where the failure might be. This procedure of using the front-panel controls is sometimes called "milking the front panel." It can save a great deal of time if done properly with a bit of logical analysis of what the front-panel controls can reveal about the problem.

These principles can be used with other equipment. When milking the front panel for symptoms, the technician can make good use of any installed indicators such as meters, LED and lamp indicators, and so forth. These are connected to specific points within the circuits and can tell a great deal about what is happening at that particular point in response to the various front-panel controls.

NARROWING THE PROBLEM TO A STAGE

Now that we know the approximate area to troubleshoot, the next step is to verify that what we think is working really is, then find the defective stage within our suspected area. For these purposes, we will need to apply signals into the circuitry and/or verify that signals generated within the circuits are being processed as they should. This procedure is called signal tracing.

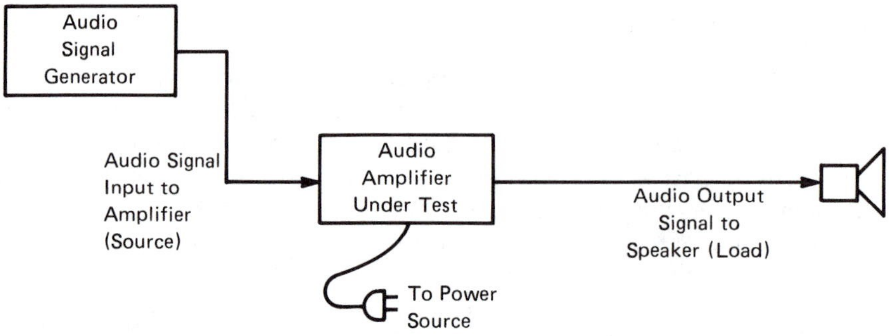

FIGURE 7-4 Application of signal to suspected equipment.

SIGNAL-TRACING CONSIDERATIONS

To illustrate what is meant by signal tracing, see Figure 7-4. A proper signal is being applied at the input of the amplifier. The output of the amplifier already has a load installed—the speaker—so we do not need to resort to the use of an oscilloscope or other "load" instrument. The speaker will do the job of informing us as to whether or not the circuit will pass a signal.

With a known good input (also called signal injection, source, or stimulus), we should be able to get an expected output (also called detection, load, or response). This is the basic idea of signal tracing. In the example in Figure 7-4, we should hear a clear tone from the speaker when the input signal from the generator is touched to the input of the circuit. If we do not, then something is wrong within the amplifier circuits.

It is important for the input signal to be compatible with the circuit. Applying radio frequencies to an audio amplifier would not produce a reasonable output because the circuit and the input signal are not compatible. The same is true for the output detector—whether it is a speaker, oscilloscope, or any other detecting instrument—to indicate whether there is a proper output. A speaker wouldn't respond to radio frequencies, but an oscilloscope might do the job very well.

In the paragraphs to follow, *source* will be used to indicate the signal being applied to the input of a stage under test, and the word *load* will refer to the detector of the output signal, an instrument or component that will give the technician an indication of circuit operation.

Source Signals

Existing Signals. Source signals can often be the signals already applied to the equipment or generated within it, such as the signals coming in from an antenna to a receiver, or the oscillator signals generated within it. These are very convenient to use when available.

A Quick Trick. In some high-impedance audio circuits, audio can be traced through equipment by touching sensitive circuits with a finger (to be done with a great deal of caution in high-voltage circuits!). This applies the AC line voltage pickup of the technician's body to the circuit input.

Caution: Introducing a signal into a circuit must be done with some care. Many operating circuits will also have DC voltages present at the points of desired signal injection. Connecting a signal generator directly into such a circuit could damage the circuit or the generator due to the DC voltage flowing into it. It is sometimes necessary to provide a DC-blocking capacitor between the generator output lead and the circuit under test. For audio circuit testing a value of about 0.1 μF should be sufficient and for RF circuits, 0.01 μF. The voltage rating required should be about 50-V DC for most solid-state circuits and up to perhaps 600-V DC for vacuum-tube circuits.

Load Instruments

Existing Indicators. Load instruments are necessary to detect whether or not the circuitry is working as it should. Installed detectors can be speakers, lights, or cathode-ray tubes (CRTs), to name a few. If the circuits under test do not have a built-in load for troubleshooting, the technician must provide one. Oscilloscopes are probably the most common instrument in analog circuits, but a VOM, a DMM, or other instrument can be used, depending upon the frequencies involved.

It is a very good idea to be sure that the connection of a load instrument will not upset the DC levels of the circuit under test. If the VOM is being used on the AC function to trace signals, it will be a good idea to use the *output* function to block any DC, which will almost certainly give erroneous readings if not blocked. The output function inserts a capacitor in series with the meter test leads to block DC, allowing the reading of AC in the presence of a DC component. DMMs usually provide an internal blocking capacitor on the AC voltage ranges.

While signal tracing through equipment it is occasionally possible to mistake one output for another. For instance, when tracing signals through a receiver it is possible to see the local oscillator signal instead of the desired signal from a signal generator. To verify that you are seeing the result of the signal input in such a case, vary or turn off the input signal. The output signal should do the same if you are looking at the right signal.

TRACING THE UNWANTED SIGNAL

Unwanted signals such as hum and "squeal" in audio amplifiers and similarly unwanted signals in radio-frequency equipment can be easily traced to their sources by shorting out the signals at the input to the various stages. In discrete transistor amplifiers, shorting transistor bases to their emitters does no harm and quickly eliminates the signal from that transistor. If the unwanted signal ceases when an input is shorted, you have identified one path required for the signal. Work back through the various stages until shorting the input has no effect. The signal is being produced within the last stage that eliminated the signal.

Vacuum-tube amplifiers may be shorted from grid to ground without harm in most amplifiers. Be careful when shorting input signals in these circuits that the DC bias voltage for the tube is not also shorted out, as this may cause excessive current flow through the tube.

Integrated-circuit audio-input circuits may be shorted to ground using about a 0.1-μF capacitor. This will prevent shorting any DC voltage present on the chip input, yet will load any incoming signal almost as severely as a short circuit. In radio-frequency circuits, a capacitor of 0.01-μF should do quite well.

SUMMARY

This chapter has given the basics of signal tracing from which the technician may learn the concept of how to analyze equipment using front-panel controls, how to introduce a signal into a stage with a source instrument, and how to detect the output with a load instrument. Specific examples of how this is done will be covered in the chapters ahead, with audio and low-frequency equipment treated in Chapter 8, radio-frequency equipment in Chapter 11, and pulse-operated equipment in Chapter 12. Depending upon the type of equipment undergoing repair—discrete components, ICs, or vacuum tubes—the reader is referred to Chapters 9, 10, or 20.

REVIEW QUESTIONS

1. What is meant by "shotgunning"?
2. What is the generally accepted way of troubleshooting looped or feedback circuits?
3. You have just purchased an expensive bench power supply. It is voltage regulated and current protected. How could you possibly damage it by trying to trickle-charge a battery?
4. Assume you have a completely "dead" radio receiver. There are no visually defective components. What is the next troubleshooting step?
5. What is meant by "milking the front panel"?
6. What is the purpose of a "load" instrument in troubleshooting?
7. What is the purpose of a blocking capacitor?
8. What special technique may be used to trace unwanted signals?

chapter eight

Troubleshooting and Signal Tracing in Low-Frequency Circuits

CHAPTER OVERVIEW

This chapter will deal with tracing signals through low-frequency circuits to identify a defective stage. For the purposes of this book, low-frequency circuits process signals from DC up into the ultrasonic frequencies of roughly 20,000 hertz. An excellent reference book for the transistors used in these circuits is Motorola's *Small-Signal Transistor Data,* #DL126. It contains information on bipolar and FET transistors. See Appendix III for ordering information.

SOURCE INSTRUMENTS

It is faster to use existing signals than to connect an external instrument. When troubleshooting audio equipment, for instance, it may be easy to use a radio tuner for an input signal.

A very convenient input signal is the 60-hertz hum picked up by a technician's body. Place a screwdriver on a sensitive input circuit and monitor the output of the circuit for a hum. Touching the screwdriver shank with a finger may force a 60-hertz signal through the amplifier. This quick trick works if the impedance of the circuit is high enough that the body hum pickup is able to produce enough input to the circuit.

In a pinch, the output of the oscilloscope calibrator might be used as a convenient source of audio signals. The calibrator signal of most oscilloscopes is about a volt or less, a good amplitude for an input test signal. Be sure to use a series-blocking capacitor of about 0.1 μF to isolate the DC that may be present from either the calibrator or the circuit under test.

Another possibility, of course, is to use a suitable signal generator to produce the signal for tracing. The technician should set the generator for about a thousand cycles (the standard test frequency in the audio range). For routine signal tracing a DC-blocking capacitor should be placed in series with the "hot" lead before applying the generator output to the equipment.

LOAD INSTRUMENTS

Installed Loads

When servicing low-frequency and audio equipment, it is usually most convenient to use the installed output devices such as speakers to do the troubleshooting.

The Oscilloscope

If a signal must be observed before the speaker output, an oscilloscope is the instrument most often available. It is particularly good for seeing any distortion of a pure sine wave input signal (Figure 8-1). The input resistance at the front-panel connector of a high-quality oscilloscope is typically one megohm. This resistance is high enough that it will not affect any but the most critical low-frequency circuits. Direct connection to the input connector is seldom done, however. It is preferable to use an oscilloscope probe. The probe offers the advantage of even higher input resistance, most commonly 10 megohms in a 10X probe. The sensitivity of the vertical amplifier must be mentally *decreased* by one decimal place (move the decimal point one place to the right) to get the sensitivity at the probe tip. For example, a vertical sensitivity setting of .1 volt/centimeter will effectively be 1 volt/centimeter at the input to the 10X probe.

The Frequency Counter

A frequency counter should be used only when the frequency of the output waveform is of importance. It will not be useful for detecting varying levels of input amplitude, so its use is quite restricted for signal-tracing purposes.

The VOM as an Indicator on Low Frequencies

A VOM can be used as an indicator of AC signals. It will read AC-signal voltages and provide relative voltage indications, and it will detect the presence or absence of signals. When using the AC ranges of a VOM for signal tracing, the VOM will also be influenced by any DC present in the circuit. When using the

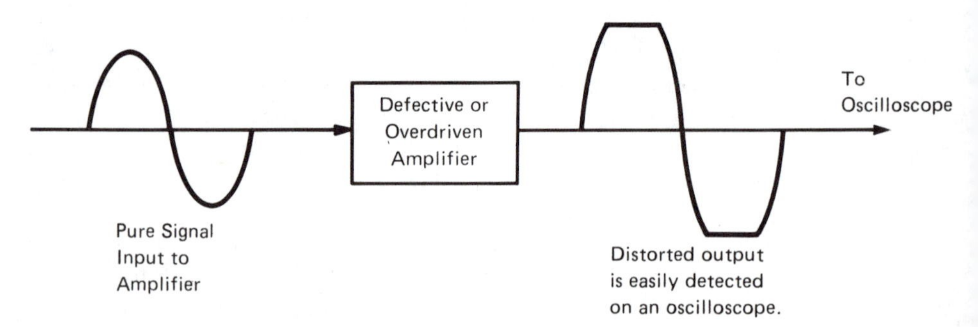

FIGURE 8-1 If distortion is severe enough, it can be observed on an oscilloscope.

VOM for tracing AC signals, the "output" function must be used. The output function places a DC-blocking capacitor in series with the positive test lead, thus preventing the DC portion of a composite voltage from showing on the meter.

Whenever a VOM is used on its lowest AC scale, the technician must remember to use a different scale on the meter face. The rectifiers within the VOM are diodes, which require a small voltage drop before they begin working. Because of this, the meter needle doesn't indicate voltage linearly near the bottom of the lowest AC voltage range (Figure 8-2). It is very easy to forget to read this scale when switching to the lowest AC range.

The needle of a VOM responds to the *average* value of an applied AC voltage. Average values are of no particular interest, so the scale behind the needle of a VOM is drawn to give rms readings *for a sine wave*. The VOM will not give accurate voltage readings for waveforms other than a sine wave. For most troubleshooting, the technician is interested in the presence or absence of signals or one voltage compared to another. Under these conditions this limitation on accuracy is not a problem.

The DMM as an Indicator on Low Frequencies

As a rule, a DMM is AC coupled on the AC voltage ranges and does not require a series capacitor to reject the DC portion of a composite AC plus DC voltage. Digital multimeters are sometimes made to respond to average values like a VOM. This is the easiest way to make them. The readings presented are skewed to indicate the rms value of a sine wave input. An average-responding DMM will give erroneous readings when used on anything other than a sine-wave input. See Chapter 12 for more information on this subject. Fluke Instruments makes at least one DMM that responds to the rms (heating effect) of a waveform. This would be the instrument to use on waveforms other than sine waves.

The DMM will indicate much smaller AC voltages than the VOM will show. This is because of the limitations of the VOM in the rectifier circuits and the lack of amplification necessary to show low-level signals.

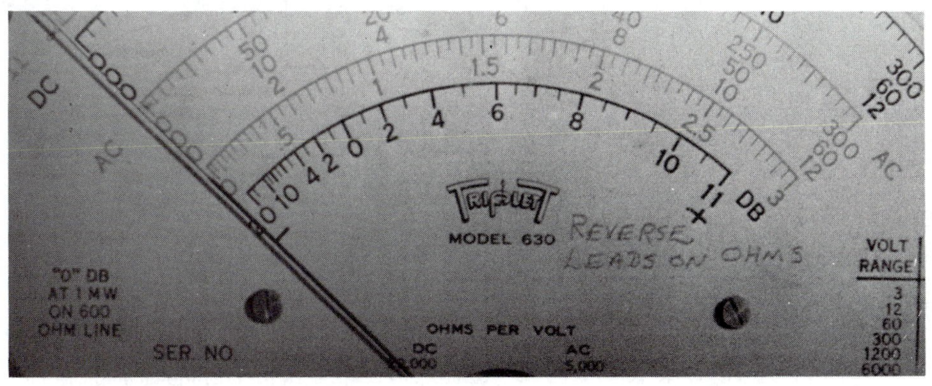

FIGURE 8-2 Lowest AC voltage range.

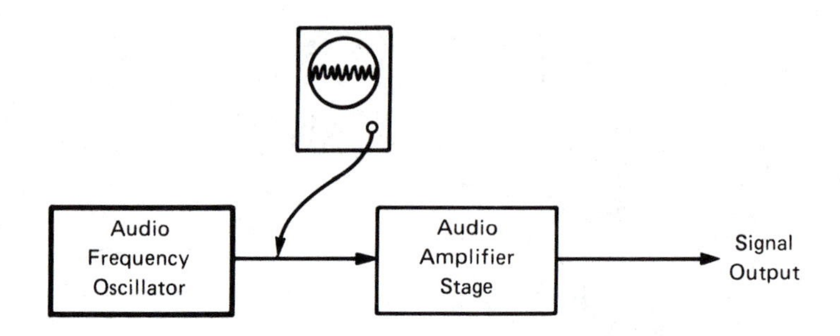

FIGURE 8-3 Testing the output of an originating circuit.

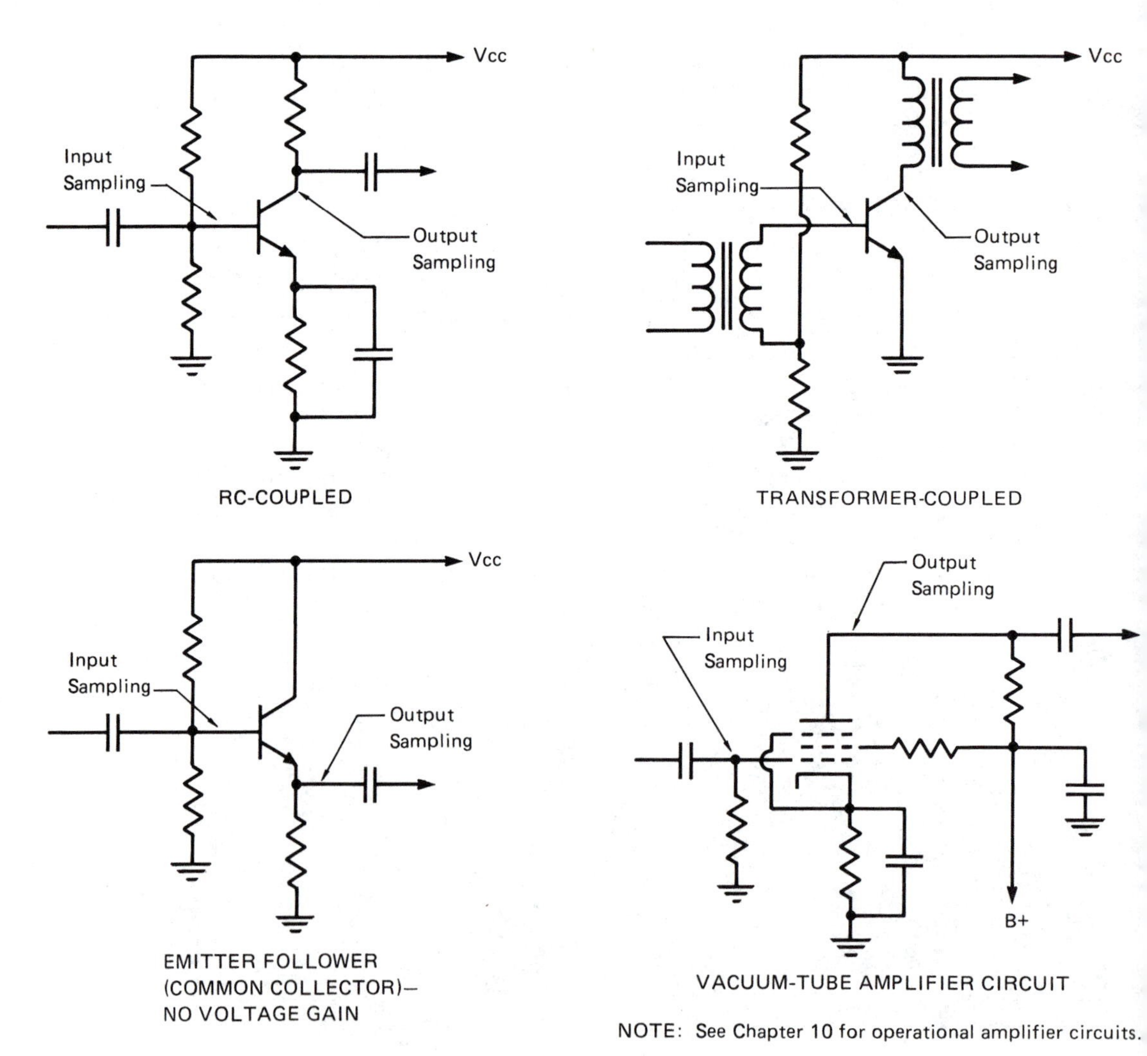

NOTE: See Chapter 10 for operational amplifier circuits.

FIGURE 8-4 Typical discrete low-frequency amplifiers and test points.

THE ORIGINATING LF CIRCUIT

The troubleshooting of the LF originating circuit is just about the simplest of all. This stage is an oscillator, and since there are no inputs and only one output, just look for an output with an oscilloscope (Figure 8-3).

THE PROCESSING LF CIRCUIT

This stage does something to the signal on its way through. Most often it is an amplifier; all that you need do is apply a signal of the proper frequency to the input and check for the proper signal out the output. An oscilloscope is about the handiest instrument to use for a load, and the signal from a fingertip is the quickest for an input (Figure 8-4).

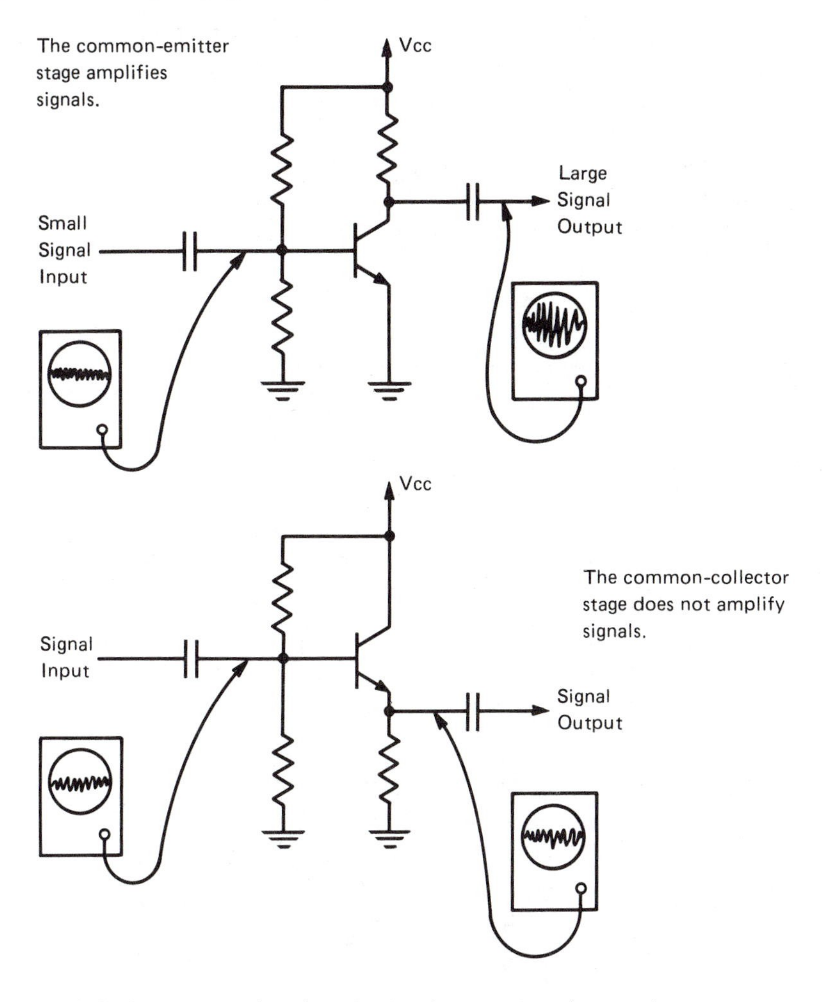

FIGURE 8-5 Examples of monitoring the operation of processing stages.

A few applications may use one of two special processing stages, the emitter-follower (common-collector circuit) or the common-base amplifier. These circuits have the transistor or tube amplifier "turned around" in the circuit. (See Figure 8-5 for examples of these stages.) Note that the emitter-follower stage does *not* provide any amplification of the input signal.

TRACING SIGNAL DISTORTION

The special case of tracing distortion through a stage bears some explanation. In this case the input waveform should be a pure sine wave. If the defective circuit is producing enough distortion that the technician can see the distortion on an oscilloscope, the stage causing the problem can be easily identified. If the distortion is difficult to see, then the technician may resort to using a special pattern called the Lissajous pattern. This pattern will show a clean diagonal line (or narrow loop if there is a little phase shift in the system) when the distortion is very low. The presence of appreciable distortion quickly becomes evident as the bending of one or both ends of the slanted line. See Figure 8-6 for the layout using this technique of distortion detection.

If the signal input to a stage is excessive, this will cause the stage to produce distortion. Be careful not to overdrive a stage while signal tracing. This can produce what looks like a major problem but is just too much input voltage from the signal generator.

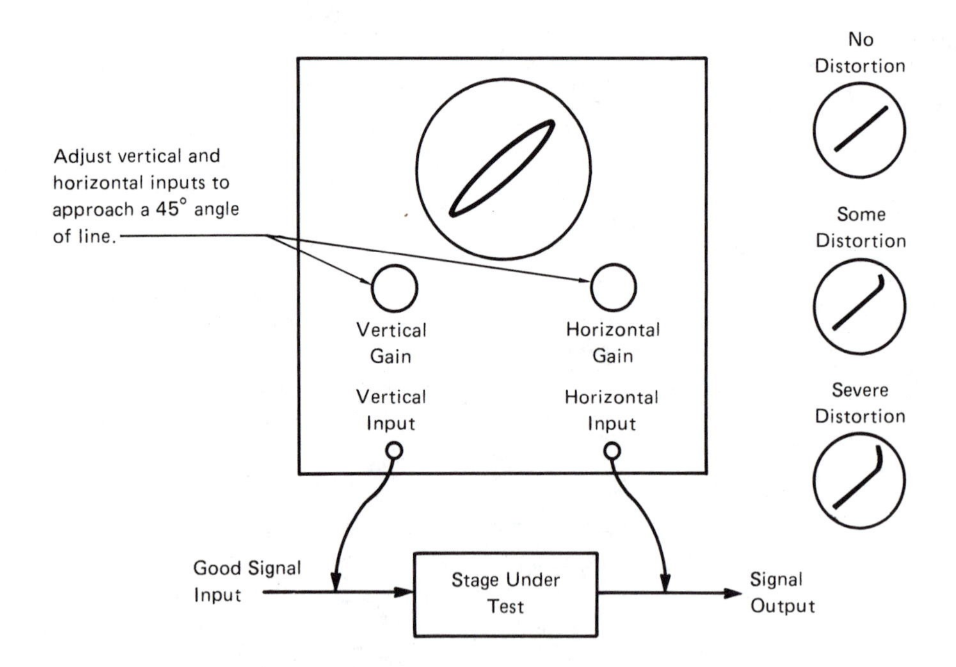

FIGURE 8-6 Setting up a Lissajous pattern to show distortion.

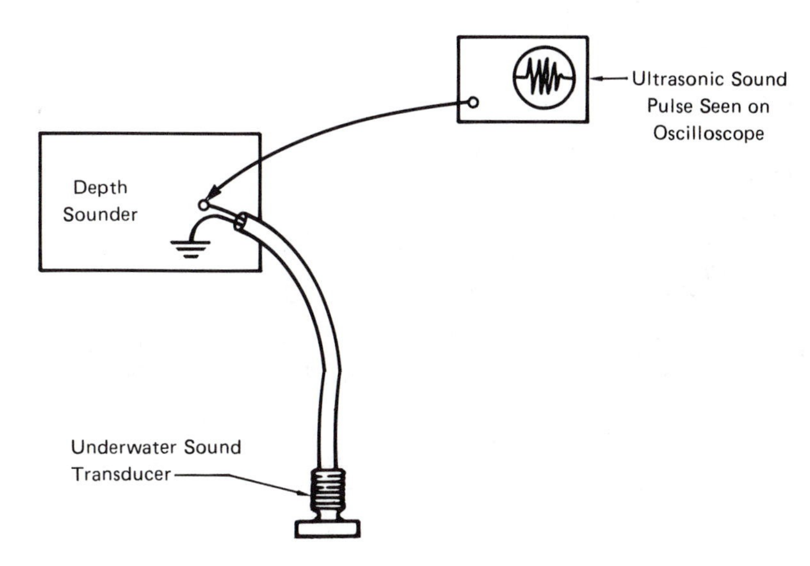

FIGURE 8-7 The oscilloscope may be used to monitor signals that cannot be heard.

THE TERMINATING LF CIRCUIT

The end result of a series of LF circuits is usually a speaker. The speaker is a good load to use as an indicator during signal tracing. A depth sounder, on the other hand, uses a transducer to couple pulses of ultrasonic "sound" waves in water. This terminating circuit doesn't have an output that a technician can hear because of the frequencies used. An oscilloscope may be used to detect the ultrasonic pulses at the input to the transducer (Figure 8-7).

SUMMARY

This chapter has covered the approach to troubleshooting audio circuits. The uses of the oscilloscope, VOM, and DMM in troubleshooting low-frequency circuits have been covered.

REVIEW QUESTIONS

1. What is the usual audio frequency for most signal tracing and testing?
2. If an oscilloscope vertical amplifier setting of 0.5 MV/cm is used and a 10X probe is connected to the input, what voltage *at the probe tip* will produce a centimeter of deflection?
3. Electrically speaking, what does the output function of a VOM do?
4. Why do most digital multimeters *not* have an output function?
5. Does an analog meter (VOM) respond to the average or rms value of an AC sine wave?
6. What is the purpose of a Lissajous pattern?

chapter nine

Special Techniques for Discrete Semiconductors in Operation

CHAPTER OVERVIEW

Discrete semiconductors are individual solid-state components such as transistors, diodes, SCRs, and components not using integrated circuit technology. This chapter presents some of the troubleshooting procedures that are unique to this category of electronic circuits.

Static conditions are assumed in this chapter. Normal operating voltages are applied to the circuitry, but no input signals are applied. This is important to note because signals may seriously affect the DC readings obtained. The VOM is effective in making these voltage tests, but the DMM, with its ability to indicate polarity without the necessity of reversing test lead polarity, is more convenient to use.

Several books are helpful when working on discrete circuits. A book of transistor specifications is handy for identifying transistor leads. It can also be used to advantage when substituting transistors, a subject covered in Chapter 22. RCA, Sylvania, and Motorola publish their own cross-references, which can also be used to identify transistor leads. Additional references may be needed for special components such as silicon-controlled rectifiers (SCRs), junction field-effect transistors (JFETs), metal-oxide semiconductor field-effect transistors (MOSFETs), and unijunction transistors (UJTs). See Appendix III for availability of these references.

TROUBLESHOOTING THE PN JUNCTION DIODE IN OPERATION

If a diode is conducting a constant DC current in the normal forward direction (positive anode, negative cathode), there should be a voltage drop of about 0.3 volt for a germanium diode or about 0.7 volt for a silicon diode. The precise voltage drop across the junction depends to some extent on the current flowing; 0.3 volt and 0.7 volt are typical of most circuits, therefore these values will be used throughout this book as representative values.

The pn junction diode in a working circuit might be conducting DC all the time, conducting intermittently, or used in such a way that it may never conduct under normal circumstances. Here are some examples:

The normal voltage drop of a conducting junction is 0.3 or 0.7 V DC. If there is substantially less voltage than 0.3 V, the diode is shorted. If there is more than about 0.75 V, the diode is open. Be sure that you have the proper voltage polarity across the diode for this test. If the diode is reverse biased (positive on the cathode end), a good diode may well have a large voltage across it.

Rectifiers in the power supply will have both DC and AC voltages across them while in full operation. Rectifiers may be tested in operation by using the AC voltage scales of a voltmeter. Look for the AC voltages across each rectifier of a bridge to be about the same, say within 10 percent of each other. See Figure 9-1 to identify the bad diode. It may be worthwhile to replace all of the associated diodes in the power supply, since one of the apparently good ones may have been stressed.

TROUBLESHOOTING THE ZENER DIODE CIRCUIT IN OPERATION

The zener diode is often used as a shunt voltage regulator (Figure 9-2). Note the path of normal current flow and the way it divides at the zener diode. The zener appears to be in the circuit "backwards," with the anode toward the negative polarity.

A zener should have no more than its rated zener voltage across itself. Excessive voltage across a zener indicates that it is open. There may be less voltage because of an overload in the load circuit. Such an overload will not harm the zener, since the zener simply sees too little voltage to conduct and opens the circuit (Figure 9-2B). If a zener diode is found defective, be sure that a proper load is present across the diode. Lack of a load could force too much current through the zener diode in a circuit not designed for this contingency.

TROUBLESHOOTING THE BIPOLAR TRANSISTOR IN OPERATION

A primary technique for troubleshooting bipolar transistors in operating circuits is the check for proper bias. Most circuit problems will greatly affect the bias voltages present on the transistor.

Troubleshooting the Class A Amplifier Stage

A bipolar transistor will probably be biased for Class A operation when used in DC and audio-amplifier circuits. The transistors will be biased so that there is half-way conduction of the collector circuit in the absence of any incoming

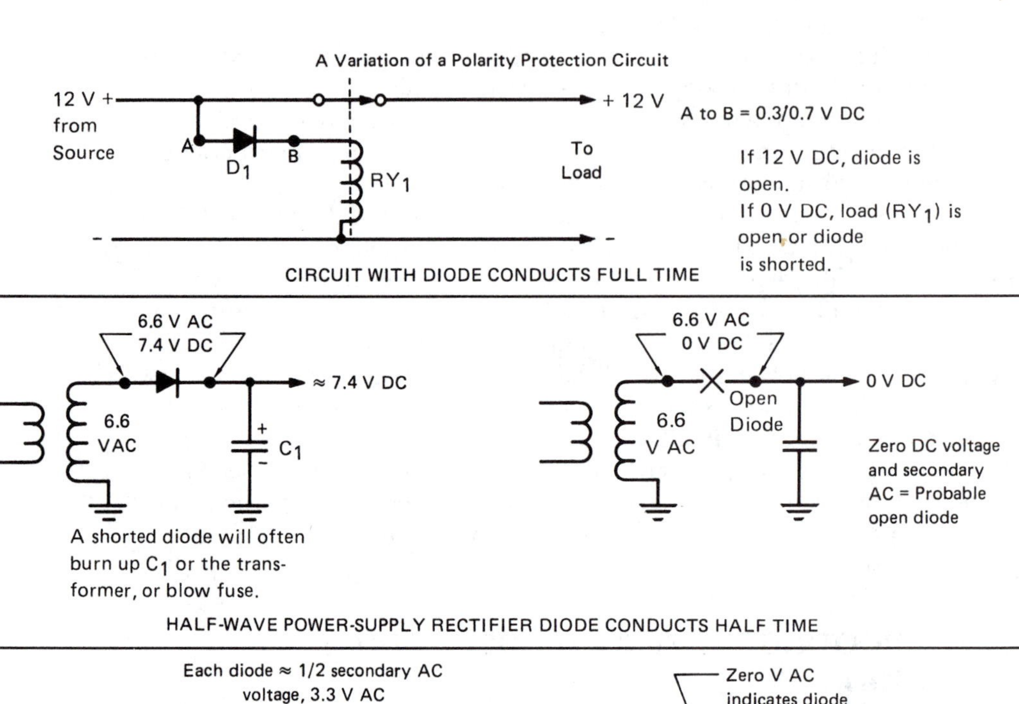

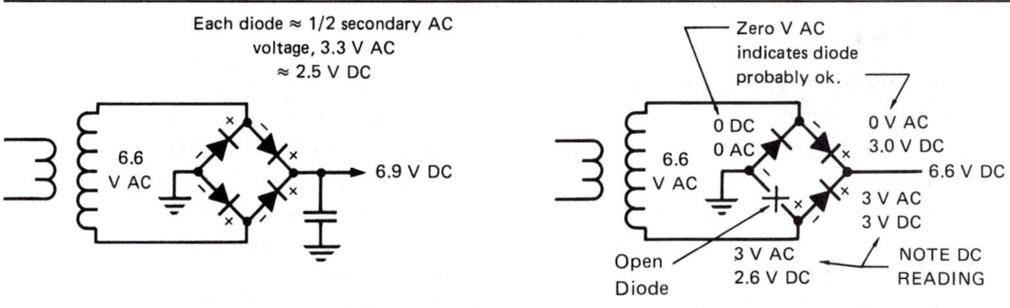

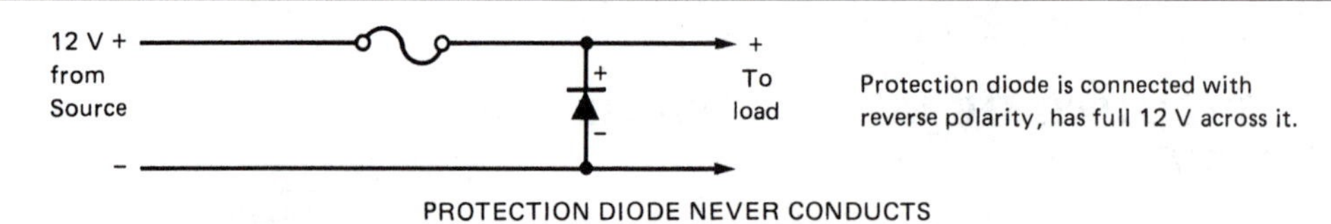

FIGURE 9-1 Sample voltages in diode circuits. Voltage readings are approximate. Use a VOM on "output" function for AC voltage readings; a DMM may read stray hum and noise and give misleading indications.

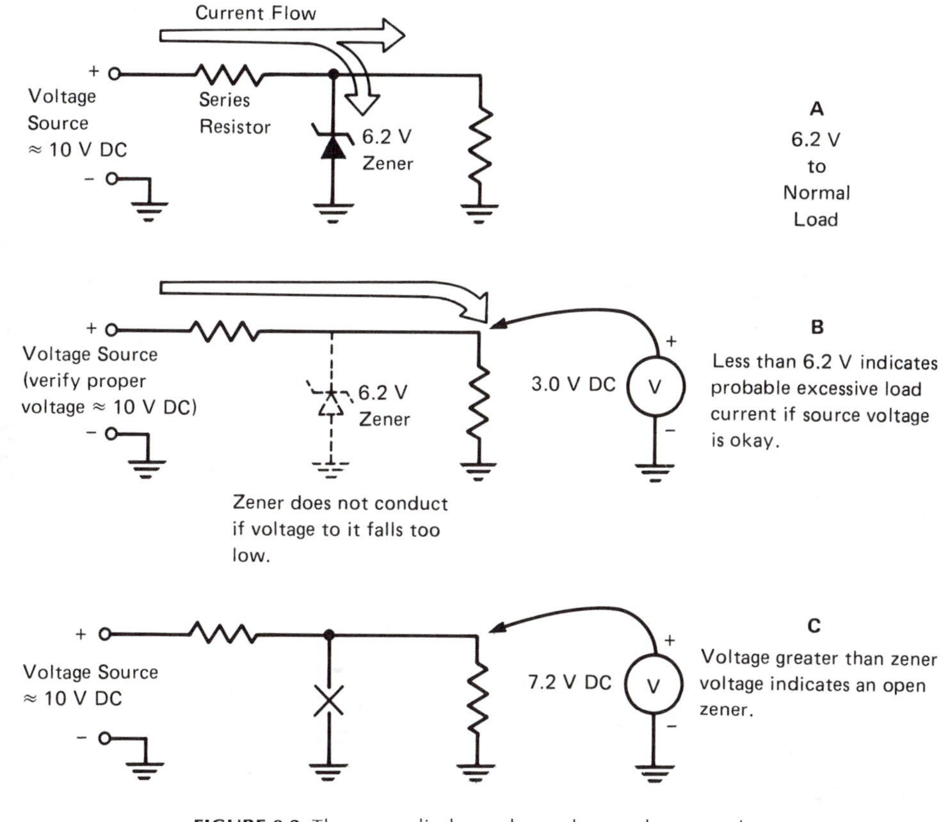

FIGURE 9-2 The zener diode used as a shunt voltage regulator.

signals. Operated in this manner, the incoming signals can either turn the transistor more on or more off, depending upon the polarity of those signals.

The Class A stage can often be identified by the presence of a resistor from the base circuit to the collector circuit or directly to Vcc, the supply voltage for the collector. There may also be another resistor from the base circuit to the emitter to assist in stabilizing the bias voltage, giving the base a voltage divider to connect into at a voltage between Vcc and ground, as it is often drawn.

Transistors can be drawn right-side up or upside down, and they can be either the pnp or npn types. This can certainly lead to confusion in trying to figure out the proper, normal polarity and amount of voltage to expect between their elements. What is needed is a simple, universal rule for examining bipolar transistor biasing that may be applied in any instance and consistently produce reliable indications. Here is that rule:

When operating Class A, the forward bias voltage at the base of a bipolar transistor (with respect to the emitter) will have collector polarity and will be either 0.3 or 0.7 volts.

Again, the 0.3/0.7-volt level is a close approximation. The actual voltage will vary slightly with the amount of biasing current through the junction. *Bias*

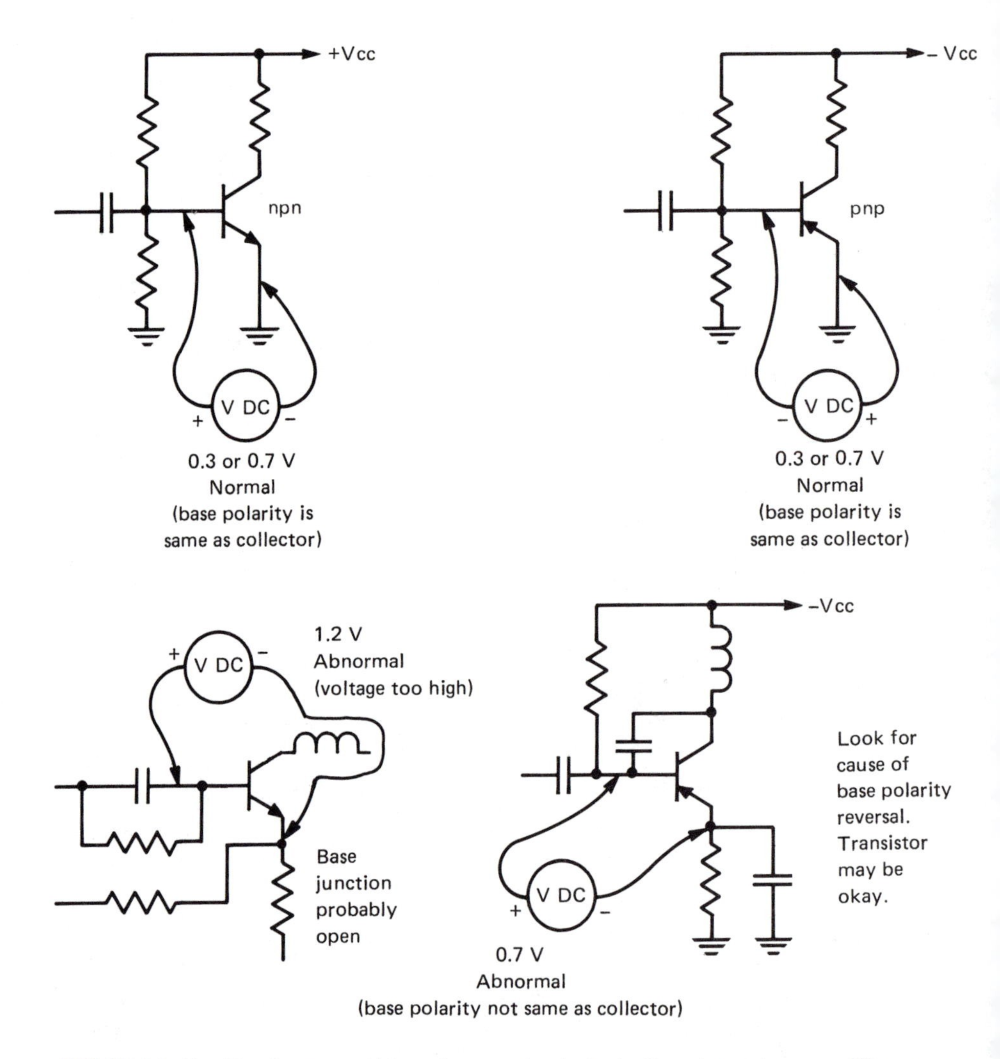

FIGURE 9-3 Checking for normal bias voltage and polarity in Class A transistor amplifier stages.

and collector voltages should be taken without signals being applied to the circuits under test (Figure 9-3).

Here are the steps to follow in troubleshooting a Class A transistor amplifier stage:

1. Identify the emitter lead and connect the voltmeter common test lead there. With the remaining test leads contact the collector lead to determine polarity and voltage there. The collector voltage of an RC-coupled transistor should be considerably less than the supply voltage, Vcc. Half is about right. If the collector voltage is almost at Vcc, look for an open transistor collector or lack of the proper forward bias voltage of 0.3 or 0.7 volts. If the

collector voltage is very low (less than a volt or so) the transistor is probably either shorted or merely saturated. Checking the bias will indicate if the transistor is saturated.

2. Check the bias voltage on the base of the transistor with respect to the emitter. If it is 0.3 or 0.7 volts *with the same polarity as the collector,* the transistor is saturated and the cause of saturation should be investigated. If the bias is not 0.3 or 0.7 volts, but considerably lower, the transistor may be shorted from emitter to base.

Sometimes an emitter resistor is provided in a Class A circuit. The voltage drop across this resistor will indicate the amount of conduction of the transistor. Too little or no voltage drop could mean an open transistor or lack of bias, or the lack of a supply voltage (Vcc) in the collector circuit. Move the test lead from the collector to the base connection. The voltage there should be of the same polarity as that of the collector. You should find 0.3 V DC there for a germanium transistor or 0.7 V DC for a silicon. If the voltage is substantially more, the emitter-base junction is open. If the voltage is very near zero, suspect a possible shorted emitter-base junction. If the voltage is reversed from proper polarity, look for the cause of bias voltage reversal, as the transistor is most likely all right. (See Table 9-1.)

This troubleshooting procedure can be used to greatest advantage in circuits that are DC coupled. Such circuits may have several transistors interconnected by DC paths in such a way that if a single transistor opens or shorts, many of the associated transistors will have improper bias readings. In these circuits, the transistors with low forward bias or with reversed bias are probably okay. Look for a transistor with a forward bias well in excess of the 0.3/0.7-volt criteria or one with a zero voltage between base and emitter (possibly a shorted transistor).

Multiple transistors interconnected in DC-coupled stages can cause one

TABLE 9-1 A Transistor Troubleshooting Chart

COLLECTOR		**BASE**		**ANALYSIS**
Polarity	**Voltage***	**Polarity**	**Voltage**	
(+)	½ Vcc	(+)	0.3 or 0.7	Normal
			<0.3	Probable E-B short
			>0.7	Open E-B junction
(+)	High	(+)	0.3 or 0.7	Open collector
(+)	Low	(+)	0.3 or 0.7	Wrong bias, see text
(+)	High	(−)	Any	Wrong bias, see text
(−)	½ Vcc	(−)	0.3 or 0.7	Normal
			<0.3	Probable E-B short
			>0.7	Open E-B junction
(−)	High	(−)	0.3 or 0.7	Open collector
(−)	Low	(−)	0.3 or 0.7	Wrong bias, see text
(−)	High	(+)	Any	Wrong bias, see text

**Compared to the supply voltage, Vcc*

another to fail when one of them starts the process; after finding one bad transistor, don't be surprised if the circuit still doesn't work. Look for other transistors that may also have failed.

Another Test

The following test is effective only if the collector circuit is fed by a resistor to Vcc, such as in the standard RC-coupled amplifier. It will not work with transformer-coupled stages. This test may be used on many Class A transistor stages under static conditions to see if a transistor will couple a DC signal through itself. Observe the collector voltage under conditions of normal bias. Then short the base directly to the emitter, thereby turning the transistor off. (Be *very* careful that the base is shorted to the emitter, *not* the collector, which could instantly destroy the transistor.) The collector voltage should increase, approaching the source voltage. This test works with resistor- and capacitor-coupled stages and many DC-coupled stages. See Figure 9-4 for an illustration of this test.

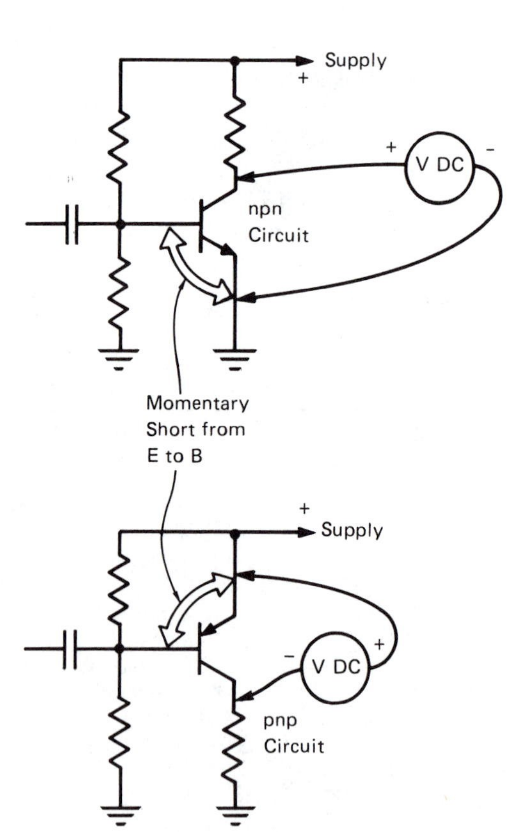

Collector voltage should vary from one half Supply E to full Supply E when base is shorted to the emitter.

This test proves that the transistor is amplifying signals.

Remember! This test applies only to Class A resistor-capacitor-coupled stages.

Do ***not*** touch base to the collector. Probable damage to the transistor will result.

FIGURE 9-4 An in-circuit transistor test for Class A stages.

Troubleshooting the Class B Amplifier Stage

The Class B amplifier is used mostly in high-power audio stages, where one of two transistors amplifies one polarity of the incoming signal and a second transistor placed in the circuit "upside down" supplies the opposite polarity. The Class B stage can be identified when the stage amplifies linear signals (audio frequencies, for instance), if there are two transistors in the stage (one is often drawn upside down), and if there is a transformer in the output circuit. This is called a push-pull stage (Figure 9-5).

These stages can also be analyzed by checking the bias voltage. The bias should be such that the transistors are turned on a little bit, thus avoiding distortion problems with very small signals. A "little bit" on means that the base junction will have the familiar 0.3/0.7-volt drop without an incoming signal. The current flowing into the base, however, will not be as much as if the transistor were operating Class A.

Emitter resistors, where provided, give the technician a very good place to monitor the current flowing into each of the transistors. Measure the voltage drop across each of the two emitter resistors to get a comparison of the current

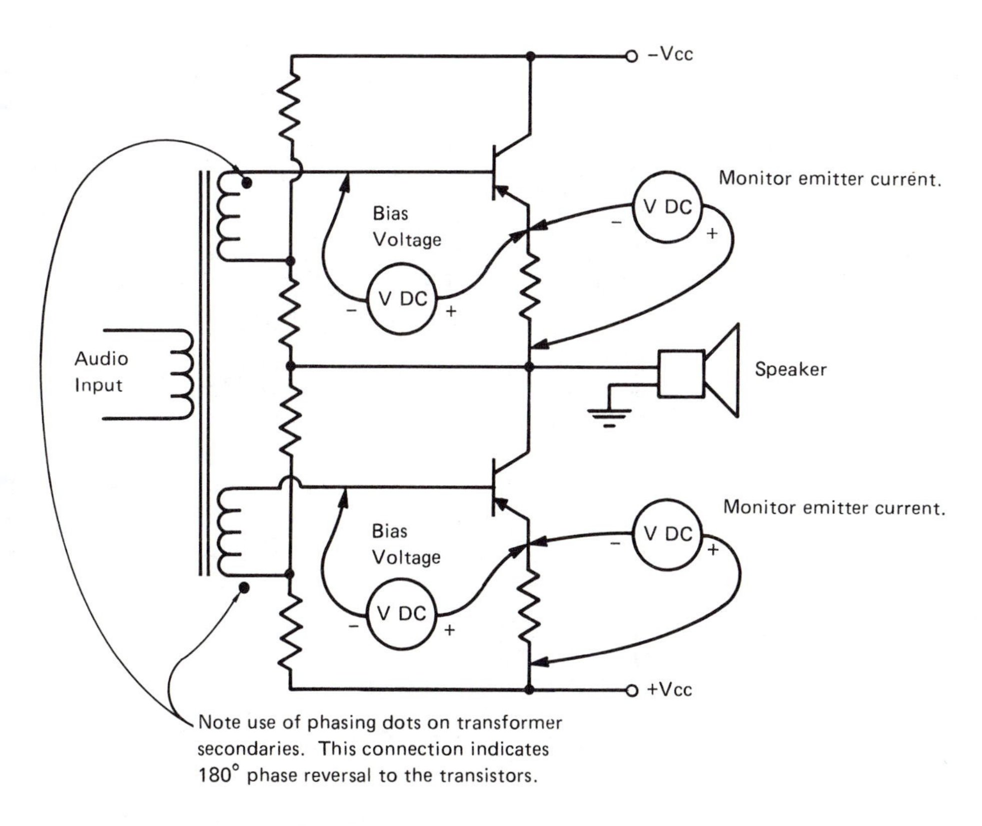

FIGURE 9-5 Testing push-pull amplifiers.

flows. An emitter resistor with no voltage drop points to a transistor with an open collector or base junction, or a transistor that does not have supply voltage.

Troubleshooting the Complementary-Symmetry Amplifier

This amplifier is easily recognizable because it has two different transistors connected to the output speaker, one pnp, the other an npn. There will also be a large coupling capacitor from between the transistors to the output, usually a speaker, if the stage is operated from a single supply voltage. This capacitor can be eliminated if there are two separate supply voltages for the stage, one negative and the other positive.

The capacitor coupling the audio to the speaker is a common source of problems (Figure 9-6). It often shorts, causing a pronounced "thump" in the speaker when power is first applied. If the speaker baffle is removed, the bass speaker will be seen to press outward or inward and remain there under the influence of the unwanted DC component to the speaker because of the defective capacitor.

Individual transistor current can be monitored in the complementary-symmetry circuit by noting the voltage drop across each of the emitter resistors if they are provided. These voltages should be very close to one another. The bias voltages should also be the familiar 0.3/0.7-volt forward bias.

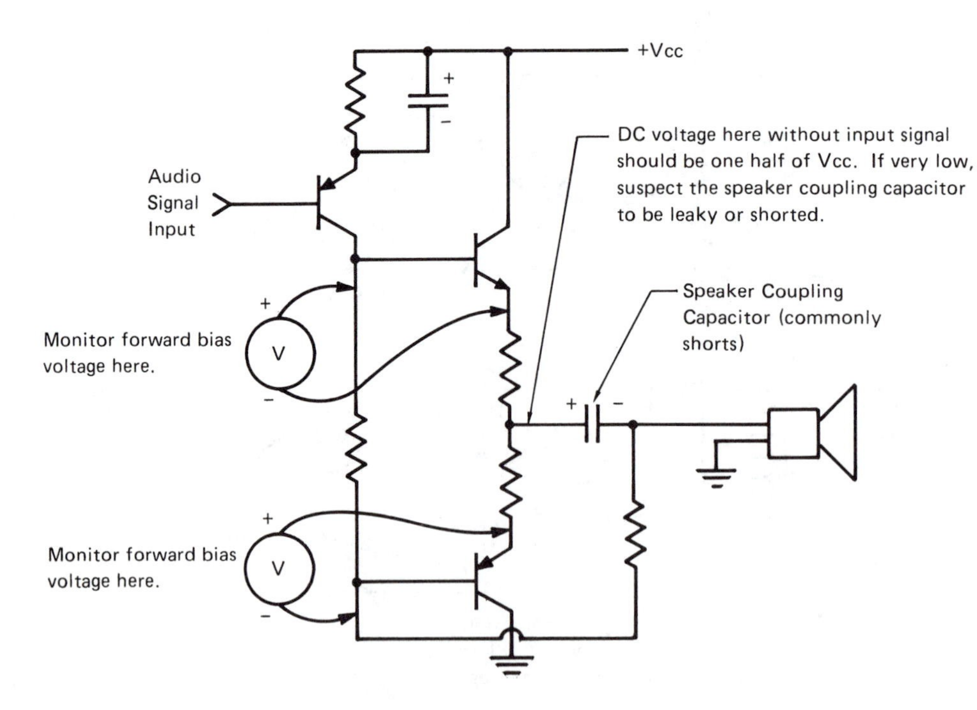

FIGURE 9-6 Troubleshooting the simplified complementary-symmetry stage.

Troubleshooting the Class C Amplifier Stage

The bipolar transistor Class C amplifier stage may have no special biasing circuitry. The incoming signal to be amplified is used to turn the transistor on by driving it from zero forward bias voltage to the point of conducting (the 0.3/0.7-volt point) and then more, causing a pulse of current in the output circuit. This class of operation is used in RF amplifier circuits, where the pulses of collector current can be smoothed into a sine wave by the flywheel effect of inductance and capacitance, and in many pulse circuits.

The servicing technician can look for the lack of biasing resistor as described for Class A amplifiers (the resistor connected between the base of a transistor and the collector circuit) as an aid in identifying the Class C amplifier stage (Figure 9-7).

A normal DC level present at the transistor emitter, if any, is an excellent indication of normal Class C circuit operation. Using a voltmeter or an oscilloscope with DC coupling, the trace of the oscilloscope should rise when the amplifier is fed with a good RF signal on the input. The DC level is the average of

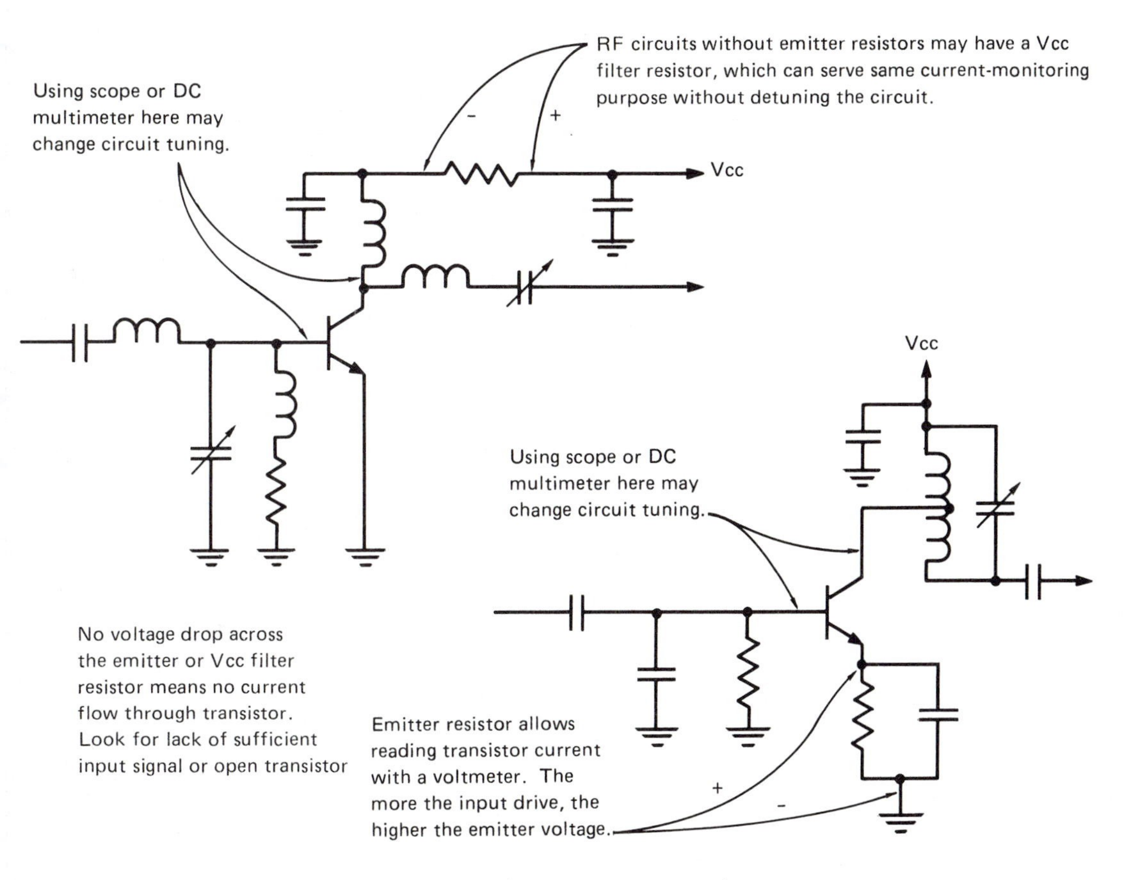

FIGURE 9-7 Monitoring Class C transistor operation.

many pulses of current in the emitter circuit, held at a DC level by the bypass capacitor across the emitter resistor.

Common problems detectable by measuring the emitter voltage are:

- Lack of any DC at the emitter—possible open transistor base, too little incoming signal, or a shorted emitter capacitor.
- Very little DC at the emitter—possible open collector circuit or lack of supply voltage at collector.
- Too much voltage at the emitter—probable shorted transistor, collector to emitter.

TRANSISTOR AND RESISTOR ARRAYS

Some circuits use *arrays* of transistors. These arrays look like integrated circuits, but instead of complex circuits within them, they have several simple transistors. Some of the transistors may be connected together internally for special purposes. See Figure 9-8 for an example of a transistor array.

Resistors can also be made to look like ICs. These arrays are convenient when, for instance, seven or eight resistors are needed of similar value. Though they can be manufactured with many different internal connections, two are in common use. See Figure 9-9 for these two wiring diagrams. In order to troubleshoot circuits using these components, it is necessary to have information on the internal wiring diagrams.

TROUBLESHOOTING FET CIRCUITS

The FET comes in two main types: the junction FET (JFET) and the insulated gate FET (IGFET).

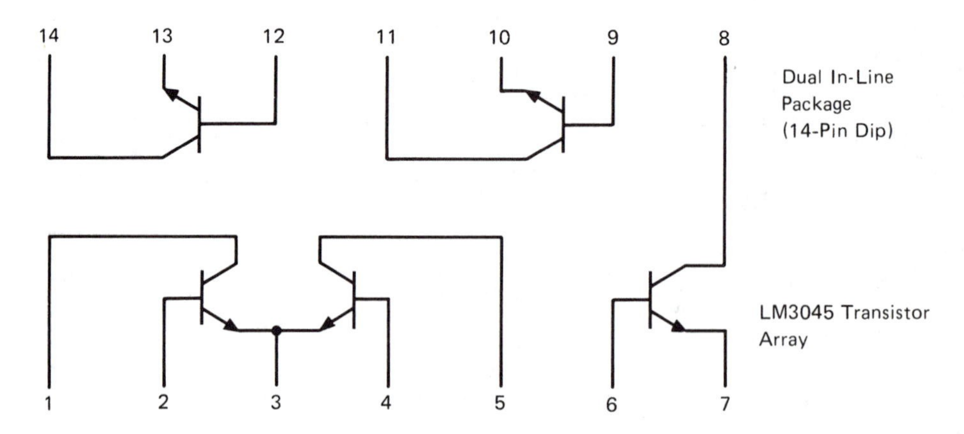

FIGURE 9-8 Schematic of a transistor array.

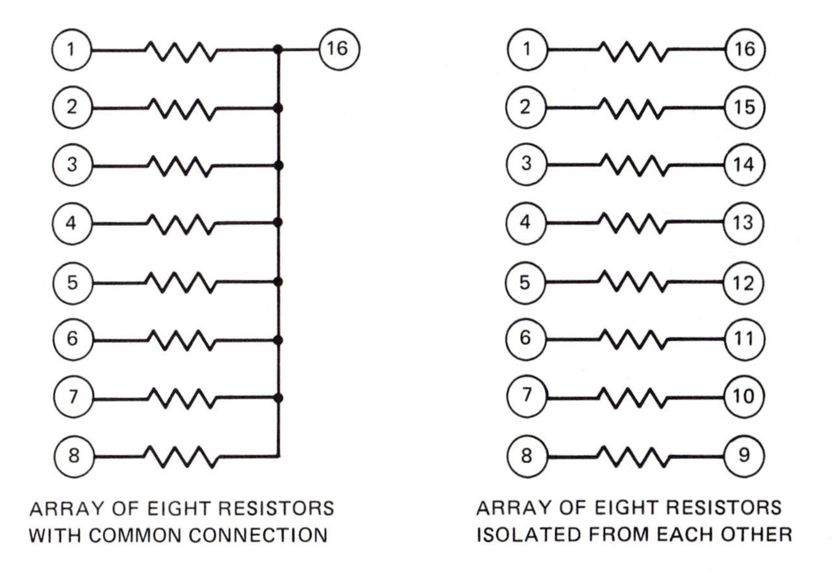

FIGURE 9-9 Sample resistor array internal connections.

The JFET Circuit

The n-channel junction FET operates with the same polarity of corresponding elements as a tube—the input or gate lead is reverse biased with a negative voltage with respect to the source. This is similar to the grid of a tube circuit. The source lead of an n-channel FET is connected to the negative side of the supply voltage, usually ground. The source lead is somewhat similar to the emitter of a transistor or the cathode of a tube. The drain connection of the FET is connected to the positive side of the voltage source. The drain corresponds to the collector of a transistor or the plate of a tube. It is the drain that produces

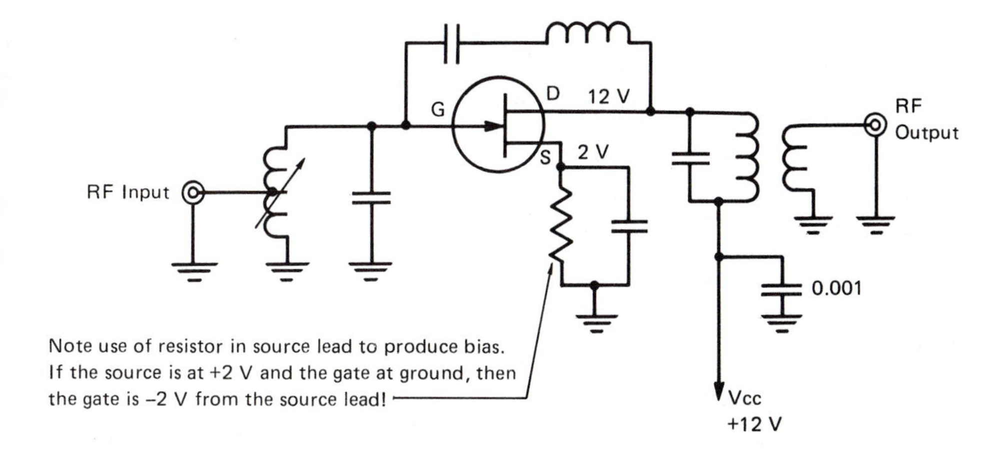

FIGURE 9-10 Simplified n-channel JFET amplifier.

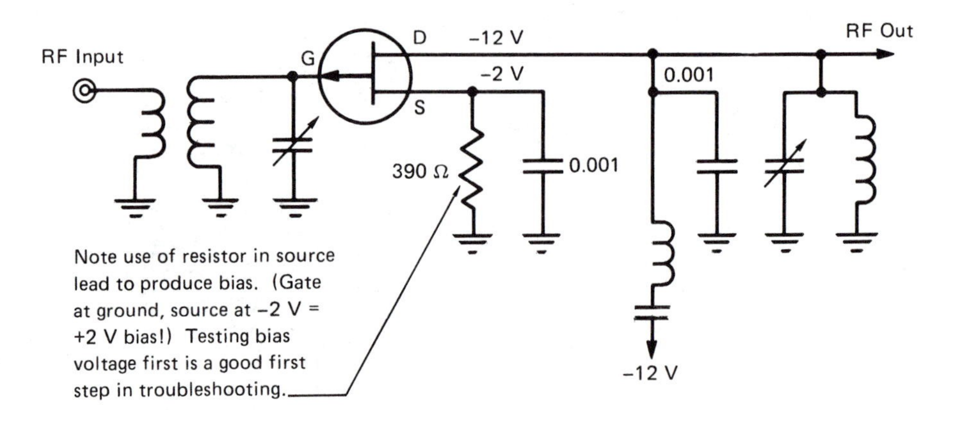

FIGURE 9-11 Simplified p-channel JFET amplifier.

the variations that become the output signal. Some technicians may choose to think of the JFET as a "solid-state vacuum tube" (Figure 9-10). The p-channel FET operates in the same manner as the n-channel but has both bias and output supply polarities reversed (Figure 9-11).

Since the input diode junction of the FET is reverse biased, the input impedance of the FET is very high. This is used to advantage in circuits that require this characteristic. The high-impedance input of JFETs is very susceptible to leakage paths. This characteristic might give a clue in FET circuits that are behaving in an erratic manner—look for circuit-board surface contamination causing a leakage problem.

The IGFET Circuit

The insulated gate FET transistor is available in either of two variations, the depletion mode or the enhancement mode (Figure 9-12). Note the difference between the depletion and enhancement mode circuit symbols. One has a solid line from source to the drain, the other a broken line. This prompts the technician as to what current flows in the output circuit *without* a bias. The solid line from source to drain of the depletion mode IGFET indicates that there *will* be current flow if there is *no* input bias. The n-channel IGFET requires a positive voltage applied to the drain connection and the negative connected to the source. This is similar to the n-channel FET and vacuum-tube polarities. The p-channel, of course, has all polarities reversed from the n-channel.

Biasing the IGFET. The depletion-type IGFET requires no special bias supply. Positive variations of the gate (with reference to the source lead) will cause more current to flow in the output circuit, and negative variations can decrease output current flow to zero. This is a sort of automatic, built-in Class A operation without the fuss of biasing circuits. Figure 9-13 shows a typical IGFET depletion-type amplifier stage.

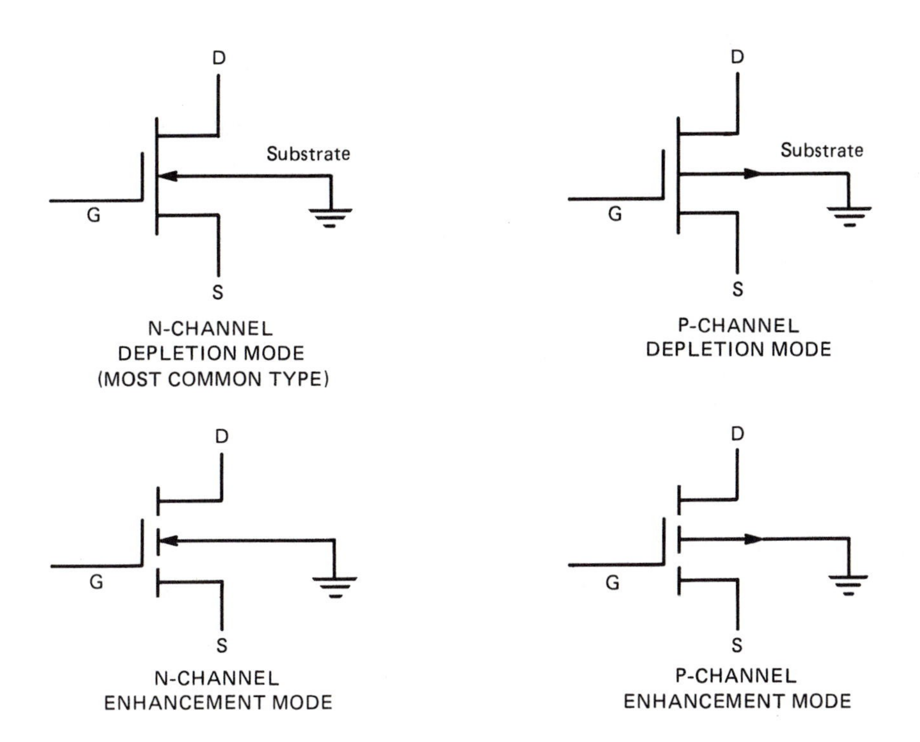

FIGURE 9-12 Insulated-gate FETs.

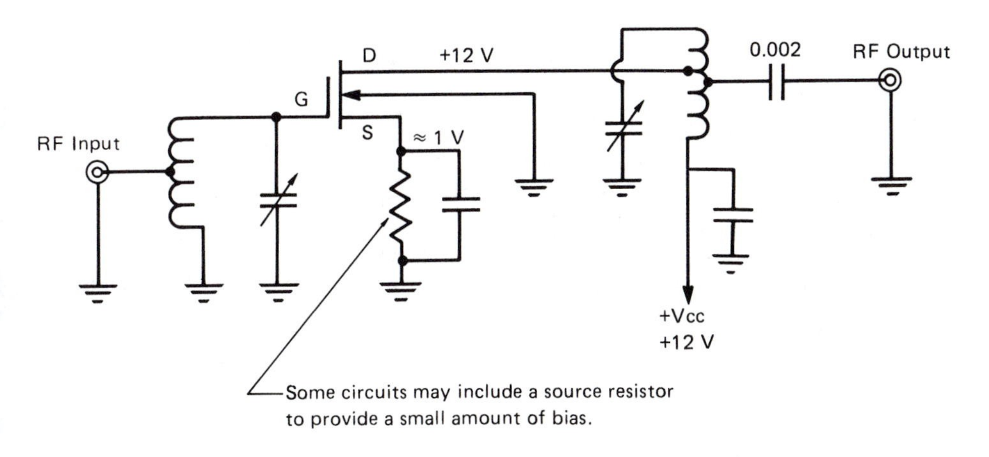

FIGURE 9-13 Sample circuit of an n-channel depletion-mode amplifier circuit.

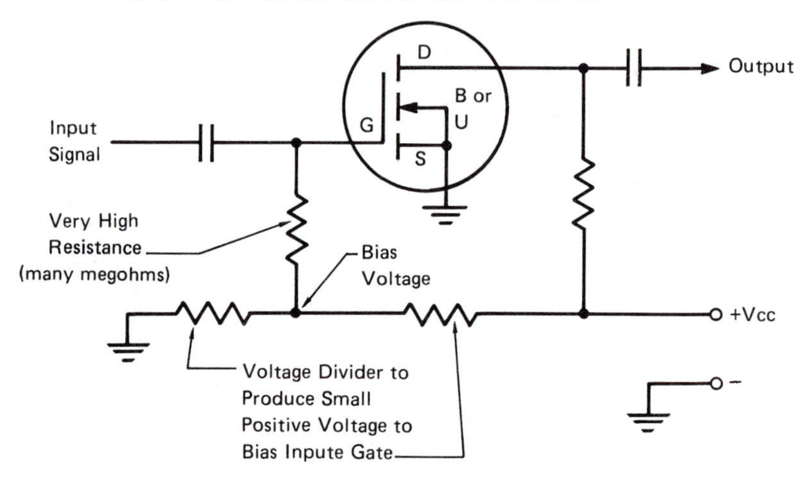

FIGURE 9-14 Biasing the enhancement IGFET.

The enhancement-type IGFET, however, requires a specific bias supply to obtain any current flow in the output circuit. This is usually derived from another available supply voltage. Figure 9-14 shows a sample circuit for an enhancement-type IGFET.

TROUBLESHOOTING UJT OSCILLATORS

The unijunction transistor (UJT) is a specialized transistor. It is used exclusively as an oscillator, producing pulses or sometimes saw-tooth waves. It is

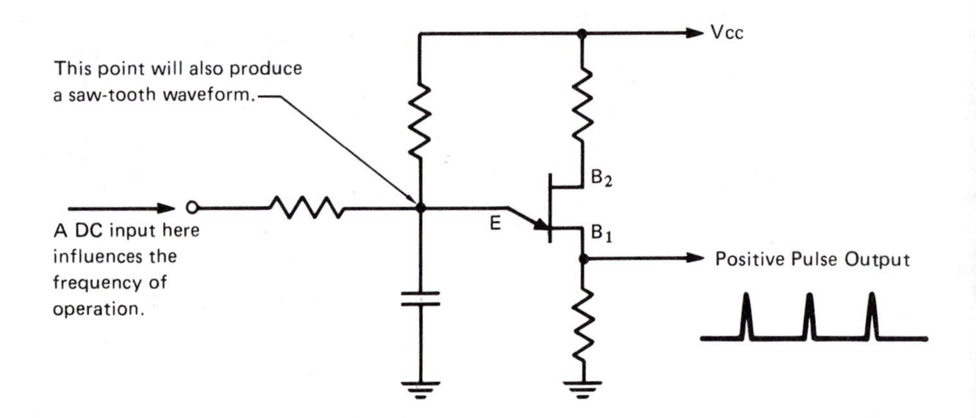

FIGURE 9-15 The typical unijunction transistor circuit.

usually used in a circuit very much like the one shown in Figure 9-15. The UJT circuit requires relatively few components, one of its chief advantages. UJT circuit operation is best determined by using an oscilloscope. If the voltage at the emitter is at supply voltage (Vcc), the UJT emitter or Base 1 circuit is probably open. Symptoms for an open Base 2 are a Base 1 voltage of 0 and a very low emitter voltage.

TROUBLESHOOTING THYRISTORS (SCRs)

Thyristors (silicon-controlled rectifiers) are often operated at high voltages, and both cathode and anode may be well above ground potentials. This makes them difficult to troubleshoot in full operation with oscilloscopes, which have one side of their input circuits grounded. Since SCRs operate mostly on AC voltages (to turn them off once conduction starts), AC voltage readings may be used to help troubleshoot these circuits. A lack of AC voltage across an SCR may be an indication of a shorted device. A shorted SCR would probably be suspected in such a case because the load, whatever it may be, is operating at full line voltage without SCR control.

It may also be possible to compare AC voltage readings between a properly operating circuit and a defective one. Be sure that both circuits are set to operate identically, then compare the AC voltage readings from cathode to anode of the two SCRs. If one of the SCRs has a substantially higher AC reading across it, you may suspect an open SCR or lack of a properly timed trigger of sufficient voltage and current capability to trigger the device.

Because of these considerations, it is often best to turn off the equipment and check the SCRs using resistance checks or a special solid-state tester as covered in Chapter 16.

SUMMARY

This chapter has presented some of the "tricks of the trade" by which defective transistors and other discrete devices might be quickly found. Biasing of diode junctions and the voltages to expect across them were emphasized because of the practical usefulness of this information. Transistor biasing was discussed in detail also because of its importance. Once a defective part has been identified in a circuit, the reader is referred to Chapter 16.

REVIEW QUESTIONS

1. What approximate voltage drop would you expect across a silicon diode?
2. What is meant by a voltage reading taken across a zener diode that is higher than the rated voltage of the diode?
3. If the load across a zener diode regulator increases to the point that the regulated voltage is half normal, will the zener be damaged?

4. What voltage test is most revealing in troubleshooting a bipolar transistor Class A stage?
5. What bipolar transistor element is the reference point when taking bias readings?
6. What is indicated if the bias on a bipolar transistor Class A stage is reversed from normal polarity?
7. How much is the Class B stage conducting in the absence of an incoming signal?
8. What is the clue in identifing a complementary-symmetry amplifier stage?
9. What is the clue in identifying a bipolar Class C stage?
10. What does the lack of a voltage reading across an emitter resistor mean?
11. How must the n-channel JFET transistor gate be biased?
12. Which type of IGFET requires no special bias voltage for Class A operation?
13. For what is the UJT used?

chapter ten

Troubleshooting in Live Analog IC Circuits

CHAPTER OVERVIEW

This chapter deals with troubleshooting analog IC circuits with power applied to them. Analog ICs are sometimes called linear ICs, because they deal with analog (varying) voltage levels rather than with the yes/no, high/low voltage levels of digital logic.

An excellent reference work on these ICs is Motorola's *Linear and Interface Integrated Circuits* Manual, #DL128. See Appendix III for ordering information. This single publication lists operational amplifiers (op-amps), voltage comparators, timers, regulators, and many other analog devices including consumer products. Consumer products are those ICs that combine many stages of common circuitry in a single chip. For instance, Motorola produces the MC1357, which combines the IF amplifier and detector circuits of a television receiver on one chip.

OPERATIONAL AMPLIFIERS

Reduced to simplest practical terms, the "op-amp" is a very high-impedance input, high-gain voltage amplifier. Its output is a low impedance of a few hundred ohms or so. It is useful for a wide range of purposes in slightly different circuit designs. As the circuit determines how the op-amp will perform, it will be necessary to consider the basic circuits that the op-amp will be used in and the troubleshooting steps for each.

Figure 10-1 shows the schematic symbol for the operational amplifier. The op-amp output voltage depends upon the *difference* between the two inputs. Often one input is held at a constant voltage while an input signal is applied to the other input.

The Inverting Voltage Amplifier

Figure 10-2 shows the op-amp in a very common circuit where the input signal to be amplified is applied to the *inverting* input. Whatever input signal is

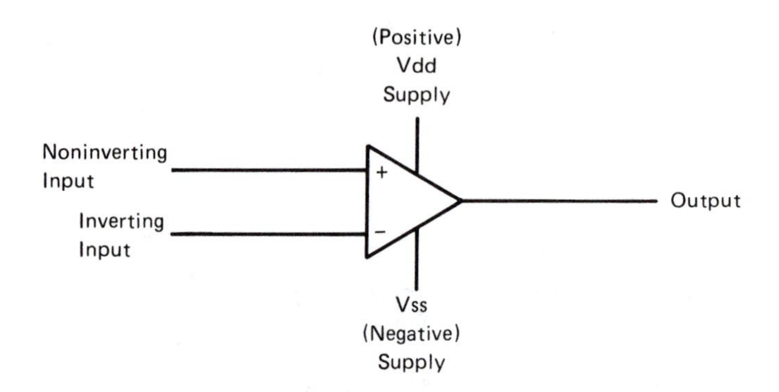

FIGURE 10-1 Schematic symbol for an operational amplifier.

applied to this circuit comes out inverted at the output, as shown by the negative mark (−) on the input terminal. The voltage gain of the circuit—and this is important to note—is, for all practical purposes, the ratio of the feedback resistor (Rfb) to the input resistor (Rin) on the *negative* input. The servicing technician should estimate the amplification factor when working with op-amps to provide an idea of the amount of amplification that should be expected from the stage.

$$\text{Op-Amp Gain} = \frac{\text{Rfb}}{\text{Rin}}$$

The remaining resistor on the (+) input is called a compensating resistor.

This circuit has a quirk that can easily confuse the beginning troubleshooter. With a good input signal being applied to the input terminal resistor and a good output signal from the IC, there will be very little if any signal noticeable at the negative input terminal (point A)! This is because the feedback resistor cancels out the input voltage. The point where no signal can be found is called a *virtual*

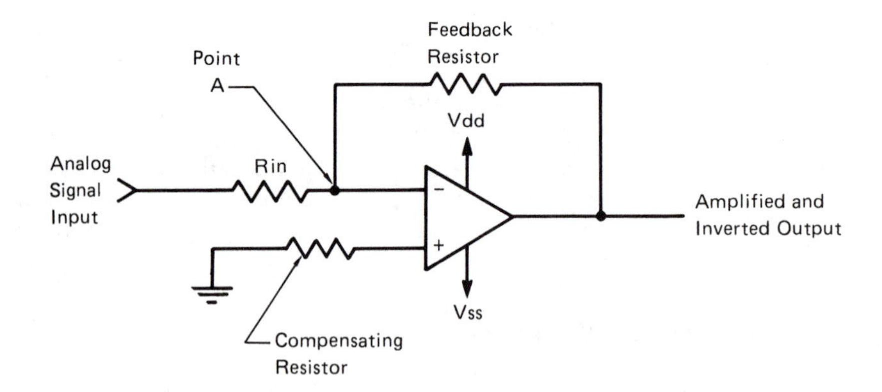

FIGURE 10-2 Basic operational amplifier inverting circuit.

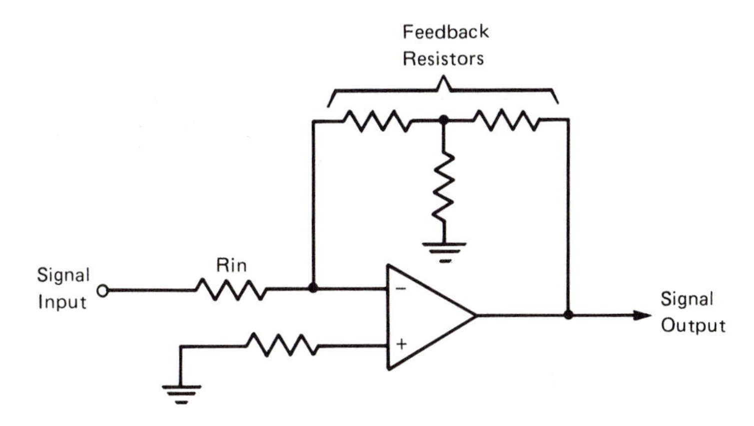

FIGURE 10-3 Inverting operational amplifier with three-resistor feedback.

ground. This is normal in a good working circuit. It may seem contrary to logic, but the presence of the same amplitude of signal at both ends of the input resistor indicates a defective IC.

In some circuits there may be a variable resistor adjustment connected to the op-amp. This adjustment is probably a DC offset adjustment and is adjusted to give a zero voltage out of the op-amp in the absence of any input signal. Check the schematic to verify the use of the variable resistor and how to set it, if need be.

Some circuits may use the three-resistor feedback arrangement of Figure 10-3 rather than a single resistor. Lower values of resistor may be used to produce high gain in the circuit. The circuit behaves as though a single resistor were used, as in Figure 10-2.

With the proper selection of resistors, this circuit may also be designed to have zero gain (a gain of 1.00) or as an attenuator of sorts, with a gain of less than one. In this case, the feedback resistor would be lower in value than the inverting input resistor.

The Noninverting Voltage Amplifier

Figure 10-4 shows a different input arrangement, the noninverting operational amplifier circuit. Again, the feedback and input resistor on the inverting (negative) input determine the voltage gain of the circuit. This circuit will always have a gain of more than one. In this circuit the input signal does not become inverted in passing through the amplifier. A good signal should be evident at the (+) input and *also at the (−) input* of the IC during normal operation! The presence of the signal at the (−) input is caused by the feedback resistor attempting to hold the (−) input at the same level as the (+) input.

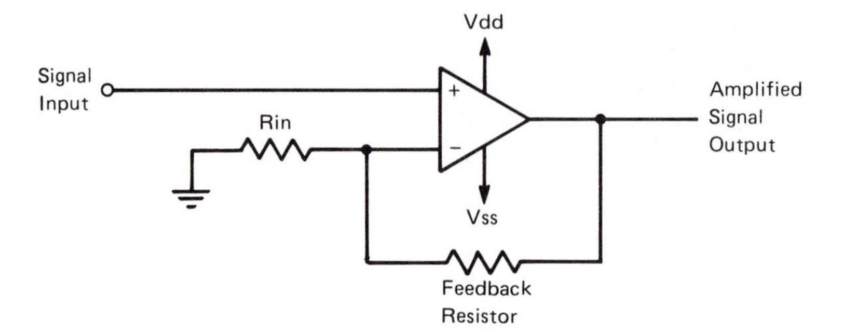

FIGURE 10-4 Noninverting operational amplifier circuit.

The Differential Amplifier

The differential amplifier is a circuit in which two inputs are provided to an operational amplifier. The signal out of the amplifier is the result of the difference between them, not their individual voltages with respect to ground (Figure 10-5). The greater the voltage difference between the two input signals, the more output is produced from the stage. If the two voltages are equal, the output of the stage is the neutral point, halfway to Vcc for a single-supply stage or zero volts for a dual power-supply stage.

The Voltage Follower

The circuit of Figure 10-6 (a *voltage follower*) has a high input impedance and a low output impedance. It is easily recognized by the direct connection between the (−) input and the output.

Operating Op-Amps from Single and Dual Power Supplies

Operational amplifiers may be operated from a pair of supplies, (+) and (−) with respect to ground, or from a single (usually positive) supply. In either

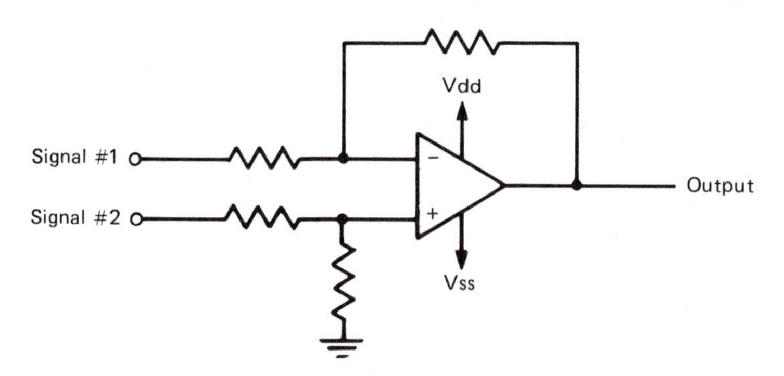

FIGURE 10-5 The operational amplifier connected for a differential output.

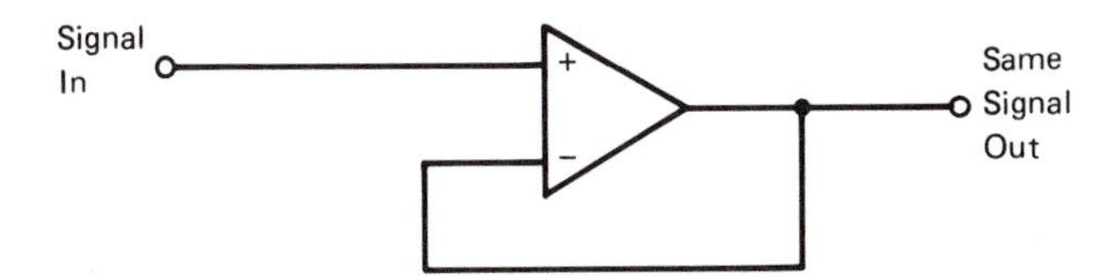

FIGURE 10-6 Operational amplifier connected as a voltage follower.

case the op-amp will amplify only the difference between its two input lines. Figure 10-7 shows an IC operated from a pair of power supplies. Its output voltage without an input signal will be approximately at ground, or zero voltage. With a signal input, the output voltage swing can approach the (+) supply voltage in the positive direction and the (−) supply in the negative direction of swing.

The maximum positive and negative voltages that a circuit may attain are called the *rails*. In a dual-supply op-amp circuit, the positive and negative supplies are the positive and negative rails, since no output signal may exceed them.

Operating from a single supply means that the quiescent output voltage from the circuit in the absence of an input signal should be about halfway between ground and the supply voltage. Connected this way, the output voltage may swing more positive or more negative. The positive swing can approach the (+) supply and the negative may come close to ground (Figure 10-8). In a single-supply circuit one of the inputs will be connected to a voltage divider to bias that input terminal to approximately half of the supply voltage. This sets the reference voltage around which the other input may operate to produce either a positive or negative input with respect to the other input. Remember, it is the *difference* between these two inputs that is amplified by the circuit.

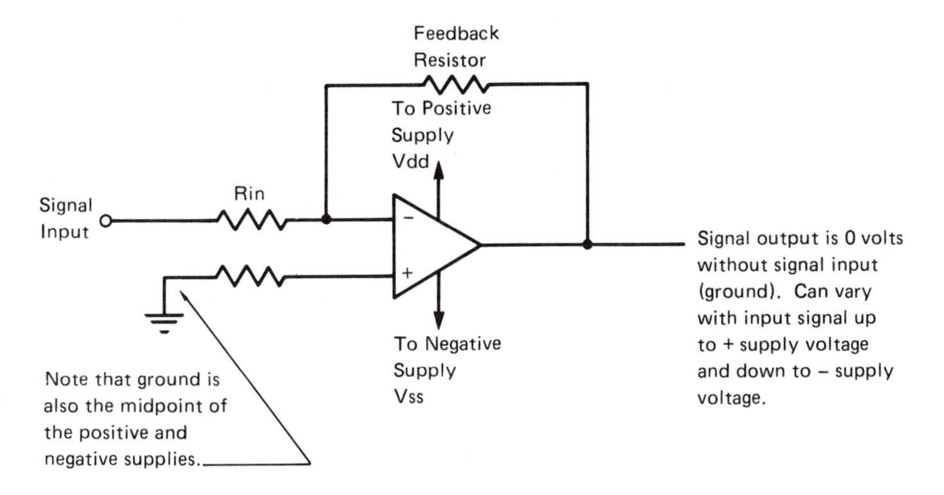

FIGURE 10-7 Operational amplifier circuit using two supply voltages.

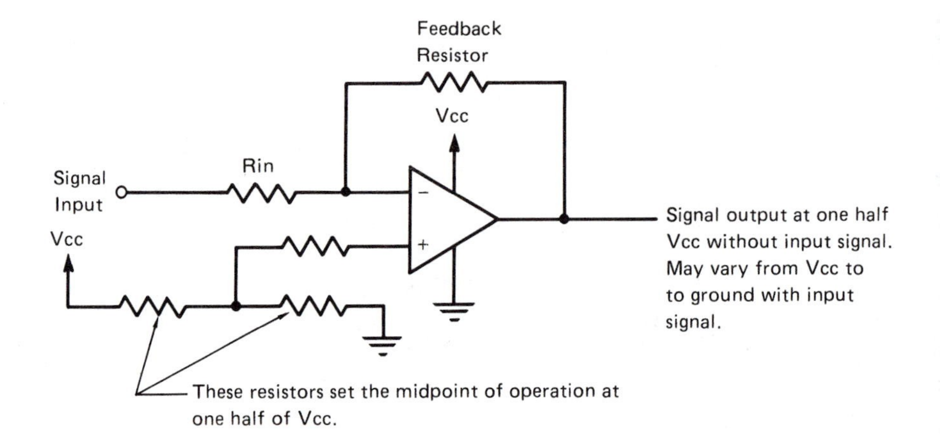

FIGURE 10-8 Operational amplifier operated from single-supply voltage.

Nonlinear Uses of Op-Amps

The Voltage Comparator. Figure 10-9 shows an op-amp without a feedback resistor. With no signal feedback to reduce the gain of the circuit, the op-amp is operating with its maximum gain. When the input voltage at one input passes the input voltage at the other, the output snaps from one extreme voltage to the opposite. Special op-amps are made that react very quickly for digital circuit uses. These are called *voltage comparators*.

Hysteresis Circuits. If the two input voltages to an operational amplifier are about equal, it is possible for the output voltage to be unstable. This can be corrected by using some *positive* feedback to add a snap-action to the circuit. Note that Figure 10-10 shows the feedback opposite to that shown in Figure 10-2. The feedback is now being applied to the (+) or noninverting input.

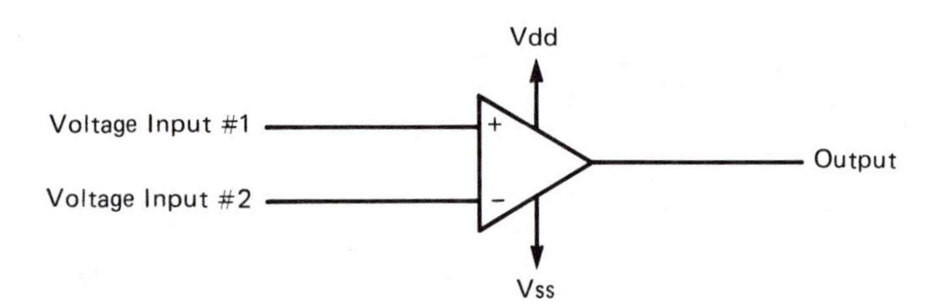

When voltage #1 exceeds voltage #2, the output is at the high rail (Vdd) and vice versa. If they change up or down together, the output does not change.

FIGURE 10-9 The voltage comparator circuit.

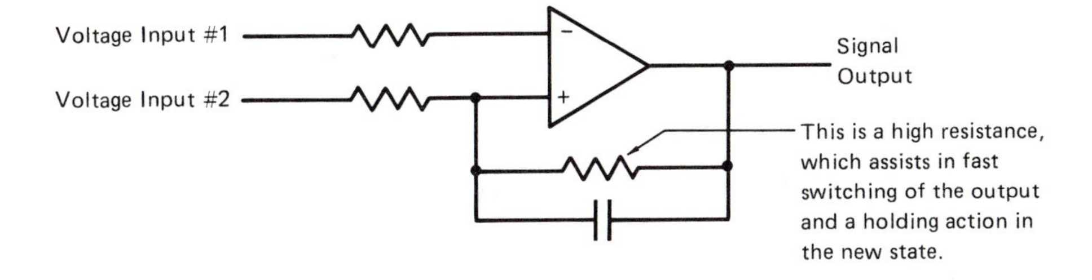

FIGURE 10-10 Adding a hysteresis or "snap-action" circuit.

The Open Collector Output. Some voltage comparators have an open collector output rather than a simple voltage output. These ICs require a resistor to pull them up to the supply voltage when the single transistor inside the IC opens (Figure 10-11). This allows the IC to operate its output circuit at a higher voltage than the chips' Vcc line in circuits requiring more output drive. This output circuitry is not restricted to voltage comparators, also being frequently used in some digital ICs.

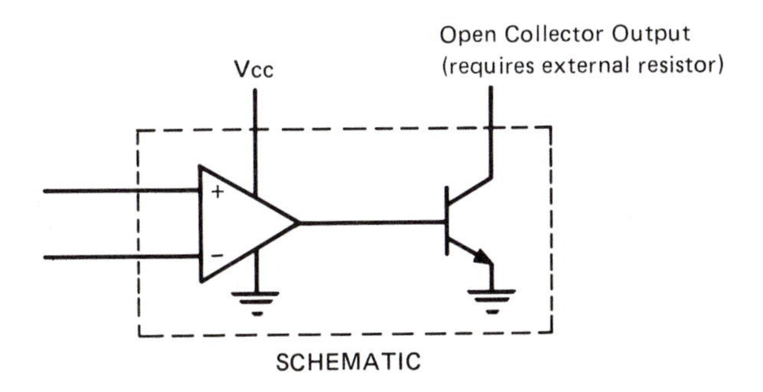

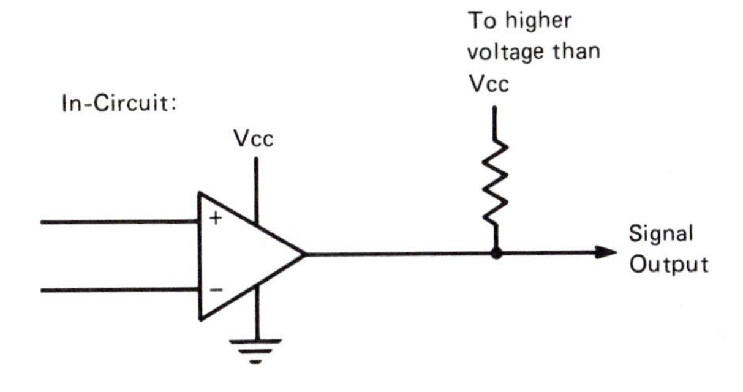

FIGURE 10-11 The open collector op-amp output stage.

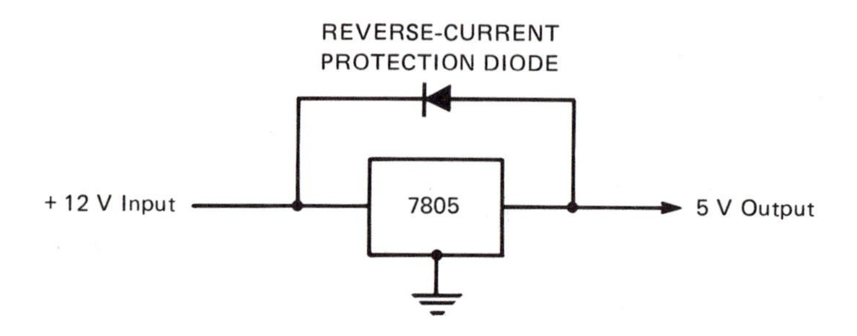

FIGURE 10-12 Protection diode for regulator ICs.

VOLTAGE REGULATOR ICs

Voltage regulator ICs are generally of the series regulator type. Figure 10-12 shows the basic circuit of the IC series regulator. (Zener shunt regulators were covered in detail in Chapter 6.) These regulator ICs have found wide use in electronic equipment. In addition to providing output voltage regulation circuitry in the chip, one small package may contain additional circuits to automatically reduce the output voltage if the regulator gets too hot and for current limiting to prevent regulator burnout if the load is excessive.

Regulator ICs can sometimes be mistaken for transistors because they are packaged in similar containers, large and small, and have the same three leads. The markings on the regulator will distinguish it from a transistor.

Checking these regulators in operation begins with checking the input and output voltages. The input voltage to the regulator must be at least 10 percent higher than the output voltage for the IC to operate. If the output voltage is too high, the regulator or the protection diode is defective. The protection diode is installed to prevent the capacitors in the output filter circuit from discharging back through the regulator when the power is turned off. If the problem with a regulator is good input voltage but low output voltage, the load should be disconnected to see if the output voltage of the regulator recovers. If it does, this is an indication that the regulator is either (1) overloaded by the following circuits or (2) unable to supply rated current and is therefore defective. If the output voltage does not recover and is still too low, the regulator is bad and should be replaced.

THE 555 TIMER CHIP

The 555 timer chip is very popular in several forms. It can be connected as a free-running, square-wave oscillator, as a monostable multivibrator, and as a retriggerable multivibrator. The best reference on this interesting chip is *The 555 Timer Applications Sourcebook* from Howard W. Sams Publishing Co. (See Appendix III.) With all of the circuits in which it can be used, the reader should become familiar with this special purpose chip.

CONSUMER IC CHIPS

There are a great many different types of consumer chips in use. They have been designed to perform a particular set of functions with a minimum of components. Figure 10-13 illustrates one of these chips. Troubleshooting these chips involves the following steps:

1. Get whatever information is available on the chip from the instruction manual for the equipment or the chip manufacturer's design specifications. Understand the use of the chip and what should be expected at each pin.
2. Verify that the proper supply voltages are available at the chip pins. Be sure that all of the supplies are accounted for, as these chips sometimes require several supply voltages.
3. Verify that the input signals are not only there, but that they have the proper DC input levels, if any, as well. An unwanted DC input level, for instance, can kill operation of a chip designed for an input without a DC offset voltage.
4. Verify that the output load of the chip is not shorted. This can make the chip appear bad when it is not.

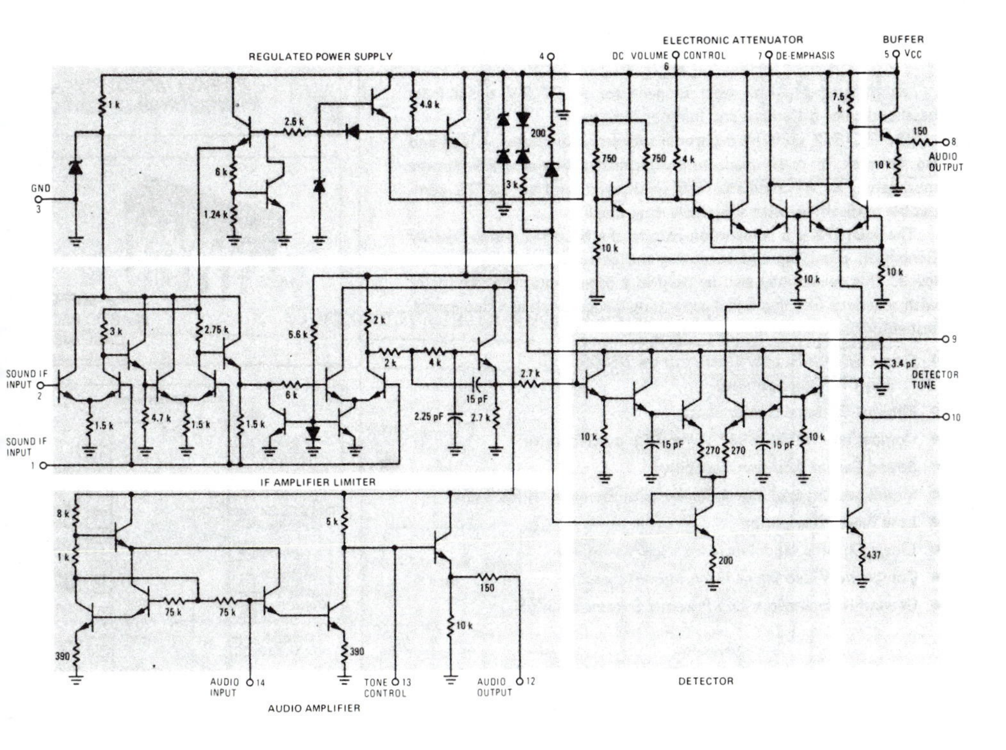

FIGURE 10-13 Internal schematic of a consumer IC chip illustrating many functions available on a single chip. (Reprinted courtesy of Motorola)

Complete audio amplifiers are often packaged as an IC. Audio-amplifier chips also often have very large electrolytic capacitors in series with their outputs. These capacitors often cause problems by shorting. This possibility should be checked first when troubleshooting these amplifier circuits. The major symptom of a shorted series coupling capacitor in an audio IC is the loud "plunk" of the speaker when it receives an initial surge of DC.

Remember, troubleshooting consumer IC chips requires the availability of proper schematics. It may be possible to obtain sufficient information for troubleshooting from the chip manufacturer. Without proper information, it is next to impossible to effectively troubleshoot these circuits.

ANALOG SWITCHES

Since the source and drain of an FET may be reversed without significant changes in the operation of the device, it follows that the FET makes a pretty

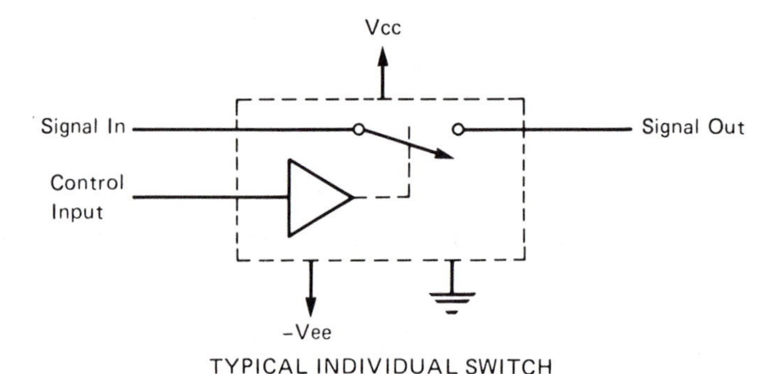

+Vcc

Chip Disable
Control Line

-Vee

TYPICAL ANALOG SWITCH CHIP

FIGURE 10-14 Typical analog switch schematic.

good switch. All one has to do is drive the gate lead of the n-channel FET switch positive with respect to the source, and the device becomes a low resistance between the source and drain connections. This is the principle of the *analog switch*.

Analog switches are ICs that contain several JFETs or IGFETs. Individual FETs are often connected so that there is a common input or output line for several of the FET channels. One or more pins are then provided to control the conduction of the internal FETs, individually or in sets. The switches may also be designed for either normally open or normally closed configuration (Figure 10-14). The resistance between the input and output pins is very low, typically from 400 to as low as 10 ohms. When open, the channels have many megohms of resistance.

Additional control lines called "enable" or "disable" may be provided on these chips. An input to either will affect the normal operation of the switch. When troubleshooting, look for these extra inputs and be sure that they are not causing an apparent problem of the switch IC because of improper voltages on them.

Troubleshooting the Analog Switch

The analog switch may be easily tested while in the circuit if either AC or DC signals are applied to the channels. Monitor the AC or DC voltage from input to output pin using a DMM (Figure 10-15). Many circuits using the analog switch will have high impedance, so the use of a standard VOM could load the circuit too much. If there is negligible voltage drop across the channel pins (less than a half-volt or so), the channel is switched on. If there is a higher voltage of several volts, the channel is switched off. Changing the actuation voltage should reverse these indications if the channel is good. Continue testing any remaining channels in the same way. A voltage that does not change with the actuation voltage indicates a bad channel. *Note:* Analog switches designed to operate from TTL signals will turn on above 2.5 V, and will turn off below 0.8 V. This is important to keep in mind when troubleshooting any TTL-associated chips.

AC signals may also be used to check the channel switching. The DMM may be used on the AC scales as above, or an oscilloscope may be used. The oscilloscope will be restricted to using ground on one side, and therefore it is easier to use it on the output side of the channel. Turning the channel on and off should turn off the AC signal at the channel output if the switch is good. Remember that a short in the output circuit may make the switch appear defective because no voltage will be present across a short.

Analog switches are often used in input circuits to operational amplifiers. Beware of the effects of a virtual ground at the input pin of an operational amplifier. No signal will be evident if the circuit is good. See the paragraphs about the operational amplifier earlier in this chapter.

OPTICAL ISOLATORS

Optical isolators are used to provide electrical separation of two circuits. Either or both sides of an optical isolator may have very high, dangerous volt-

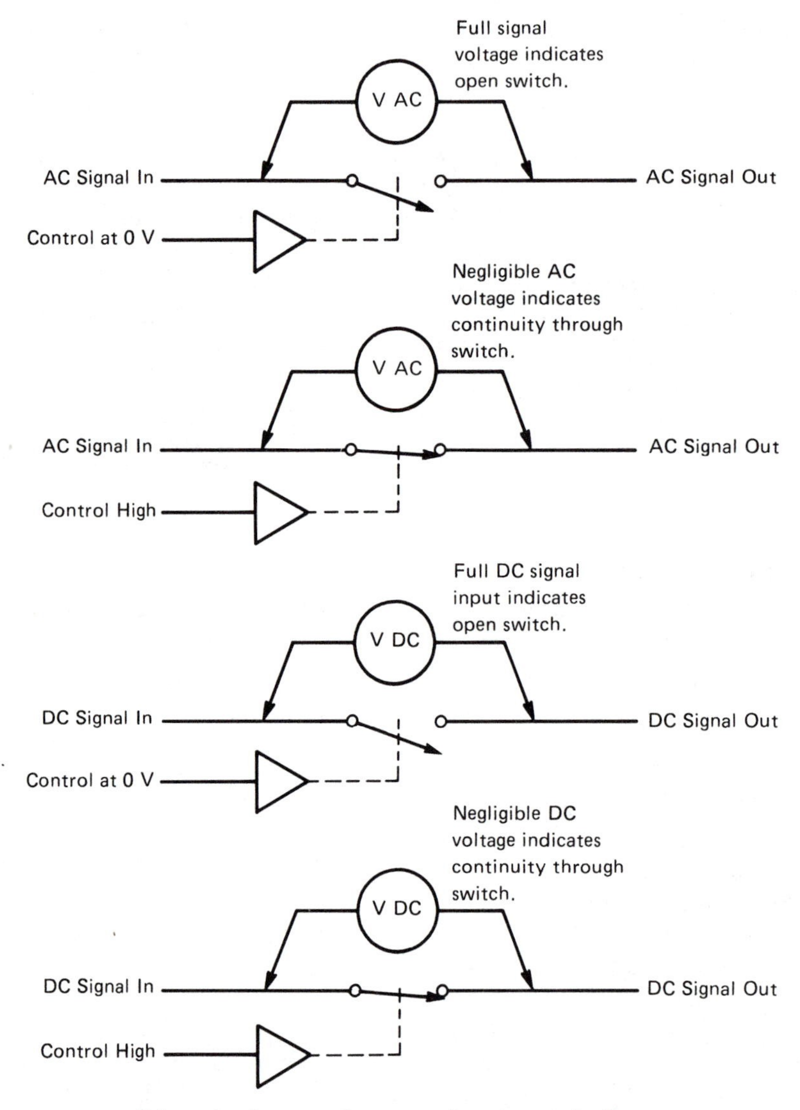

FIGURE 10-15 Testing analog switch with DMM.

ages present! In these circuits the indiscriminate use of grounded test equipment such as an oscilloscope can cause circuit damage.

Optical isolators usually consist of a diode and a transistor on the same IC chip (Figure 10-16). They are widely separated from each other, the only connection between them being the light beam that flows from the diode (an LED) to

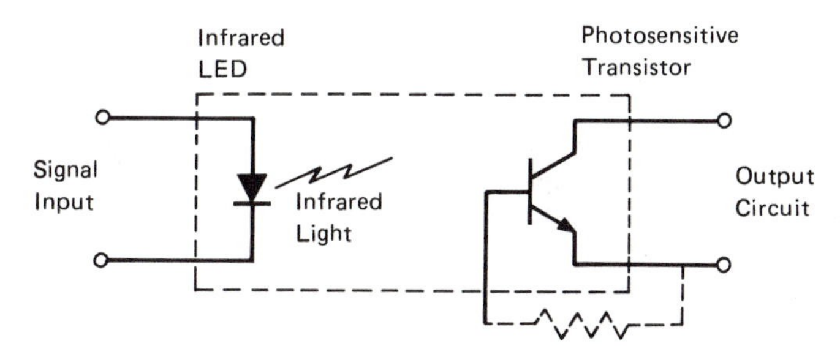

FIGURE 10-16 Inside an optical isolator.

the transistor (a photosensitive transistor). The separation of even a small six-pin optical isolator may be rated as high as 1500 volts of reliable isolation.

The internal diode of an optical isolator is often an infrared LED. It will have a typical forward voltage drop of about 1.0 volt. This is a DC value. An AC signal applied to the diode could operate it normally with a lesser average voltage as shown on a voltmeter. The output transistor of the isolator may be operating in a linear mode or it may be a switching transistor, designed for full-on or full-off operation. The specifications for the isolator may be consulted to determine the type of operation. In either case, the output transistor should be following the *current* waveform that the diode is providing.

Some optical isolators have a *Darlington-connected* output circuit rather than a single transistor. This indicates that the circuit is designed for full-on or full-off switching applications. The Darlington output stage may be considered, for practical troubleshooting purposes, as a single transistor (Figure 10-17). With a proper forward voltage drop on the LED, the output transistor should be conducting. Troubleshoot the LED side as a diode and the output transistor as a separate transistor as explained in detail in Chapter 9.

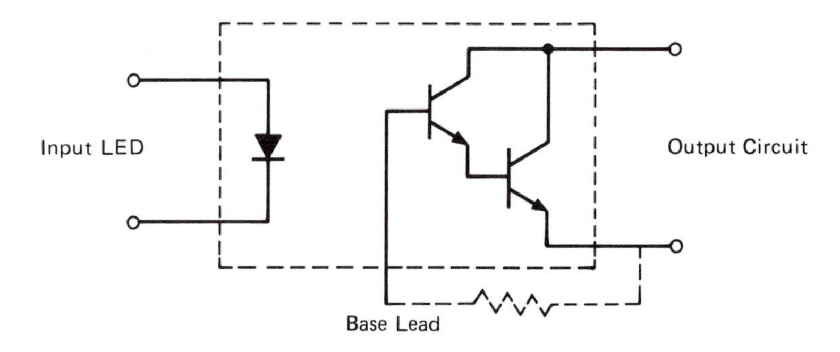

FIGURE 10-17 An optical isolator with Darlington transistor output.

SUMMARY

This chapter has covered many of the troubleshooting techniques used with analog ICs. The surprising concept of a virtual ground, where the absence of an input signal is normal circuit operation, was explained. Chapters 11 and 12 will cover troubleshooting of radio-frequency and pulse circuits.

REVIEW QUESTIONS

1. If a positive signal is applied to the (−) terminal of an operational amplifier, what would you expect the output terminal to do?
2. What would be indicated by a good output from an operational amplifier but no input signal present at the (−) terminal? The (+) terminal has a fixed voltage applied.
3. What is a DC offset adjustment, and how is it made in an operational amplifier circuit?
4. What voltage is amplified by a differential voltage amplifier?
5. What might a resistor to Vcc on the output of an IC indicate?
6. If the output of a regulator is very low, what test should be made first?
7. What two components are contained in an optical isolator?
8. What is a Darlington-connected transistor?
9. What is the normal forward voltage drop of an infared LED?

chapter eleven

Troubleshooting and Signal Tracing in RF Circuits

CHAPTER OVERVIEW

This chapter deals with circuits operating at any frequency above audio, approximately 20,000 Hz. Due to the high frequencies involved, entirely different circuits and special techniques are needed from those used at the lower frequencies.

An excellent reference work to have on hand is the Motorola *RF Data Manual,* #DL110. This reference book also covers tuning diodes and RF FETs. Many sample circuits are also included.

WHY RF CIRCUITS ARE DIFFERENT

When the length of circuit wiring becomes an appreciable part of a wavelength of the frequencies involved, strange things begin to happen. Wires can become inductors or capacitors. Shorts can appear open, and opens can appear as shorts. Currents can reflect back and forth along a wire like the waves in a watering trough.

Besides these effects, radio frequencies will radiate from circuitry. This effect is most troublesome when using signals that are very small, in the microvolt region. During signal tracing, the circuitry can be very sensitive to the mere placing of hands too close to the circuit board! In order to control these effects, RF circuits are kept physically small. If radio frequency must be sent from one place to another, it is confined within special wires to prevent radiation and to insure efficient transmission from source to load. For the best efficiency the source, the cable, and the load must all be designed to work together. This has resulted in a communications industry standard: the 52-ohm standard. Transmitters are designed with 52-ohm output circuits. Coaxial cable is designed with the same 52-ohm standard, referred to as a characteristic impedance of 52 ohms. Antennas for use with these cables must "look like" 52-ohm loads when connected to the ends of coax cables. Some antennas are already close to being a good 52-ohm load on the coax. Special tuning circuits are sometimes required to make other antennas appear to be the proper 52 ohms. These circuits are called *antenna matching networks* (Figure 11-1).

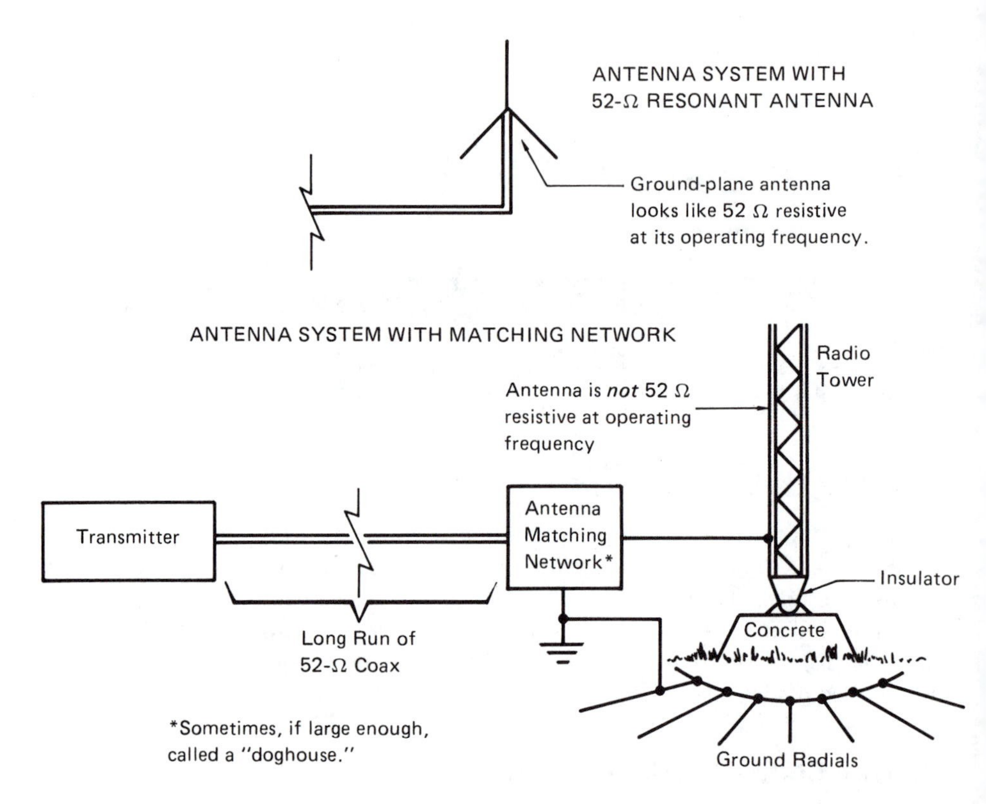

FIGURE 11-1 Antenna systems with and without matching network.

Video circuits are associated with radar and TV-screen displays. The frequencies involved are roughly from 15 Hz to 5 MHz. Video circuits have become standardized at a slightly different value of 72 ohms. Video signals use coaxial cables and terminating circuits of 72 ohms rather than 52 ohms.

SPECIAL TOOLS NEEDED

When adjusting circuits operating at high frequencies, the addition of only a picofarad or two may be enough to change the circuit's operation. A small metal screwdriver can cause such effects. The technician should have available a small assortment of the special alignment tools needed. This may consist of a few plastic Allen wrenches and a few small, plastic screwdrivers with tiny metal inserts in the tip for making adjustments. Some intermediate-frequency (IF) transformers require the use of a narrow-width, thick-bladed screwdriver.

It is wise to never force an adjustment. It is very easy to split the adjustment slugs of IF and RF transformers, making them expand like a brake shoe. This makes them very difficult to remove for replacement.

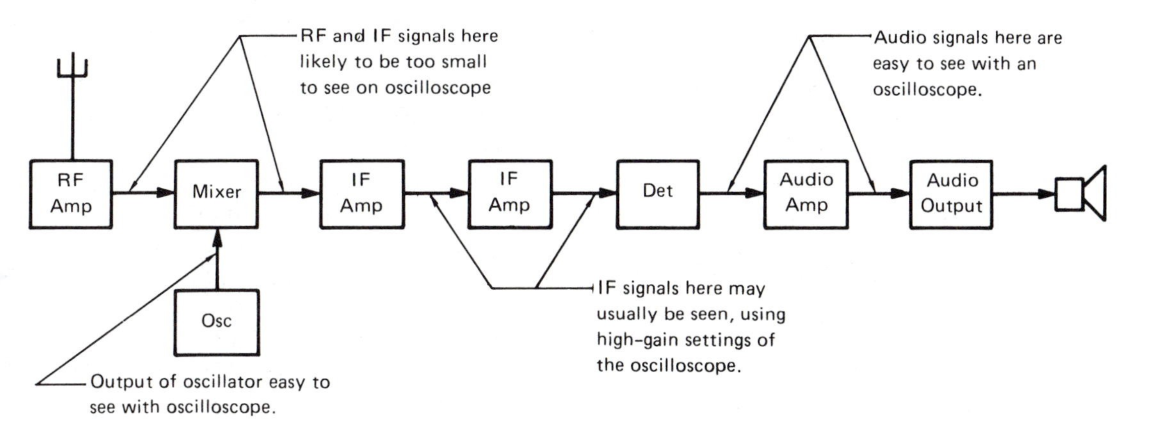

FIGURE 11-2 Relative signal amplitudes in a common AM radio. Some RF signals may be too small in amplitude to see with an oscilloscope.

SOURCE INSTRUMENTS

Existing Signals

Signals to trace through radio-frequency circuits should be the signals normally used by the equipment. An antenna will provide a small signal that may be traced in the later stages of radio receivers. The signal will be too small in amplitude to trace, even using the gain of an oscilloscope, in the earlier RF stages. The middle IF stages may provide enough gain to see the signals from that point onward (Figure 11-2).

Internal Signals

Internally generated signals, those produced by internal oscillators such as a receiver's local oscillator, are easily traced since they are of sufficient amplitude to see on an oscilloscope at any point.

The Signal Injector

The signal injector can provide signals for low radio frequencies and is especially handy for 455-KHz IF amplifiers and through the audio stages. Above about a megahertz, the signal injector may not have sufficient output to provide a reliable signal to trace through the equipment.

The Grid Dipper

In years past the grid dipper was an inexpensive, popular source of RF for signal tracing, but it has fallen out of use due to its frequency instability. Just touching the instrument can change the output frequency; thus, grid dippers are seldom used today as signal sources for signal tracing.

The Signal Generator

The usual signal source for tracing signals is the signal generator. The modern signal generator is stable in frequency and will also provide a reading of the amount of signal being fed into the circuit under test. The output frequency of the signal generator should be precise for the circuit under test. The signal generator frequency can be compared to harmonics of an internal crystal oscillator to verify exact frequency setting. The details of this calibration should be covered in the instrument instruction book, and the calibration procedure should be performed frequently.

A signal-generator output frequency can also be determined by turning up the output signal to near maximum and feeding the signal generator directly into a frequency counter. This is a good way of getting a precise frequency radiated into the working area, where direct connection to the receiver under test may not even be necessary. This method has two cautions: First, if the signal generator output is then reduced (below the point at which the counter will operate), the signal-generator frequency may change slightly, depending upon the signal generator's characteristics. Less-expensive generators may change their output frequency substantially as the output attenuator is changed. Second, the frequency may only be counted without any amplitude modulation on the signal generator.

The output amplitude control of the RF signal generator is often a calibrated dial attenuator on the instrument. Some models of signal generator require setting the amount of RF going into the final output attenuator by using a meter or other method. This also is a procedure that the technician should become very familiar with and perform whenever the signal generator is changed substantially in frequency. The newest models of signal generator automatically keep the output amplitude at the proper level; the frequency is always precise due to digital techniques of frequency generation.

The output of a signal generator should be fed into the circuits under test using a shielded cable, usually a 52-ohm coaxial cable. This is important to prevent stray-noise pickup on the way to the circuit and to help provide the calibrated amount of RF at the end of the cable. Using wires rather than coax at high frequencies can totally destroy any calibration of the amount of RF at the end of the wires.

There is an alternative way to introduce signals into circuit without direct connection between the generator and the circuit under test. Use a loop of several turns connected to the signal generator to inductively couple the signal into circuits. See Figure 11-3.

LOAD INSTRUMENTS

Normal Outputs

The normal output of the equipment can sometimes be used for detecting signals for signal-tracing purposes. The audio output of a receiver, for instance,

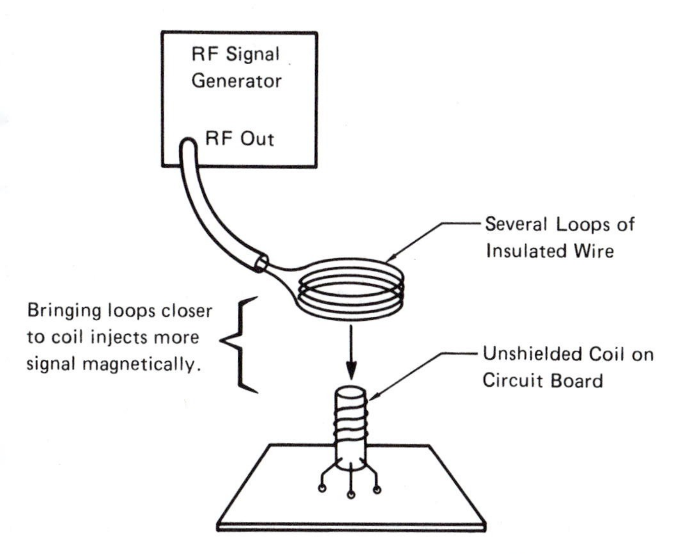

FIGURE 11-3 Inductively coupling RF signals into circuit coils.

can be used for this purpose. This avoids hooking up more test equipment. Another example is the picture tube of a TV receiver—it is a built-in indicator of equipment performance.

The VOM/DMM and the RF Probe

Another method for monitoring RF for signal tracing is to use an RF probe on a VOM or other DC meter. The RF probe just rectifies the RF into a DC, which the meter can then read. The VTVM was commonly supplied with this accessory. The RF probe has been largely replaced by modern oscilloscopes with their high-frequency vertical amplifier response. Attempting to use a VOM, FETVM, or DMM directly on RF circuits without a probe is useless. The meters simply don't respond at all, or they go crazy with strong RF fields and slam their needles on the end pegs, or the digits jumble meaninglessly.

The Oscilloscope

The oscilloscope is a very versatile instrument. It will show RF and can be used to indicate amplitude, frequency, and distortion. But it is not a perfect instrument. It must be used with some knowledge of its limitations.

Vertical Amplifier Frequency Response. An oscilloscope can be used to show RF voltages only up to a certain usable frequency limit. Frequencies above that limit will show less and less amplitude on the screen until nothing at all will be shown other than a baseline sweep. Some method for determining a useful frequency limit had to be standardized. This is known as the "3DB point" of vertical frequency response. To find this limit, an input frequency was applied to a scope and increased in frequency slowly while its input amplitude was held constant. When the vertical amplitude dropped to 0.707 of the real amplitude being applied, the frequency of the input signal was noted. This became the standard frequency used to compare oscilloscope vertical amplifiers. It means, roughly speaking, that at this frequency and beyond the oscilloscope will be more and more deficient in providing calibrated amplitude measurements.

Signal Tracing with the Oscilloscope. The oscilloscope may be used for signal-tracing purposes as an output indicator at reasonable frequencies. "Reasonable" frequencies are those not far above the upper frequency limit of the vertical amplifier in the oscilloscope. An oscilloscope vertical amplifier with a response of 20 MHz, for instance, may be quite usable for signal tracing up to as much as perhaps even 40 MHz, but the amplitude indications would be far less than they really are. Below this example of 20 MHz the oscilloscope would provide more accurate voltage readings. The technician is only rarely interested in the precise value of an output voltage for signal tracing, more often being concerned with the simple presence or absence of a signal.

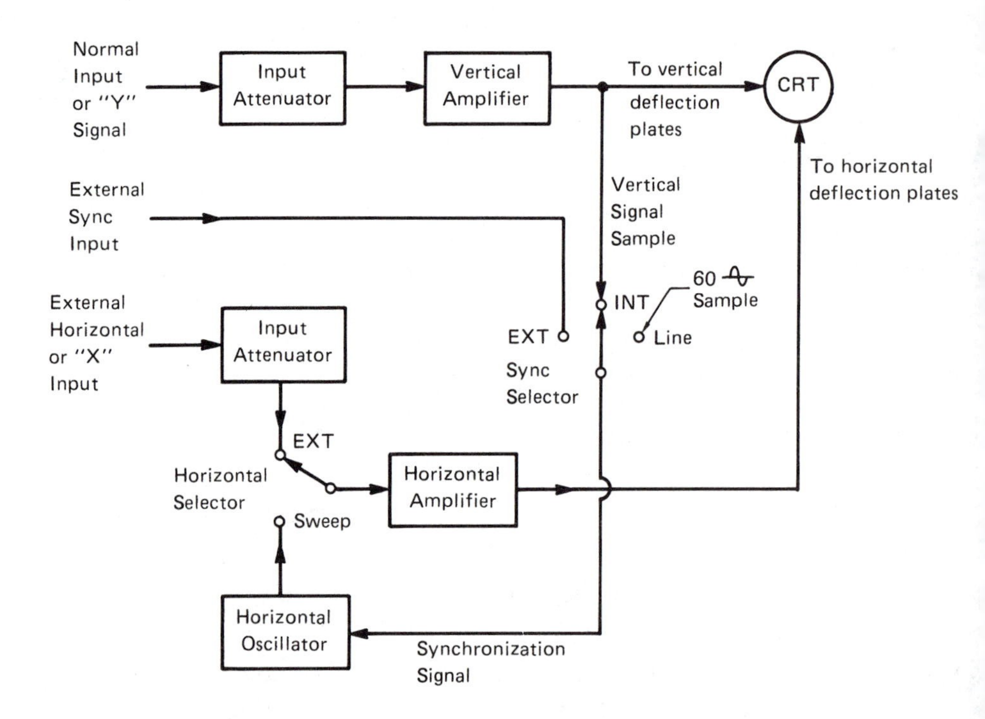

FIGURE 11-4 Basic oscilloscope block diagram.

It is important to be very familiar with the block diagram of a basic oscilloscope (Figure 11-4). It is helpful in understanding the functions of the controls, of which there are plenty.

Oscilloscope Probe Limitations. A high-quality probe will not appreciably change the waveform on the input signal as it is presented to the oscilloscope. In order to accomplish this, the probe will probably have an adjustment, located at either end of the cable, to properly balance the capacitance of the probe and cable to the input circuits of the oscilloscope. Whenever a probe is used for *any* measurements, it should be checked for a "flat" response. This is easily done by touching the probe tip to the "calibrator" output of the oscilloscope and adjusting the probe for a flat waveform (Figure 11-5).

Oscilloscope probes provide some very real advantages over connecting an oscilloscope to a circuit with a wire or a simple shielded cable. Open wires to an oscilloscope will pick up tremendous amounts of noise and "garbage." Shielded wires have undetermined amounts of capacity to load high-frequency signals. Probes are available in standard attenuation factors of 1X, 10X, 100X, and more. The 1X probe is often used because there is no conversion factor to consider when using it—what is seen on the screen is basically what is in the circuit. The 1X probe, however, has one serious disadvantage: It will pass on to the circuit

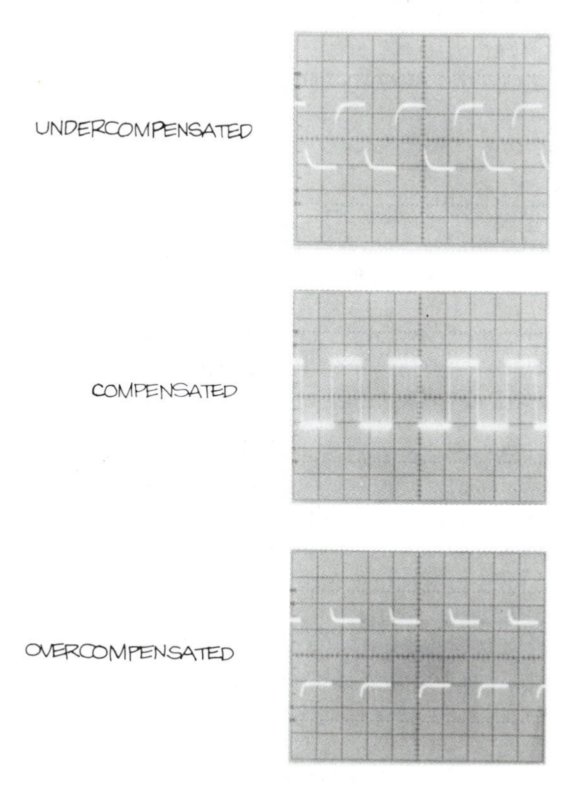

FIGURE 11-5 Waveforms for probe adjustments. (Courtesy of Tektronix, Inc.)

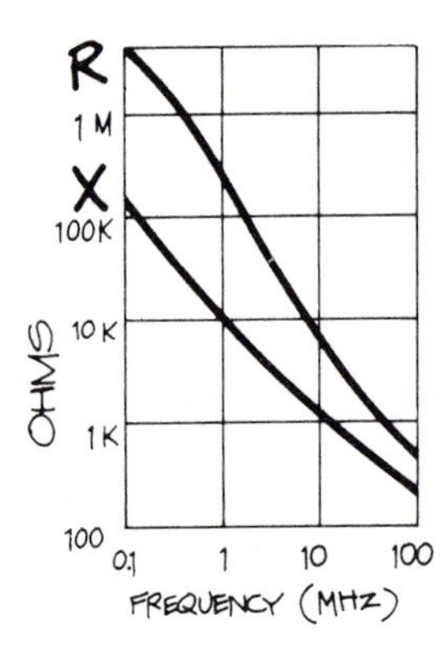

FIGURE 11-6 How a probe loads a circuit as the frequency increases. (Courtesy of Tektronix, Inc.)

under test all of the inherent capacity to ground of the oscilloscope's input circuit, the connecting coax cable, and the probe tip itself, right into the circuitry under test. This can amount to a considerable amount of capacity, as much as 100 picofarads or more. In DC applications this may be of no particular consequence, but in RF circuits it is murder! A much better solution is to use an attenuator probe. A 10X probe, for instance, will have only one tenth of the capacity loading effect of a 1X probe, perhaps 10 picofarads. A 100X probe will have even less, but the attenuation factor becomes a problem for viewing low-amplitude signals. When making RF measurements be sure to ground the working end of the probe to the circuit under test with as short a ground clip as possible.

When using the oscilloscope as a signal detector, keep in mind that although the resistance of the usual test probe is a high-impedance 10 million ohms, the capacity of the probe to ground may be 1 pF or so. This capacity might be enough to upset the tuning of critical circuits. Using the oscilloscope in high-impedance circuits at very high radio frequencies may disturb the circuits excessively. For instance, even a good 10X probe can load a 100-MHz circuit with less than a thousand ohms! (See Figure 11-6.)

Frequency Determination with the Oscilloscope. The approximate frequency of unknown signals may be determined by using an oscilloscope. Count the number of *whole* cycles (counted from the same point on each cycle) in a given number (and fractions) of horizontal divisions. Don't try to estimate portions of cycles to make the measurement come out an even number of horizontal divisions. The best accuracy is obtained when many divisions are used (Figure 11-7). Use the formula below to determine frequency within approximately five percent. The accuracy will depend upon the accuracy and linearity of the oscilloscope's horizontal oscillator and the skill of the operator in reading the screen.

$$F = \frac{C}{D \times S}\text{, where}$$

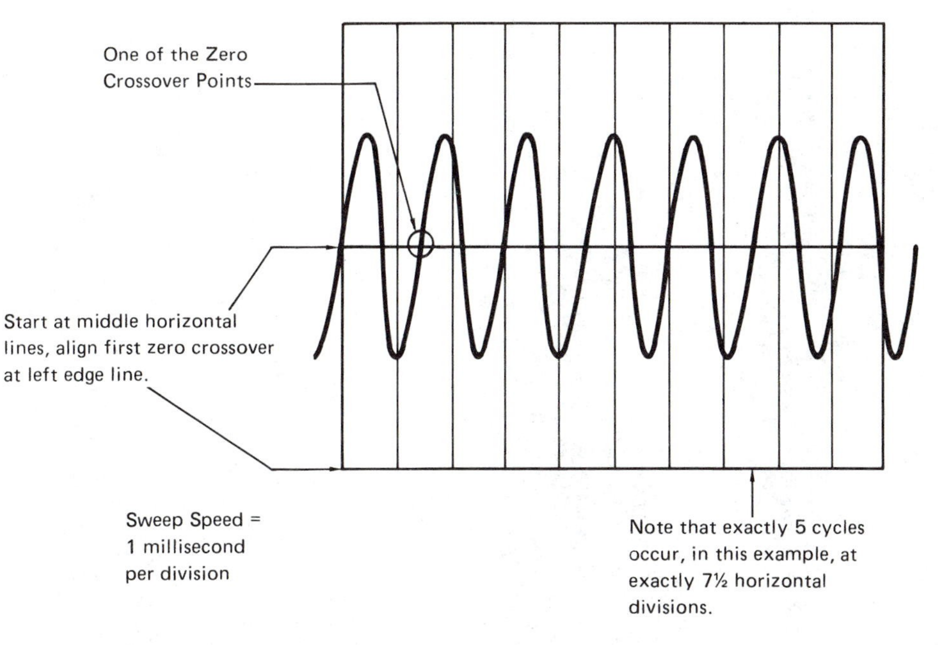

FIGURE 11-7 Use of divisions in determining frequency with an oscilloscope.

F = frequency
C = whole cycles
D = divisions and fractions
S = sweep speed per division
(convert to seconds; 1 ms = 0.001; 1 μs = 0.000001)

The frequency of the signal in Figure 11-6 would be

$$S = 1 \text{ ms} = 0.001 \text{ second}$$
$$D = 7.5$$
$$C = 5$$
$$F = 5/(7.5 \times .001) = 5/0.0075 = 666.67 \text{ Hz}$$

Be sure to convert the sweep time per division to seconds.

The RF Wattmeter

RF output circuits generating appreciable power must be terminated in a proper load. Without a terminating load, the RF amplifier could be damaged. For most transmitters this load has been standardized at 52 ohms. The power

FIGURE 11-8 Terminating RF (Termaline®) and Thruline® wattmeters. (Photos courtesy of Bird Electronic Corp.)

output of an RF power amplifier may be monitored by dissipating the power in a proper load and measuring the RF voltage across that load. With a known load resistance a voltmeter can be calibrated to read directly in watts. This meter is called a terminating RF wattmeter and is accurate only when used with the right load resistor. A similar arrangement can be connected together using a separate terminating 50-ohm load and a Thruline® wattmeter set to read the power flow into the load. The wattmeters shown in Figure 11-8 are calibrated to read the power of a continuous carrier. Special steps must be taken to measure peak-envelope power (PEP), a specification unique to single-sideband transmitters (SSB). This topic is covered in Chapter 18.

The Standing Wave Ratio (SWR) Meter

This instrument is installed in a coaxial transmission line and measures the voltage on the coax line due to the power flowing into the antenna. It also measures the voltage resulting from any reflections from the antenna back toward the transmitter. By comparing these two voltages, a measure of the efficiency of the antenna system may be obtained. As a signal-tracing instrument, it provides a good indication of the presence of RF in the antenna circuit. To decide whether a problem is in the antenna circuit or not, consider the ratio of these two voltages: the ratio should be very high in a good antenna system—very little voltage should be produced by reflections as compared to the voltage flowing toward the antenna.

Using a Separate Receiver as a Detector

There is one completely nonloading method of signal detection available. A sensitive receiver set to the proper frequency can detect signals without any direct connection at all. There are a few things to keep in mind when using a receiver as a detector: The receiver must be set to precisely the right frequency; the receiver must be capable of processing the type of modulation used in the test, such as an FM output for FM signals or a tone output for a continuous pure carrier input (commonly called CW reception); and the signals searched for must be of reasonable amplitude in order to be sufficiently radiated to the receiver. Direct connection to a receiver input should be made only from circuits that are producing very low power levels, comparable in amplitude to the output of a signal generator.

When working on transmitters, it is not unusual to have other receivers available for use as test receivers. For instance, the amateur radio equipment technician would likely have several vhf receivers on hand to use in testing a vhf transmitter. Amateur technicians sometimes have two or more pieces of equipment of the same frequency range that can be used in this manner.

Another approach to detecting signals is to use a communications receiver. A communications receiver often covers the HF range, 3 to 30 MHz, and sometimes below that frequency range down through the broadcast range to about 550 KHz. Tuning a communications receiver to an exact frequency may be a problem with older, analog-tuned receivers. Only very expensive analog receivers have anything near precise frequency indications. A few of the more recent surplus communications receivers may fill this requirement. It may be necessary to use a modern, digital communications receiver for detecting signals in larger shops. The frequency readout of a digital receiver is precise. It is also possible to set up a good communications receiver to receive continuous RF signals (called CW or continuous-wave signals by some), modulated AM or SSB signals, or even FM signals in the HF range.

The Frequency Counter

The frequency counter can be used only for giving the frequencies of signals. It cannot be used to indicate relative amplitude of signals. When using the frequency counter, be sure that the input of the counter will not load the circuit under test. An oscilloscope's 10X probe may be used with the frequency counter. This will reduce the loading on the circuit. When using such a probe,

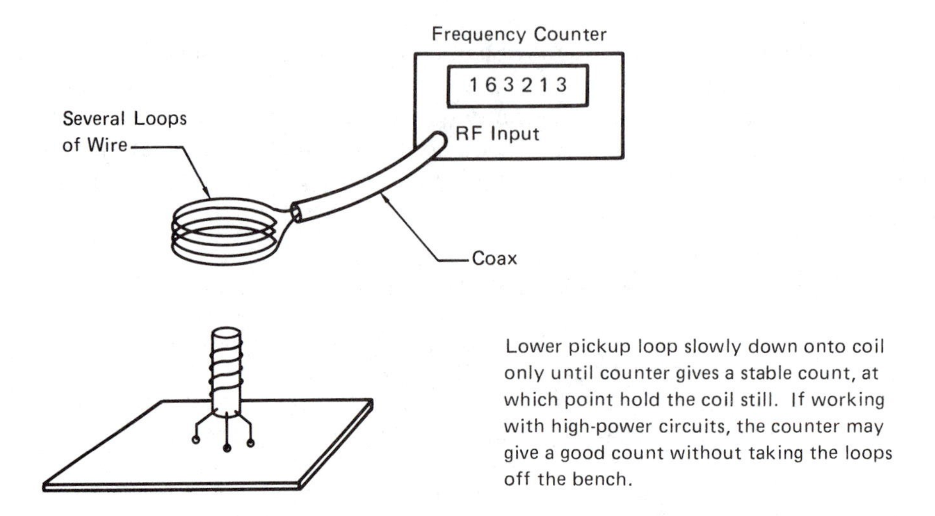

FIGURE 11-9 Using magnetic pickup loop with frequency counter.

the circuit under test must have a large enough amplitude to provide an input to the counter sufficient to give a stable count. This is particularly important to note if a 10X probe is used.

A small loop of several turns of wire can also be used to detect RF without the direct connection necessary when using a probe (Figure 11-9). Connect the loop directly to the end of a piece of coax.

Another consideration when using the frequency counter is that the input signal must have a fairly constant amplitude. Attempting to count a waveform, such as an amplitude-modulated waveform, will produce a useless, unstable count. The counter will be counting only those cycles that are of sufficient amplitude to trigger the counter input, ignoring all others.

SIGNAL TRACING IN RF CIRCUITS

The Originating RF Circuit (Oscillators)

Signal tracing these stages consists merely of verifying that they have a signal output of sufficient amplitude for the following stages and that the output frequency is proper. To determine frequency, the calibrated communications receiver approach may be best (there is no loading on the circuit); the frequency counter is most accurate but loads the circuit, and the oscilloscope is the least accurate method of frequency determination.

Some quartz crystal oscillator circuits must be properly adjusted for them to operate. The adjustments available in a failed oscillator circuit should be changed slightly to see if the circuit suddenly begins working. To use the oscilloscope on RF oscillators, try to select a pickup point that is at low impedance.

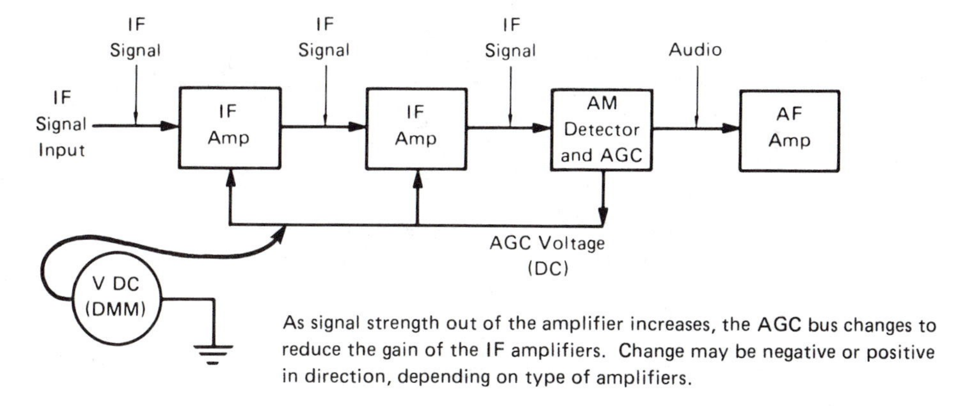

FIGURE 11-10 Monitoring IF amplifier performance by watching AGC bus voltage.

This could be the emitter of a transistor, if it is not bypassed with a capacitor, or the output winding of an interstage transformer.

Processing RF Circuits—Small-Signal Amplifiers

Small-signal amplifiers have small input signals. These amplifiers will be operating much like similar circuits designed for low frequency. A small input signal produces a larger output signal. Some of these amplifiers have provision for control of their gain by the application of a DC signal external to the stage (Figure 11-10). This is a common situation in intermediate-frequency (IF) amplifiers of AM receivers. This control voltage should be checked to be sure that it is proper before stage gain is considered insufficient. This is treated more thoroughly in Chapter 18.

Source signals for signal tracing small-signal amplifiers are mostly either normal signal inputs or signals from a signal generator. Loads for signal-tracing purposes in small-signal amplifiers are most often oscilloscopes. Remember the RF-loading effects of the oscilloscope, and be sure that the frequency of operation is reasonably within the vertical amplifier frequency limitations of the oscilloscope.

Detectors and Converting Stages

These stages process incoming RF signals and convert them to signals other than RF. Receiver detector circuits have an RF input and produce an audio frequency output (Figure 11-11), which may be traced using the techniques as covered in Chapter 8.

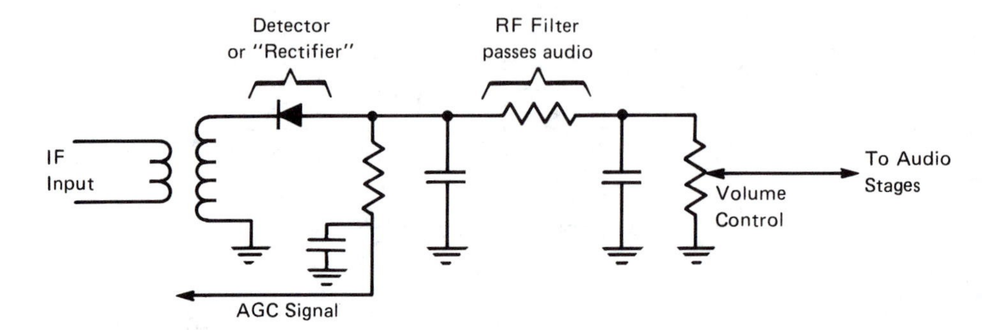

FIGURE 11-11 A typical amplitude-modulation detector.

A few circuits produce a DC output from RF signals for measurement purposes (Figure 11-12). The inputs, as above, are analyzed as RF stages, and the output as DC circuits as covered in Chapter 5.

Processing RF Circuits—The Large-Signal Amplifier

Large-signal amplifiers, such as the last amplifier stages of transmitters, amplify an initially large signal. The output circuit includes an inductor and a capacitor. Because of the flywheel effect of the L (inductance) and C (capacitance), the amplifier stage is able to operate in efficient Class C operation. The L/C output circuit smoothes the pulses, producing a smooth sine-wave output voltage. Since it takes less power to produce RF with Class C, this method is often used in designing RF stages, especially those operating at a single frequency or those stages that an operator can be expected to put into proper tune during normal operation.

These Class C stages also make very good frequency multiplier stages, because of the short, sharp pulses of current in the output circuit. The output L/C may be tuned to two, three, or more times the input frequency and produce a respectable output at the higher frequency.

Because of Class C operation, the bias of a large-signal RF amplifier can be estimated. The bias voltage will be well below cutoff so that no current will be flowing at static DC levels. Insufficient drive voltage feeding a Class C stage will

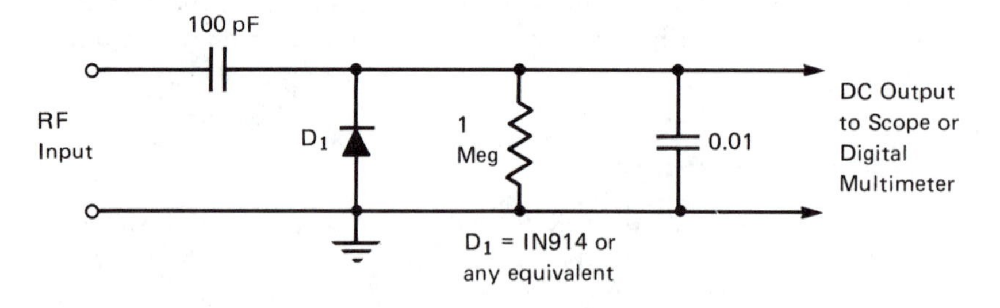

FIGURE 11-12 Circuit for RF demodulator probe.

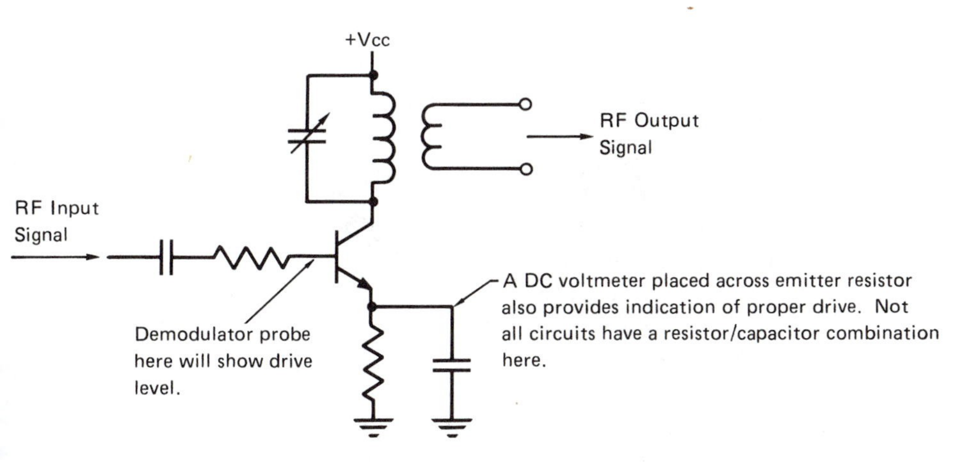

FIGURE 11-13 Typical Class C RF power-amplifier stage.

result in no output. A high input voltage is required to drive the input far enough into conduction to produce current pulses in the input circuit. (See Figure 11-13.)

Radio frequency of proper strength entering the input of a bipolar transistor or a tube Class C amplifier meets a diode action. In the transistor, the emitter-base junction, and in the tube, the cathode to grid section, make a rectifier. Normal drive to these stages produces a negative DC voltage with respect to the emitter or cathode, which can be used as an indication of the proper amount of drive to that stage. Sometimes special test points are provided that, with a voltmeter, are used for this purpose. The more drive there is, the more DC voltage present in the input circuit.

Drive signals for power RF amplifier circuits must produce a power input, since both voltage and current are required to drive them. A simple signal generator may not be powerful enough, even at maximum output, to properly drive a power output stage. Internal signals must therefore be used to signal trace in RF power amplifiers. The load instrument for signal tracing through RF power amplifiers is usually a wattmeter with a dummy load.

TERMINATING RF SIGNALS

The antenna is the usual termination for operating RF equipment. During signal tracing the amplifier should always be terminated in a dummy load to prevent possible radiation of interfering signals. A dummy load is a special resistor of the proper value (usually 52 ohms) built in such a way that it has very little extraneous inductance and capacity to ground. Figure 11-14 shows a pair of dummy-load resistors for use on low-powered transmitters. Both of these provide a sample of the RF signal at the other end (they may be used in either direction) and can be used as power attenuators. Never operate a transmitter without a proper load, either a dummy load or an antenna. To do so may damage the final amplifier because of the excessive voltages that may develop. If the

FIGURE 11-14 Two resistive "dummy" loads used to absorb RF energy. These two also provide an attenuated sample of the RF.

termination is a terminating wattmeter or a dummy-load resistor with an SWR meter or directional wattmeter, the meters will provide the indication of operation that is necessary for signal tracing.

MIXING RF CIRCUITS

The RF mixer used in radio circuits takes two frequencies simultaneously and produces several at its output, only one of which is the desired one. The proper signal is selected by output circuit tuning, while the unwanted ones are attentuated by that tuning to virtual nonexistence.

One of the two input signals must be a large one. A large signal is required to operate the mixer stage into the "rectifying" portion of its output curve and enables the whole stage to "work." It is this larger input that is probably generated in the equipment. In receivers it is called a local oscillator signal. The local oscillator signal can be detected with a communications receiver in most cases. There should be plenty of radiation from the local oscillator for this test. The presence of this signal may also be checked with an oscilloscope.

The second signal may be very small in comparison. It could be the signal coming from an antenna, with or without amplification. This signal may be too low in amplitude to see on an oscilloscope if it is a signal originating from an antenna. It may be necessary in this case to provide a signal from a signal generator through the antenna connector to verify a good signal path through to the mixer input. The output amplitude and frequency of the mixer may be checked with an oscilloscope.

THE RF LOOP CIRCUIT (FREQUENCY SYNTHESIZER)

The frequency synthesizer is sometimes called a phase-locked loop, or PLL for short. Radio-frequency synthesis circuits operate by generating a DC "steering" voltage to correct the output frequency of a voltage-controlled oscillator (VCO) until it is on exactly the right frequency (Figure 11-15).

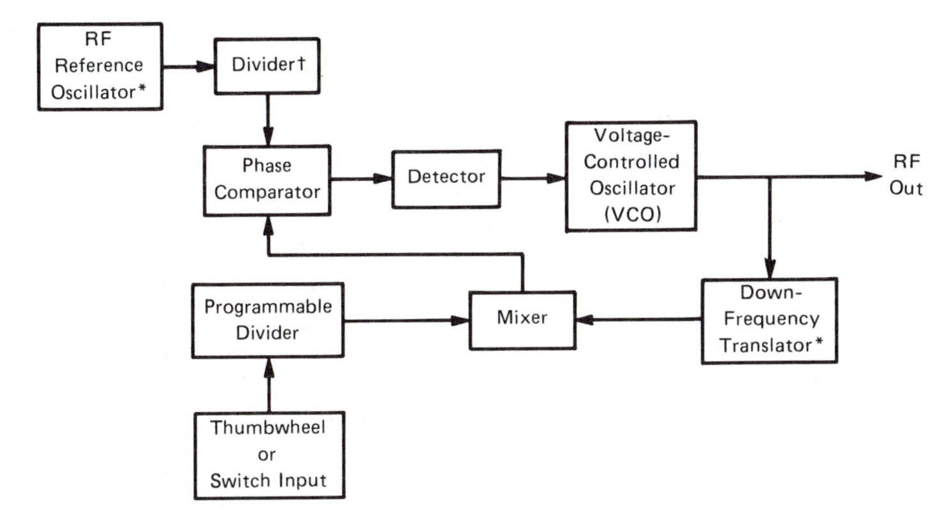

FIGURE 11-15 Basic phase-locked-loop frequency synthesizer.

Troubleshooting the phase-locked loop begins with determining a few easily checked facts. Test each of the fixed-frequency oscillators to be certain not only that they are in operation, but also that they are on exactly the right frequency. These are indicated in Figure 11-15 with an asterisk (*). A frequency counter or a communications receiver can be used. Neither of these instruments will give an indication of the amplitude of these oscillator signals, but if the signals are present and at the right frequency, it is reasonable to go on and check other sources of possible problems before worrying about the precise amplitude of these signals. This step alone can identify many of the problems that occur in the PLL circuit. The frequency counter is particularly needed for working in frequency synthesizer circuits, since precise frequency rather than signal amplitude is the major issue.

The next step is to check the output of any fixed-frequency dividers. In our example of Figure 11-15, this is shown with a dagger (†). A frequency counter or an oscilloscope may be used to make this test.

The output frequency of the voltage-controlled oscillator (VCO) will be determined by the voltage applied to the oscillator frequency control element, a varactor. Use an oscilloscope to see if the VCO has any RF output. If there is no output from the VCO, none of the following circuits can possibly have a valid output. If the VCO has an RF output, proceed to the next paragraph. If the VCO does not have an RF output, it may be necessary at this point to break the PLL circuit where the filtered DC control voltage is applied to the varactor. A well-regulated DC voltage should be substituted to see if the VCO will operate with a control voltage somewhere between the supply voltage and ground. If the VCO produces an output of approximately the right frequency, the VCO is probably all right and the problem is elsewhere in the loop circuits. If the VCO will not generate RF with this substitution of a control voltage, the VCO is defective.

The mixer circuit following the VCO may be traced through by the procedures already covered in this chapter. Both input signals have enough amplitude to be seen easily on an oscilloscope.

Having determined that the RF oscillators are all working, examine the programming circuitry for a valid code. Some PLLs are programmed so that out-of-band operation results in a shutdown of the oscillator. Since the programmable frequency dividers are digital, it should be reasonably easy to determine if they are working. Set up an easy-to-test combination of the thumbwheels, and check for proper coding from them into the divider. Then look with an oscilloscope at the frequency going into the divider, count the cycles in a convenient horizontal space, and then look at the output of the circuit. Make another count there. For example, if the divider circuit is divided by ten, there should be exactly one-tenth the number of cycles on the output side as on the input.

Note: After careful examination of RF circuits, it is possible that the technician may still not be able to identify the cause of a malfunction. It would be well to consider the possibility of a signal that is off frequency far enough that the circuits following cannot operate properly. It would be well in this case to use a frequency counter to verify the actual frequencies within the equipment.

SUMMARY

This chapter has introduced the reader to the considerations in troubleshooting and tracing signals through radio-frequency circuits. When the bad stage is located, the reader may proceed to Chapter 16.

REVIEW QUESTIONS

1. What accessory will enable the use of a voltmeter in RF circuits?
2. Could a 10-MHz oscilloscope be used to signal trace in a 20-MHz circuit?
3. Name three advantages of using a proper oscilloscope probe rather than a 50-ohm coaxial cable connection to the scope input.
4. What is the frequency of a waveform that completes 5 cycles in 5 divisions of an oscilloscope set for a calibrated 10 ms/cm?
5. Why cannot a frequency counter be used to measure the frequency of AM signals?
6. What might you expect to see if you connected an oscilloscope to the antenna circuit of a radio receiver?
7. Name the two outputs of an AM detector used in a receiver.
8. What are two purposes of the dummy load?
9. What collector current would you expect in a Class C bipolar transistor amplifier stage in the absence of an incoming signal?
10. One of the two signals' input to a radio receiver mixer is larger in amplitude. Which of the two inputs is it?
11. What determines the output frequency of a VCO?

chapter twelve

Troubleshooting and Signal Tracing in Pulse Circuits

CHAPTER OVERVIEW

This chapter will deal with the special techniques of tracing signals through circuits with nonsinusoidal signals (signals that are not sine waves). The sudden application and removal of voltage and current leads to some very different effects from those of circuits using only DC or sine waves. Pulse circuits are used in applications such as television, radar and loran. They are not to be confused with digital circuits, which are treated as an entirely different subject for troubleshooting purposes in Chapter 13.

A special concept should be understood when dealing with pulsed circuits that charge the magnetic field of an inductor and then discharge it through a load: The polarity of the voltage across the inductor depends on both the direction of current flow *and* whether the current is increasing or decreasing. When the current flow through an inductor changes from an increasing current to a decreasing current—or when the current flow is suddenly interrupted—the voltage across the inductor reverses.

TERMS UNIQUE TO PULSED WAVEFORMS

Duty cycle is a term that is used frequently in dealing with pulsed circuits. The duty cycle of a pulse waveform is the time that the signal is "on" divided by the time of the whole cycle, expressed as a percentage (Figure 12-1). The names of various pulse parameters are given in Figure 12-2. The servicing technician should be familiar with these.

Another point peculiar to pulse circuits is the polarity indicator on transformers used in these circuits. When the marked lead of a transformer primary has a specified polarity applied, the same polarity will appear at a marked lead of the secondary. The mark is a small dot or other coding on both the schematic and the transformer. This marking of the leads is necessary to properly phase the transformers in building and maintaining the circuits (Figure 12-3).

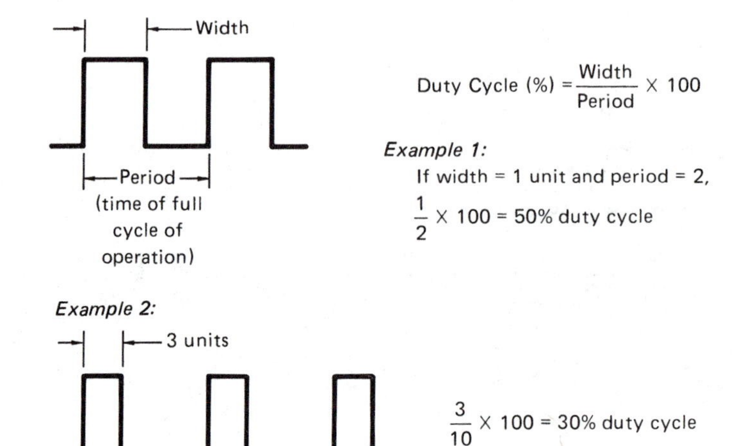

FIGURE 12-1 Duty cycle explained.

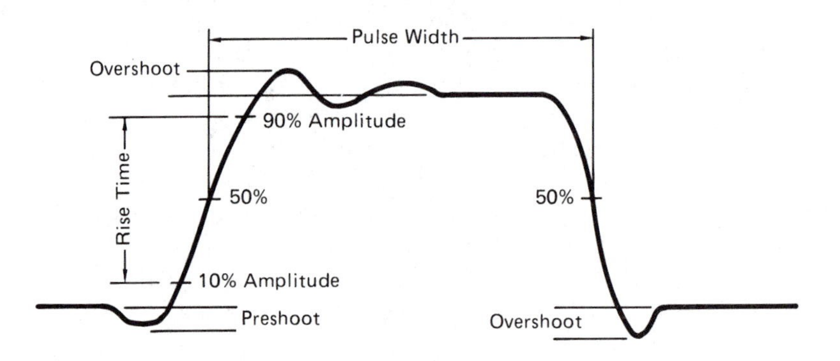

FIGURE 12-2 Pulse waveform parameters (parts).

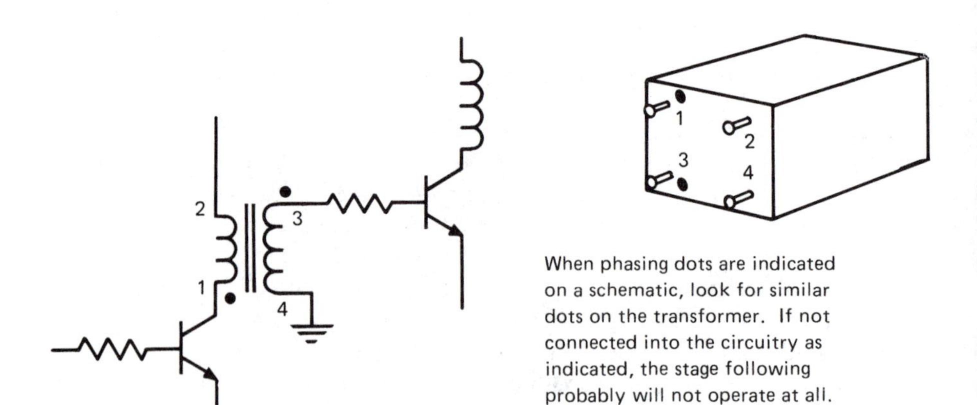

FIGURE 12-3 Use of dots to identify phasing of transformer leads.

Biasing in Pulsed Circuits

Pulsed circuits almost always operate Class C. As such, they have no output current flow in the absence of an input signal. Bipolar transistors by their nature are cut off until driven into conduction, so no special bias supply is necessary. A separate bias supply must be provided for vacuum-tube circuits to keep the stages in the cutoff condition. Failure of the bias supply may cause excessive current flow through many stages, likely causing an overload of the power supply.

Troubleshooting pulsed circuits should begin as it does for other circuits such as DC, LF, and RF: Check the power-supply voltages to be sure they are correct. In tube pulsed circuits, be sure to locate and check the bias supply, too. When a defective stage is identified, check the bias and Vcc right at the stage, to be sure they are correct at that point.

An input signal to a Class C stage must be sufficiently high in amplitude and have enough current capability to overcome bias supply, if any, and to then bring the stage into full conduction. Remember that it takes up to 0.7 V of forward bias to make a bipolar transistor conduct. Watch for this as a possible problem and pay attention to input pulse amplitudes as compared to the bias of the stages.

SOURCE INSTRUMENTS

The signals normally produced within equipment to be repaired are the primary signals used in tracing signals through pulse circuits. External sources of pulse signals may come from one of two generators for this purpose, the *pulse generator* and the *function generator.*

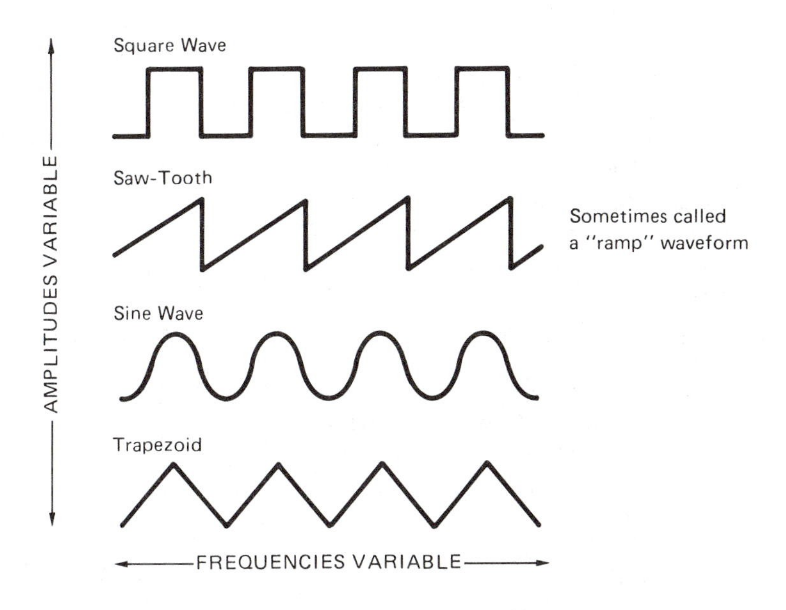

FIGURE 12-4 Waveforms often available from a function generator.

Depending upon the individual equipment, pulse generators may be adjustable for setting the pulse amplitude, duration, and how often the pulses occur. The frequency of recurrence is called the *pulse repetition rate* (prr). ("Frequency" is a term used mostly for sine waves.) For instance, if pulses occur 1000 times a second, they have a prr of 1000.

The function generator may have the capability of producing square waves, sine waves, trapezoidal waves, and saw-tooth waves. It may be used as a signal source in some pulsed signal tracing applications. See Figure 12-4 for an example of these waveforms.

The pulses coming from the function generator must provide sufficient amplitude and current to drive the stage to which it is connected if it is used as the source of signals for signal tracing . A pulse of too low an amplitude or current capability will not trigger the stage into operation and may make the stage seem defective when it is really not. For most troubleshooting, the internal signal sources within the equipment are quite sufficient, and a function generator will not be needed.

LOAD INSTRUMENTS

Often a technician can use the indicators installed in the equipment under test to monitor proper operation. Installed cathode-ray tubes (CRTs) are commonly used. In the absence of a suitable output indicator the usual load instrument will be the oscilloscope. The oscilloscope will be discussed in some detail because of some special considerations that must be kept in mind when using it with pulsed circuitry.

The Oscilloscope

Before discussing the oscilloscope itself, the oscilloscope probe should be considered as it relates to pulsed circuits. Some pulsed circuits generate hundreds of volts because of the nature of these circuits. Take for example the ignition circuit of a car, where the interruption of the ignition-coil primary current produces voltages greater than 20,000 volts from a 12-V source. Oscilloscope probes have voltage limitations, commonly 500 volts maximum. Maximum means the *peak* value that the probe can withstand without possible damage. Be sure that the circuit you are working on will not damage the probe before using it. Surprisingly high voltages can be produced from low-voltage sources in pulsed circuits.

Whenever an oscilloscope is used for any application where the vertical amplitude of the signal is of importance, the probe must be compensated for use with the oscilloscope's input circuit. This is done by placing the probe tip on the oscilloscope's calibrator output and adjusting the probe (the adjustment may be at either end of the probe cable) for a proper square wave. This was covered in detail in Chapter 11, page 145.

Triggering the oscilloscope in pulsed circuits is best done using the *system* trigger of the equipment under test. Radar and other pulsed applications are synchronized with a common trigger, which is often provided at the front panel specifically for troubleshooting purposes.

When triggering the oscilloscope for troubleshooting most pulsed-circuit problems, the oscilloscope should *not* have a trace if the trigger is removed momentarily. If there is still a trace with the trigger disconnected, the oscilloscope is very likely not triggering on the signal it is supposed to, and the signals observed, if any, will be unsynchronized. The signals will appear to run or drift across the face of the CRT. Be sure to use the oscilloscope's *external trigger* input jack to apply the trigger when using an external input such as a system trigger. Switch the *trigger source* control of the instrument to "external." Adjust the trigger level so that the trigger input reliably triggers the sweep, but so that the sweep ceases when the trigger is removed. In order to accomplish this, the "auto trigger" position, if available, should not be used.

The repetition rate of a pulsed waveform will have a major effect on the intensity of the oscilloscope display. If the pulsed waveform has a high repetition rate, the pulses will occur many times per second and the oscilloscope picture will also be traced many times per second, producing a bright picture. If, on the other hand, there is a very short pulse that occurs only once in a relatively long time before repeating the cycle, the trace will probably be very dim and may be difficult to see in normal room lighting. Low-duty cycle signals are best shown using a storage oscilloscope. The standard high-frequency oscilloscope may be used for many pulse applications. Most pulse waveforms occur often enough to be quite visible on a standard oscilloscope.

The oscilloscope's high-frequency capability is a specification that should be kept in mind when using it on pulsed waveforms. An oscilloscope of insufficient high-frequency capability will round off the leading and trailing edges of a pulse that should be steep and sharp at the corners (Figure 12-5). To determine if a specific oscilloscope is suitable for use with a waveform of a given *rise time*, it will be necessary to calculate the approximate rise time of the oscilloscope in question and compare the oscilloscope's rise time with that of the signal under test. This is done by dividing the bandwidth of the oscilloscope's vertical amplifier (given in megahertz) into a constant of 350. This will give the approximate oscilloscope rise time in *nanoseconds*. As a rule of thumb, the rise time of the

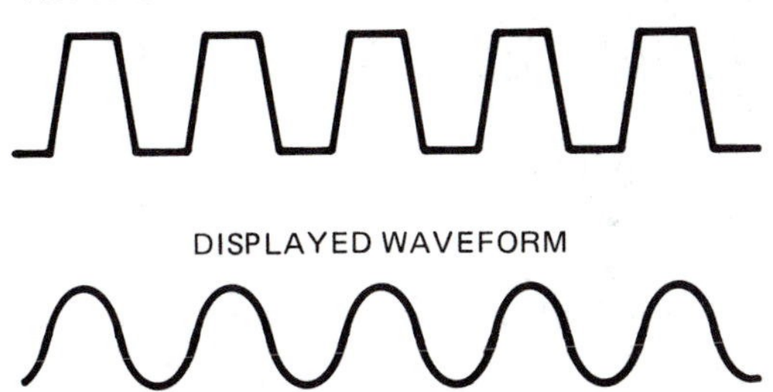

Note loss of sharp corners and amplitude.

FIGURE 12-5 Effect of insufficient oscilloscope high-frequency response on display of high-frequency waveform.

oscilloscope should be less than one fifth the rise time of the signal to be observed. This will result in no more than a two-percent error in the observed signal over its actual wave shape.

An example: An oscilloscope with a vertical amplifier rated at 20 Mhz would have a rise time of about 17.5 nanoseconds (350/20). This scope would be good for viewing any waveform with a rise time of 90 nanoseconds or more (17.5 × 5).

Another control requiring some explanation for pulsed applications is the *holdoff* control of the oscilloscope. This control will introduce an additional variable delay at the end of each sweep, which *may* help stabilize the observed pattern with *some* complex waveforms. It is a useful control when the triggering signal is not a simple, repetitious signal, but consists of several possible points upon which to trigger (Figure 12-6).

The Storage Oscilloscope

The storage oscilloscope can be of great help when viewing low-duty cycle waveforms. With proper setup of the triggering circuits, the oscilloscope can be "just waiting" for the pulse to occur. When it does, the scope will sweep at a fast

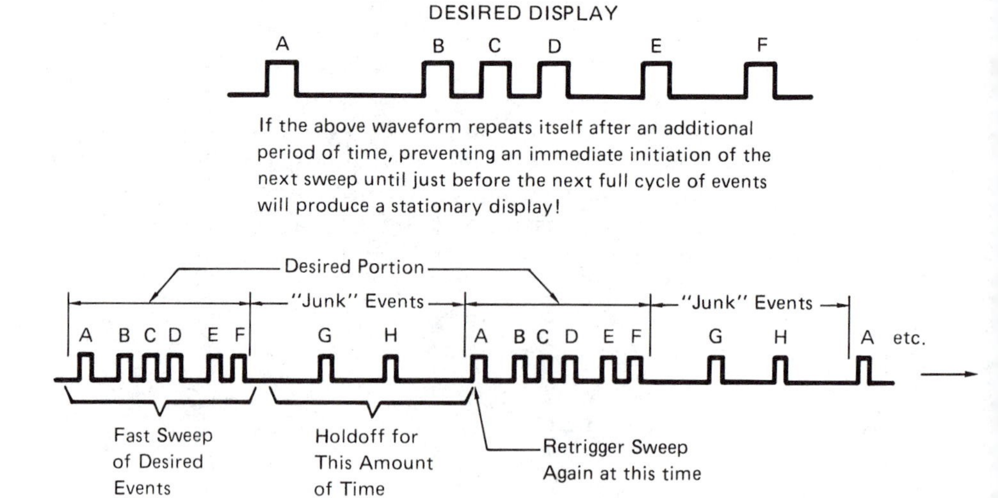

FIGURE 12-6 The oscilloscope holdoff control explained.

rate just as a standard scope does, but the difference is that after the sweep occurs the trace remains at full intensity. Some storage oscilloscopes can also hold the image after the scope is turned off!

The triggering and sweep circuits of the storage oscilloscope are the same as those of the standard oscilloscope. The main difference is in the additional controls that govern the time the image is held before fading, whether or not the image is to be stored, and a means of switching off the storage function to use the instrument as a normal oscilloscope.

The Frequency Counter

Frequency counters are very useful in pulse circuitry to indicate repetition rate. When the counter is connected to a circuit, care must be taken not to exceed the input voltage limitations of the counter. An oscilloscope 10X probe can be used in most cases to help avoid this danger to the counter input circuits. Some counters have an input voltage limitation as low as 2 volts, so be sure that you know the limitations of your particular instrument.

The Multimeter

The VOM responds to the average value of an AC waveform when using the AC scales. The scale behind the needle is marked to indicate the rms value. The VOM AC scales are accurate only when measuring sine waves with the AC scales. The VOM can, however, be used to indicate relative amplitudes of other waveshapes. In other words, a 50-percent duty-cycle square wave of 2 volts indicated on a VOM is larger in amplitude than another 50-percent duty-cycle square wave of only 1.5 V.

Most digital multimeters are constructed to respond to the average value of an AC waveform, yet read out a different value, the rms value. Such a meter is calibrated to do so with sine waves only. These meters are called *average-responding* meters. Remember that DMMs are reading out rms though not responding directly to the rms value. There is a multiplication factor of 1.11 times the average value to get the RMS value of a sine wave built into the meter.

Digital meters can also be made so that they respond to the rms value of a waveform. These are called *true-rms* meters. These meters may be used on any waveform and will give a true indication of the rms voltage or currents involved. For example, they can be used to determine the heating effect, which can be an indication of the proper fuses to use. In summary, the true-rms indicating meters are usable on waveforms other than sine waves. Average-responding meters give erroneous values. Figure 12-7 gives a table of correction values that may be used to get a close approximation of a true-rms reading on different waveforms.

INPUT WAVEFORM	DISPLAY MULTIPLIER FOR MEASUREMENT CONVERSION			
	PK-PK	0-PK	RMS	AVG
SINE	2.828	1.414	1.000	0.900
RECTIFIED SINE (FULL WAVE)	1.414	1.414	1.000	0.900
RECTIFIED SINE (HALF WAVE)	2.828	2.828	1.414	0.900
SQUARE	1.800	0.900	0.900	0.900
RECTIFIED SQUARE	1.800	1.800	1.272	0.900
RECTANGULAR PULSE D=X/Y	0.9/D	0.9/D	$0.9/D^{1/2}$	0.9D
TRIANGLE SAWTOOTH	3.600	1.800	1.038	0.900

FIGURE 12-7 True-rms conversion factors for average-responding multimeter. (Reproduced with permission of the John Fluke Mfg. Co., Inc.)

THE ORIGINATING PULSE CIRCUIT

A few common pulse-generating circuits are shown in Figure 12-8. The first step to pursue when troubleshooting pulse oscillators is to examine oscillator outputs with the oscilloscope for sufficient amplitude, approximate frequency, and proper waveform. The exact repetition rate may be checked with a frequency counter if necessary.

THE PROCESSING PULSE CIRCUIT

Pulse amplifiers, pulse shapers, and pulse inverters are most often operated Class C and brought sharply into conduction by the incoming pulse. Common ways to transformer-couple pulse stages are shown in Figure 12-9. Pulses can also be coupled from one stage to the next by resistor-capacitor networks, as shown in Figure 12-10.

Line drivers are a special type of pulse amplifier. Their output voltage is about the same as the input voltage, and there is no voltage amplification. There is, however, considerable increase in the current-driving capability of the circuit (Figure 12-11).

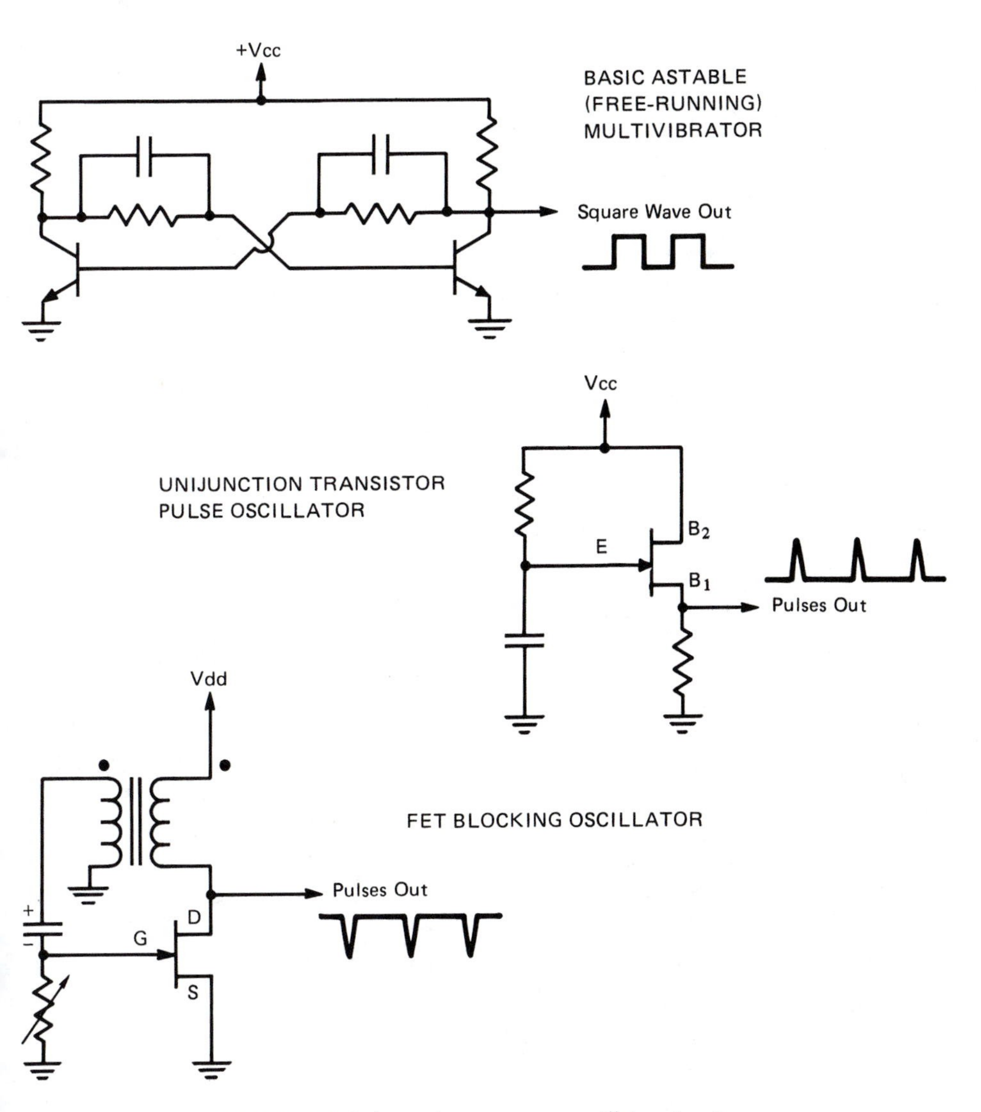

FIGURE 12-8 Pulse and square-wave oscillator circuits.

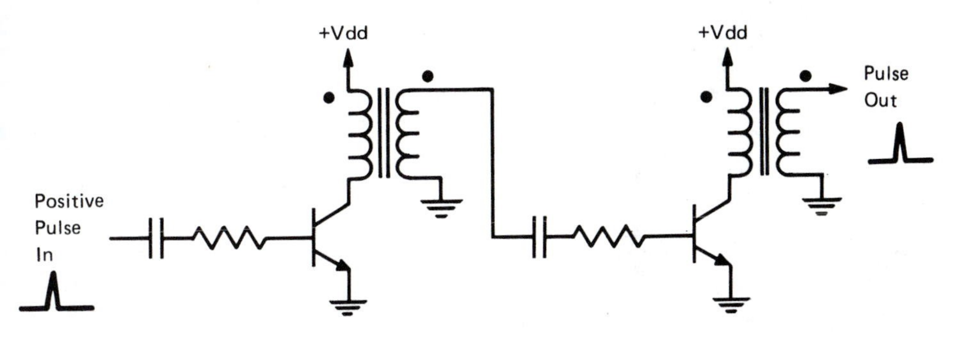

FIGURE 12-9 Transformer coupling in pulse circuits.

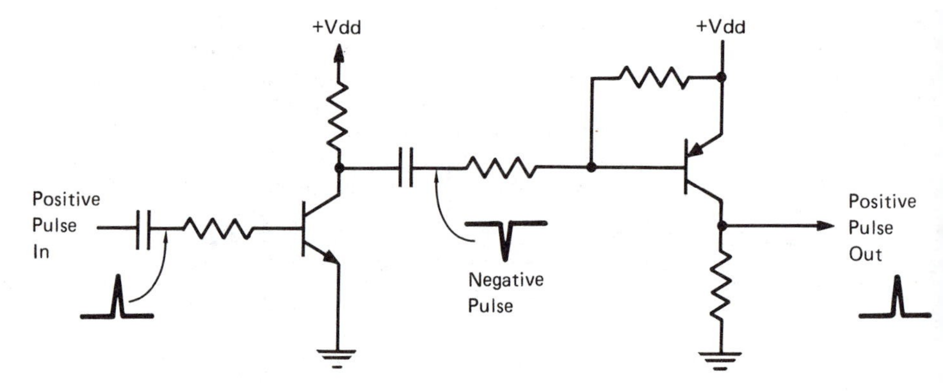

FIGURE 12-10 Resistor-capacitor-coupled pulse amplifiers.

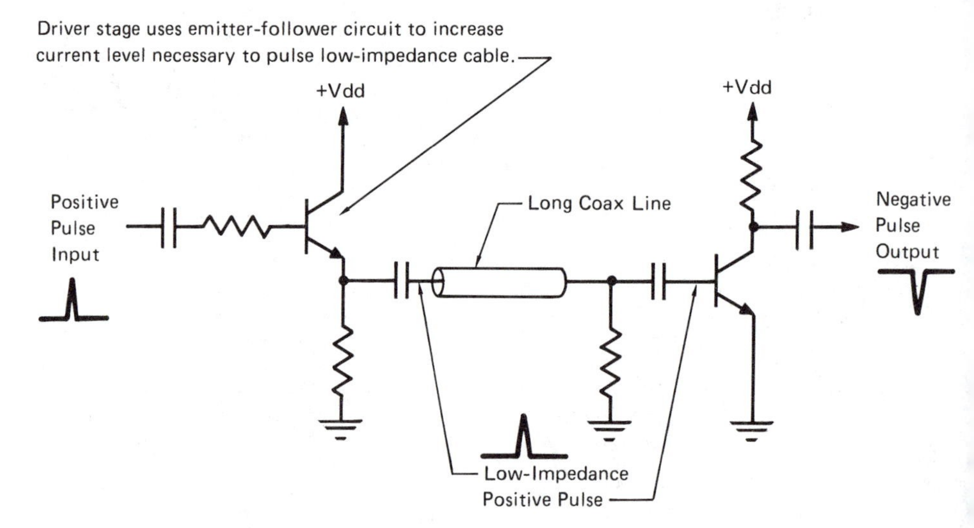

FIGURE 12-11 Simple line-driver circuit.

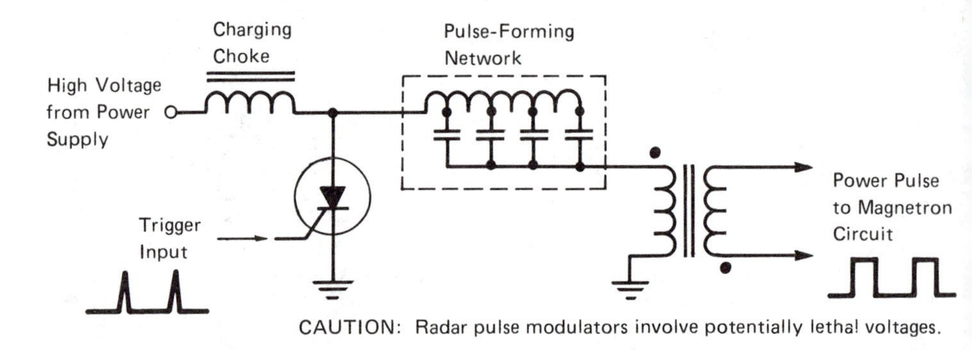

FIGURE 12-12 Simplified radar modulator.

THE TERMINATING PULSE CIRCUIT

Power pulses are often applied to a transformer winding. The transformer may also be used to produce a pulse of voltage to another circuit, which adds LF or RF to it, thus making a "transmittable" signal. In radar circuits this becomes the transmitted pulse. To trace pulses through a radar pulse-modulation circuit, it is possible to monitor signals up to a point in the circuitry where the voltage and power levels can be extremely dangerous (Figure 12-12). High-voltage pulse circuits should be monitored using an appropriate probe, perhaps a 100X probe with an oscilloscope. Be aware of, and do not exceed, the voltage limitations of the probe.

The output pulse of the comparatively low-power depth sounder may be seen at the output of the modulation transformer. The pulse of ultrasonic high-frequency "sound" can be seen directly on an oscilloscope (Figure 12-13). The amplitude of the envelope of ultrasonics is related to the amplitude of the modulating pulse. The manufacturer of the equipment should provide the necessary information as to the proper voltage at this point.

Another common termination circuit is the dual-purpose circuit of the flyback transformer. This circuit takes the 15,750-Hz horizontal oscillator signal

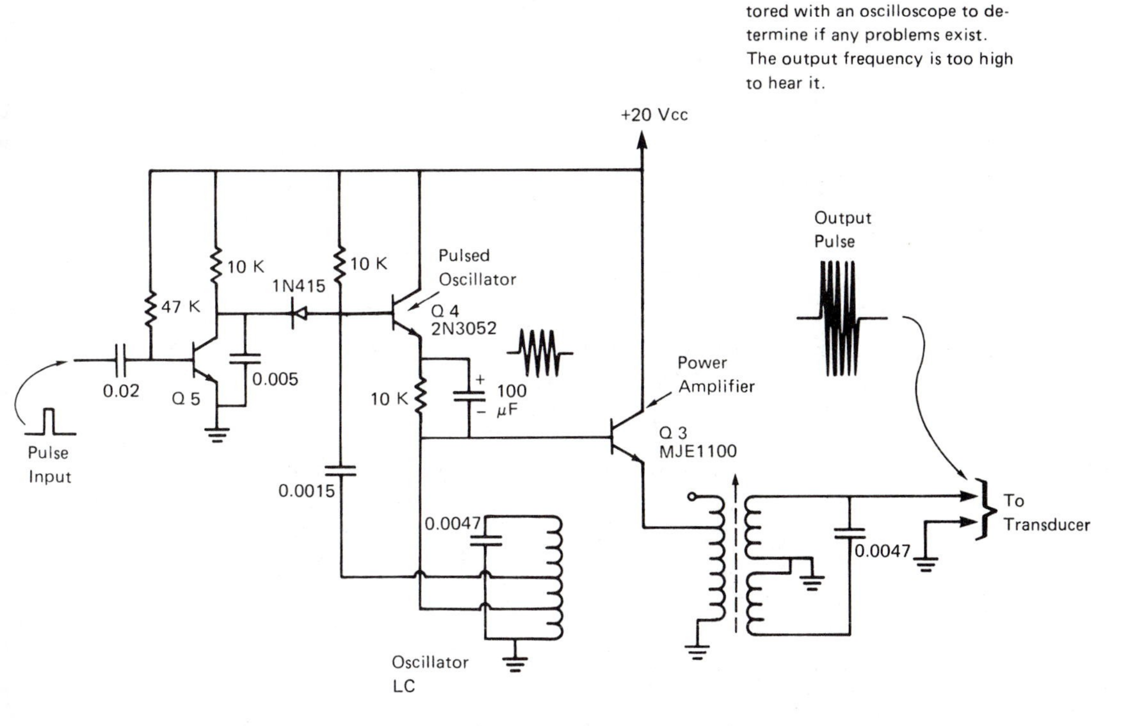

FIGURE 12-13 Simplified depth sounder power output stages. (Courtesy of Ross Laboratories Inc., Seattle, WA)

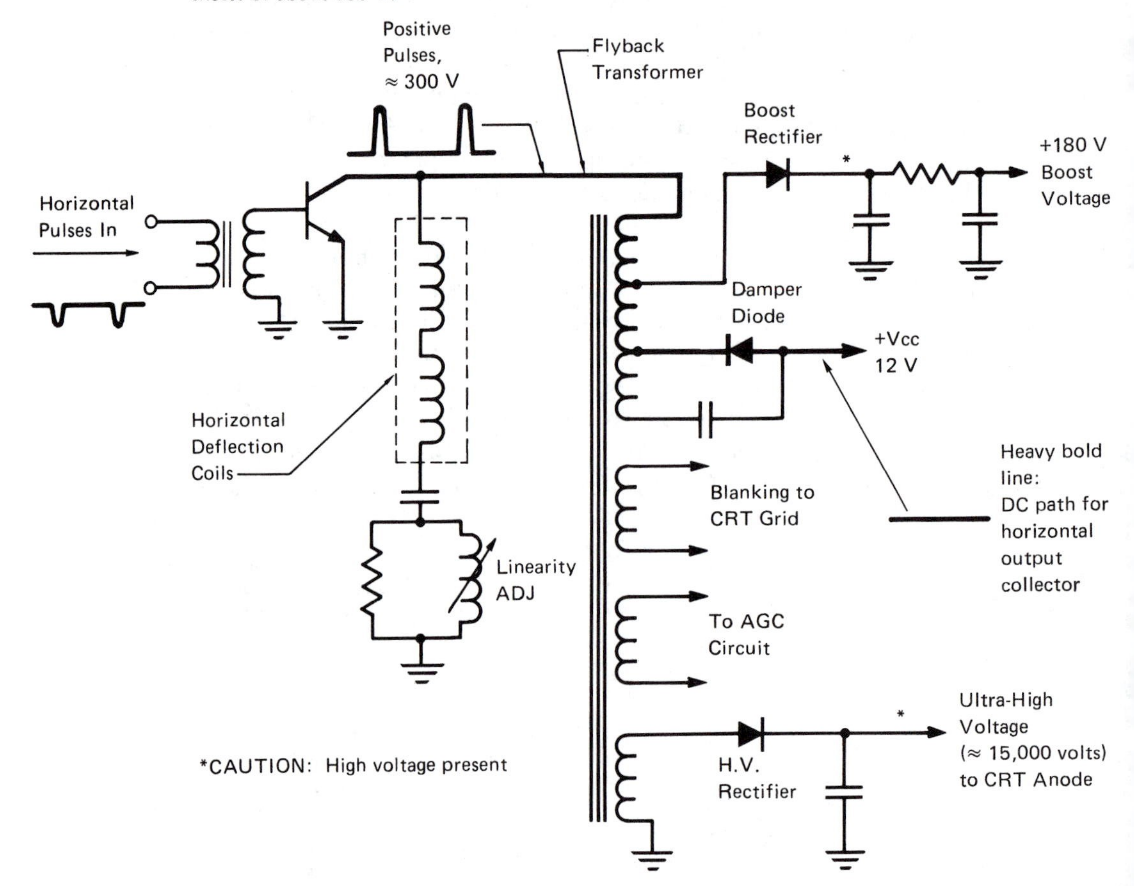

FIGURE 12-14 Simplified TV horizontal output circuitry.

and produces the horizontal sweep currents for the tube *and* the very-high-voltage DC required for the acceleration anode of the CRT at the same time (Figure 12-14). When measuring the high-voltage DC produced by this circuit, a special very-high-voltage probe must be used. Since this circuit can produce in excess of 25,000 volts, a standard meter will be damaged if not provided with additional multiplier resistance. Figure 12-15 shows a probe designed specifically for measuring these ultra-high voltages.

THE MIXING PULSE CIRCUIT

Mixing circuits in pulsed equipment are more often called *gates*. Figure 12-16 shows some examples of gating circuits used.

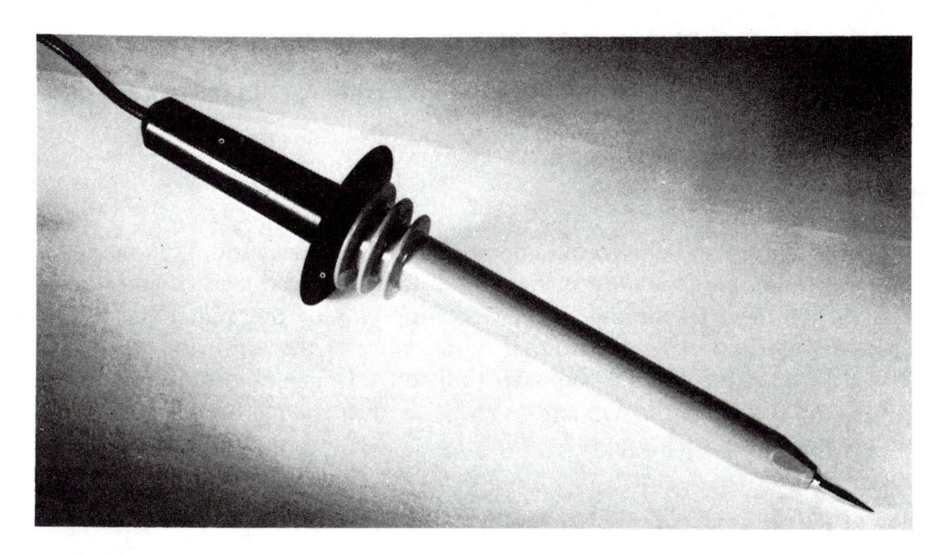

FIGURE 12-15 Probe for measuring very high voltages. (Reproduced with permission of the John Fluke Mfg. Co., Inc.)

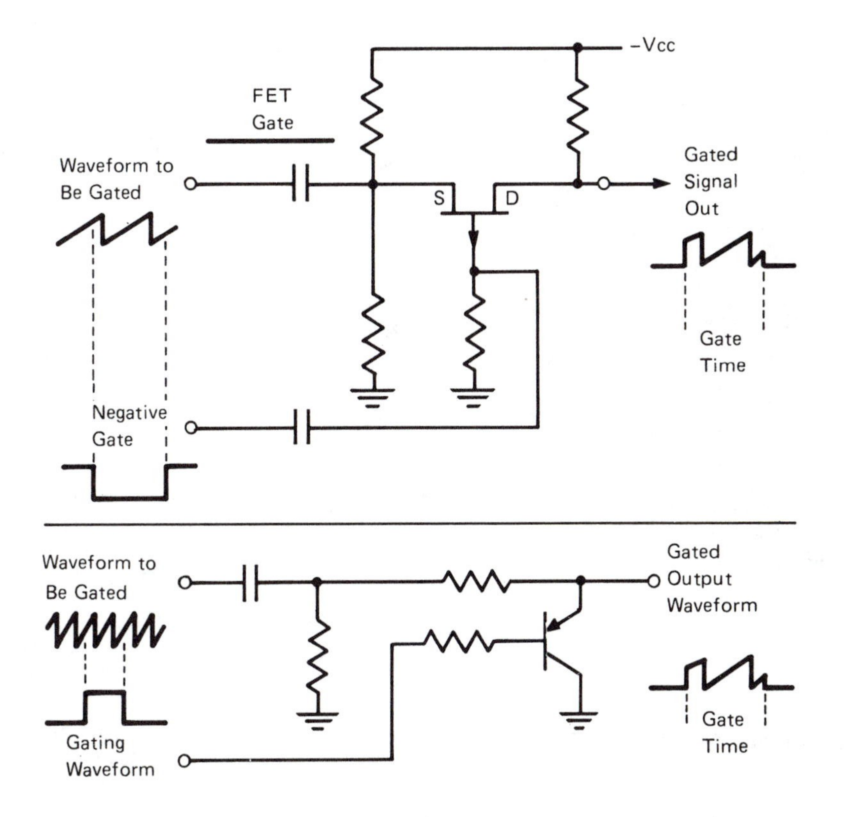

FIGURE 12-16 Sample gating circuits.

HOW SIGNALS CHANGE IN PULSE CIRCUITS

Capacitors and inductors can cause drastic changes in pulse waveforms as signals progress through the circuitry.

DC-Level Changing

A capacitor can be used to remove a DC component, add it, or change it. A common example is the output stage of an RC-coupled transistor amplifier. The capacitor is used to remove the DC component of the signal because that component would upset the normal functioning of the following stage (Figure 12-17). A capacitor can also be used to insert a DC level into a circuit (Figure 12-18). And of course the capacitor can be used to change from one level of DC component to another (Figure 12-19).

Signal Rectification

When alternating signals, LF, RF, or pulses are coupled into a bipolar transistor, vacuum-tube grid circuit, diode, or any other rectifying element

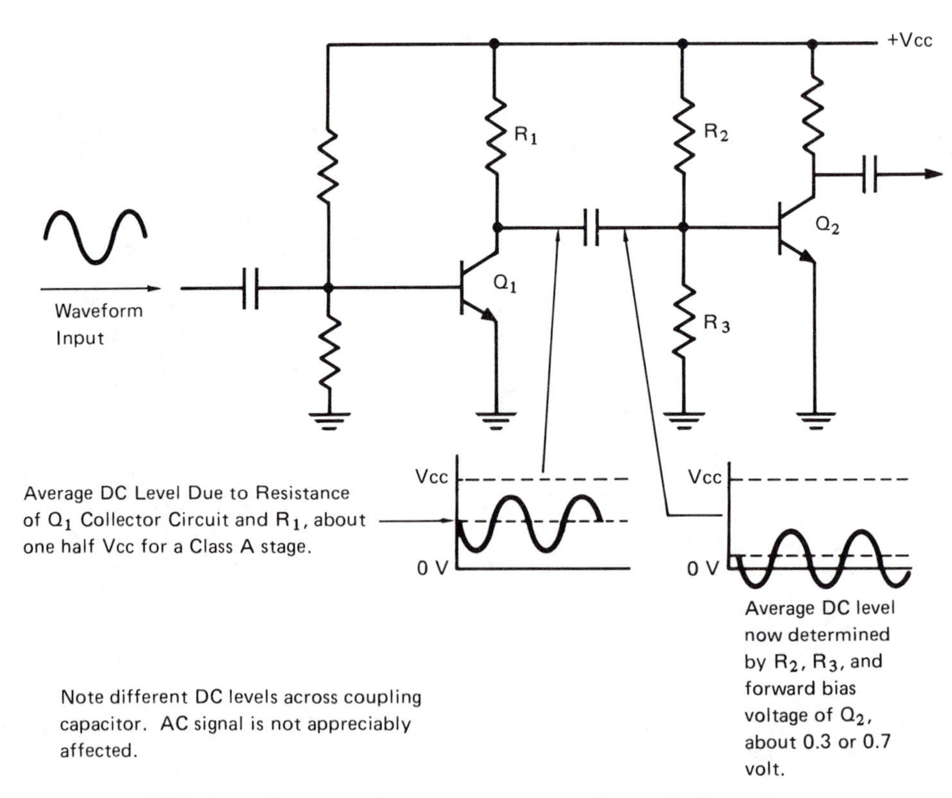

FIGURE 12-17 A capacitor can remove DC levels of applied waveforms.

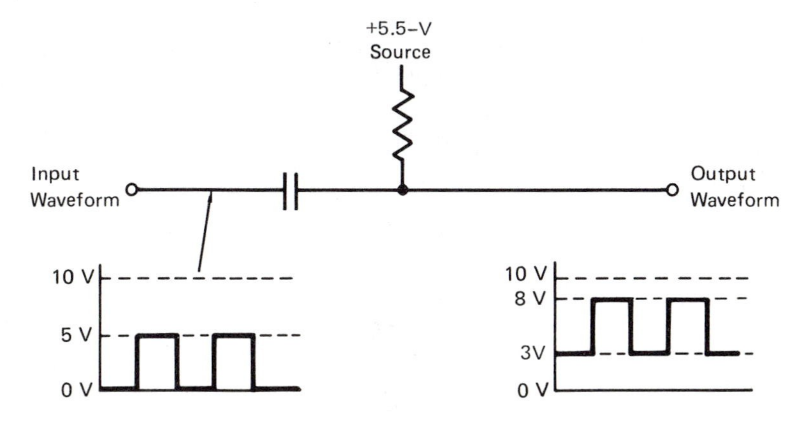

FIGURE 12-18 A capacitor can add DC levels to applied waveforms.

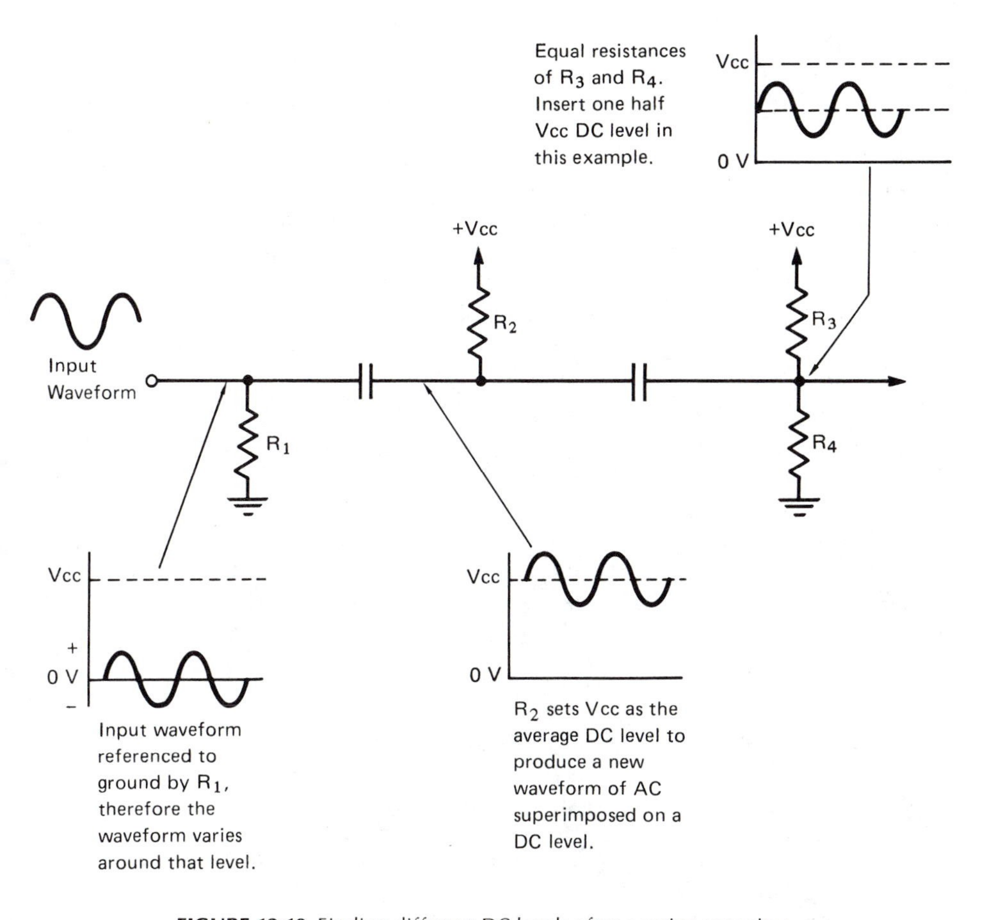

FIGURE 12-19 Finding different DC levels after a series capacitor.

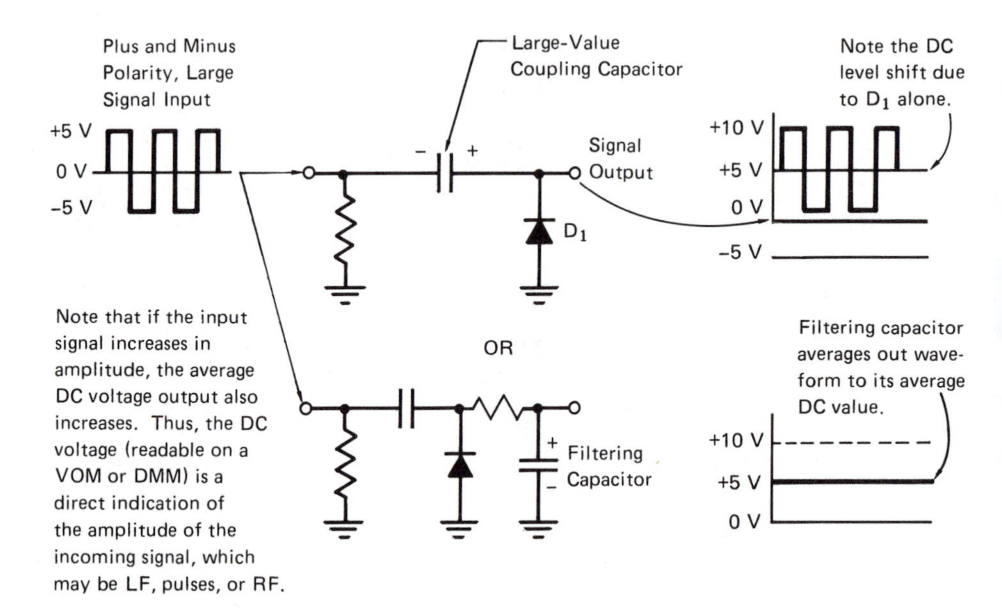

FIGURE 12-20 A rectifying junction will produce DC from AC.

through a capacitor, a special effect occurs. This effect has been called grid rectification in the past and is probably best known now as the "driving level" (Figure 12-20). When the incoming signal goes into the positive half of the cycle (for this type of transistor), the voltage at the base cannot exceed about 0.7 volts due to the forward voltage drop across the base to emitter junction. Additional voltage must be dropped across the capacitor. Upon reversal of the input signal, the charge on the capacitor adds to that of the signal, producing a very negative average voltage at the base of the transistor. This voltage provides a very good indication of the amount of input-signal voltage. A DC voltmeter can be used in this case to monitor the amount of RF driving the stage. This fact is often used by manufacturers, who provide special test points at circuit inputs for this purpose.

Knowing that signals can be converted in some inputs to a DC voltage, it should come as no surprise to the technician when a large pulse signal fed into a vacuum-tube grid or transistor base suddenly becomes a large negative voltage with apparently no direct source. The input to the horizontal output tube or transistor of a television is a good example of this.

THE TIME CONSTANT

The Resistor-Capacitor Combination

A resistor and a capacitor form a natural timing circuit. It takes time for a capacitor to charge or discharge to a new voltage level through a resistor. Figure 12-21 shows a basic timing circuit to illustrate this point. If the DC source

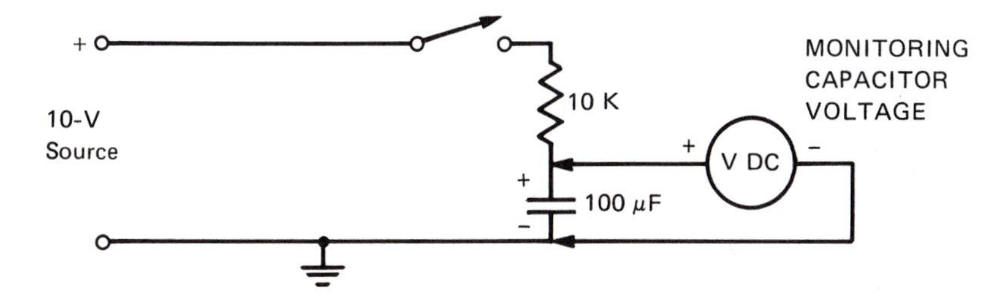

FIGURE 12-21 Basic RC timing circuit.

voltage does not change, the capacitor will reach 63.2 percent of the source voltage in a number of seconds determined by the simple formula:

$$\text{Time Constant (seconds)} = \text{Resistance (ohms)} \times \text{Capacitance (farads)}$$

Thus, the capacitor in Figure 12-21 will reach 6.32 volts of charge in

$$10{,}000 \times 0.0001 = 1 \text{ second}$$

After one second the capacitor will reach 63.2 percent of the source voltage. After another second 63.2 percent of the remaining voltage would have been reached, and so on. After five time constants, the capacitor is considered to be fully charged.

Time constant is a very important concept in dealing with pulsed signals. In operating circuits, *the relationship of the time constant to the period of the incoming signal determines how the incoming signal is changed coming out the other end of the RC circuit.* To examine how the time constant relates to the period of the incoming signal, consider the circuit of Figure 12-22. With this circuit, called an *integrator,* the output waveform becomes more and more like the *average* value of the incoming waveform as the time constant gets longer. Averaging waveforms is a principal use for this circuit. Note also that the output circuit will retain any DC present in the input signal and superimpose the average value of the AC signal upon it.

Now consider the effect of reversing the position of the components and the effect of time constant on the output waveform (Figure 12-23). The technical name for this capacitor-resistor combination when used in pulse applications is a *differentiator.* Notice that as the time constant gets *shorter,* rapid changes get through the circuit, but the incoming DC levels do not. The more rapid the input voltage changes, the higher the voltage output of the stage. A major use for this circuit is to make spikes out of square waves. Note that this circuit also blocks any DC present on the input from appearing in the output signal.

As the time constant gets shorter compared to the period of the incoming signal, notice also that the capacitor has time to reach a full charge. Because this capacitor charge adds to the voltage of a suddenly reversing input, the

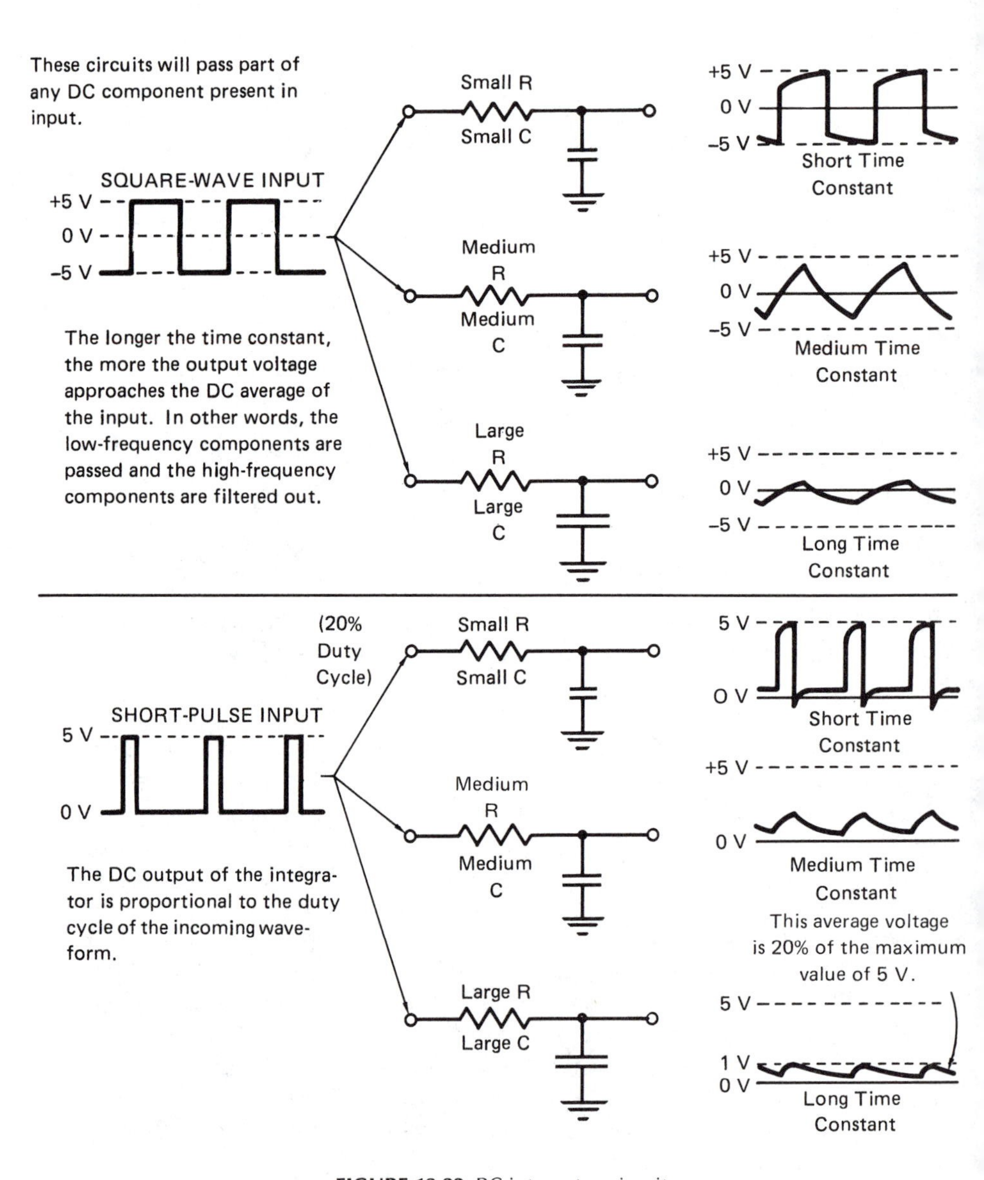

FIGURE 12-22 RC integrator circuits.

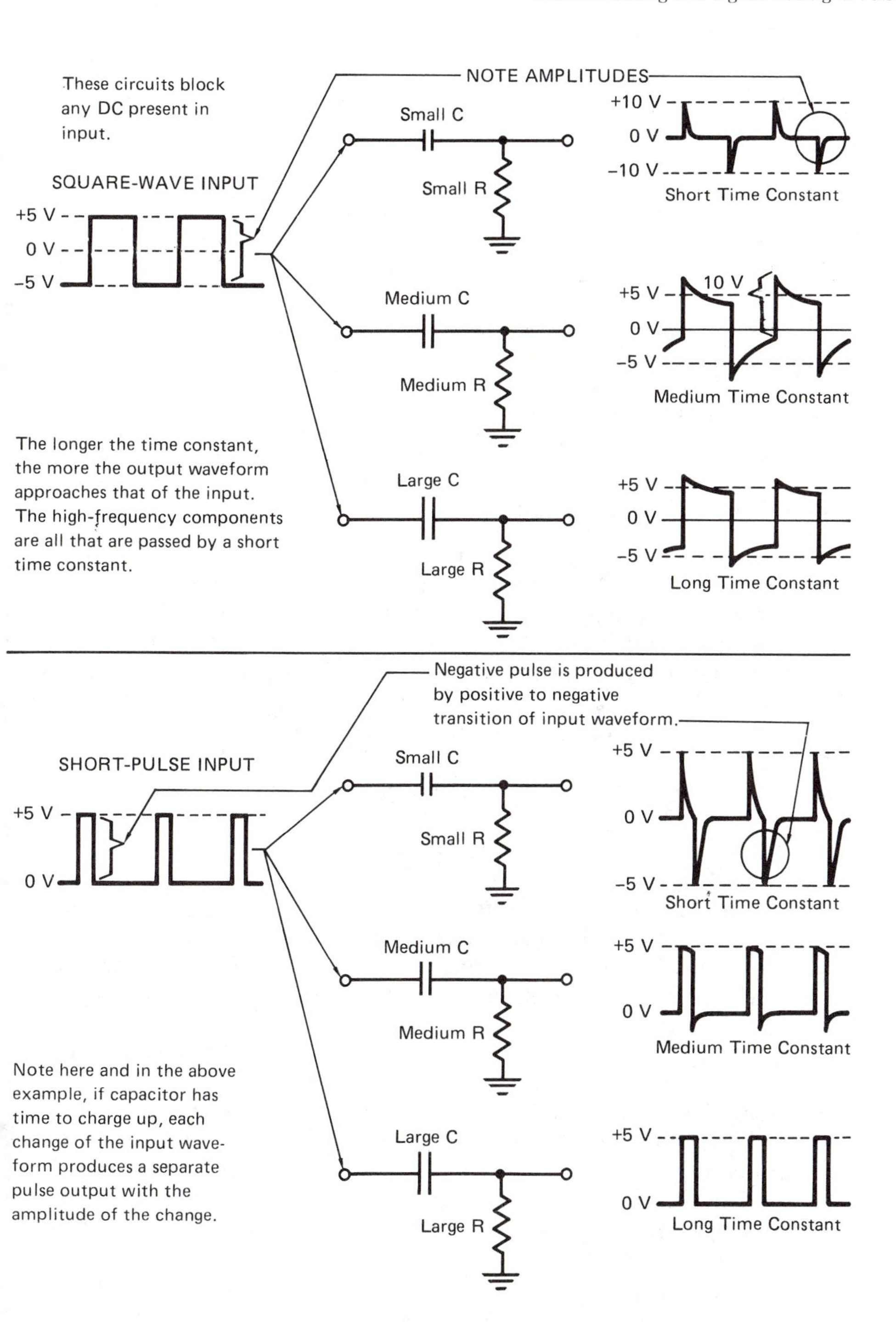

FIGURE 12-23 RC differentiator circuits.

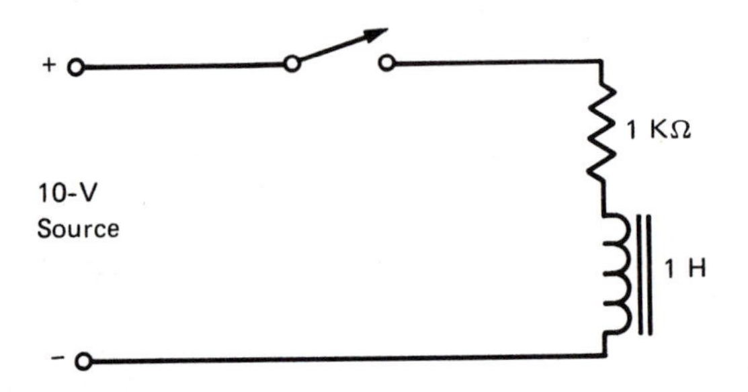

FIGURE 12-24 Basic RL timing circuit.

output voltage from the differentiator results in output pulses that momentarily have *twice* the maximum input voltage.

The Inductor-Resistor Combination

A resistor and an inductor also form a natural timing circuit. It takes time for an inductor to build up or decay a magnetic field to a new level through a resistor. Figure 12-24 shows a basic timing circuit to illustrate this point. From the instant that the switch is closed, the inductor will reach 63.2 percent of the maximum magnetic field strength in accordance with another simple relationship:

$$\text{Time Constant (seconds)} = \frac{\text{Inductance (henrys)}}{\text{Resistance (ohms)}}$$

Thus, the inductor of Figure 12-24 will reach its first time constant in

$$\frac{1}{1000} = 0.001 \text{ second, or } 1 \text{ millisecond}$$

After one millisecond the inductor will reach 63.2 percent of the full field strength possible with the voltage applied. This process repeats until after five time constants, at which point the inductor is considered to be fully magnetized. Current at that time is the maximum that the total resistance of the circuit will allow.

Note that if the resistance *decreases,* the time to reach full magnetization *increases*. In other words, if the resistor of Figure 12-24 were changed to a 10-ohm resistor instead of a 1000-ohm one, the inductor would take 0.1 second to reach full magnetization. This is the opposite effect of the resistor in a resistor and capacitor combination.

To examine how the time constant of an RL circuit relates to the period of the incoming signal, see Figure 12-25. With this circuit, also called an *integrator,* the output waveform becomes more and more like the average value of the incoming waveform as the time constant gets shorter. Averaging waveforms is a principal use for this circuit. Note that the positions of the L and the R are reversed from those of the capacitor and resistor integrator. This circuit will pass along any DC present in the input signal to the output waveform due to the low

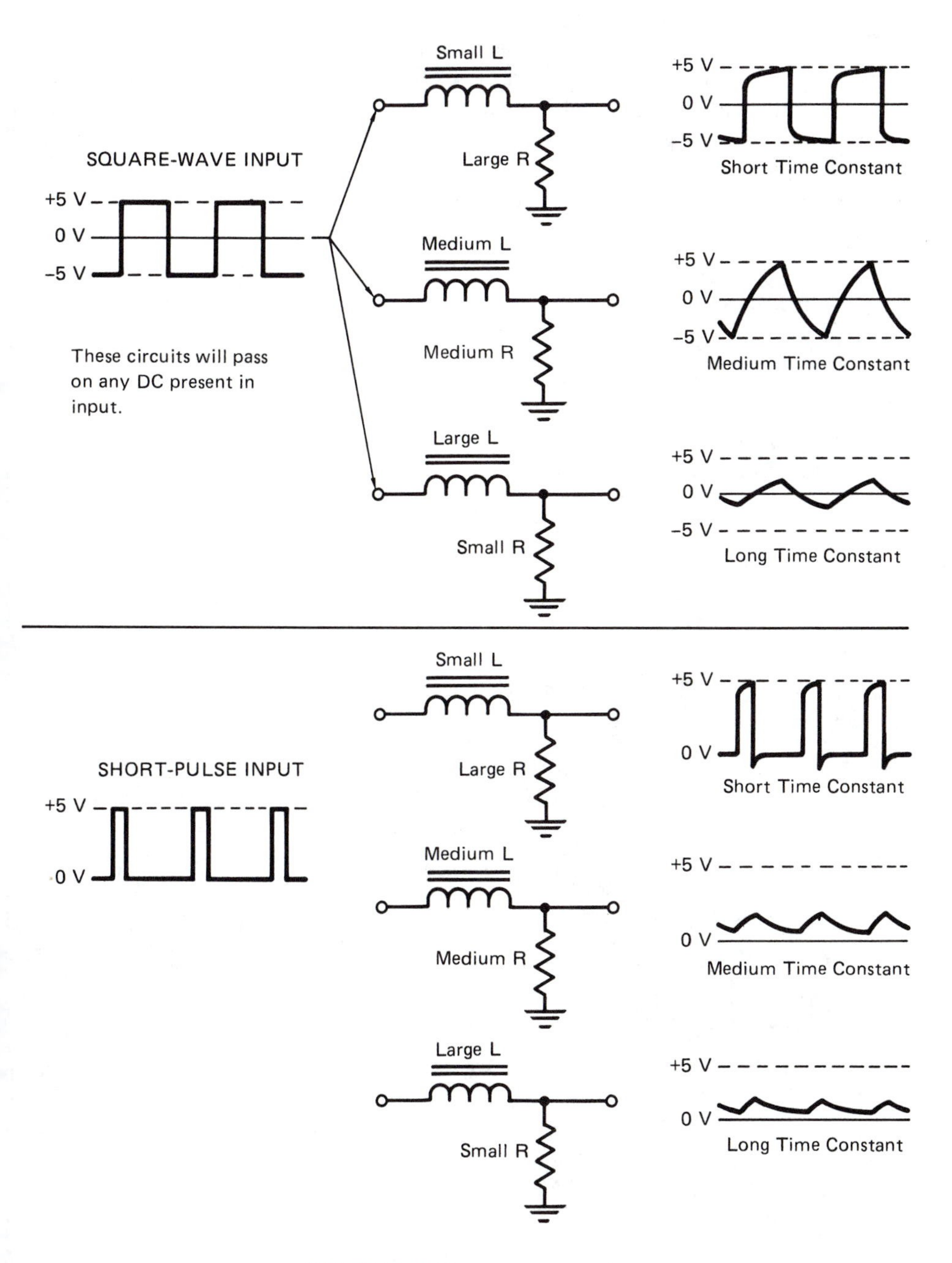

FIGURE 12-25 RL integrator circuit.

resistance of the inductor winding. In addition, any changes in current flow through the inductor are opposed by the inductance. Thus, this circuit is also widely used to smooth out DC power sources such as the pulsating output of an AC rectifier.

Now consider the effect of reversing the position of the R and L components and the effect of time constant on the output waveform (Figure 12-26). This inductor and resistor combination is also called a *differentiator.* Notice that as the time constant gets *shorter* the more rapid changes get through the circuit, but the incoming DC levels do not. A major use for this circuit is to make spikes out of square waves. Note also that any DC present in the input signal is shorted

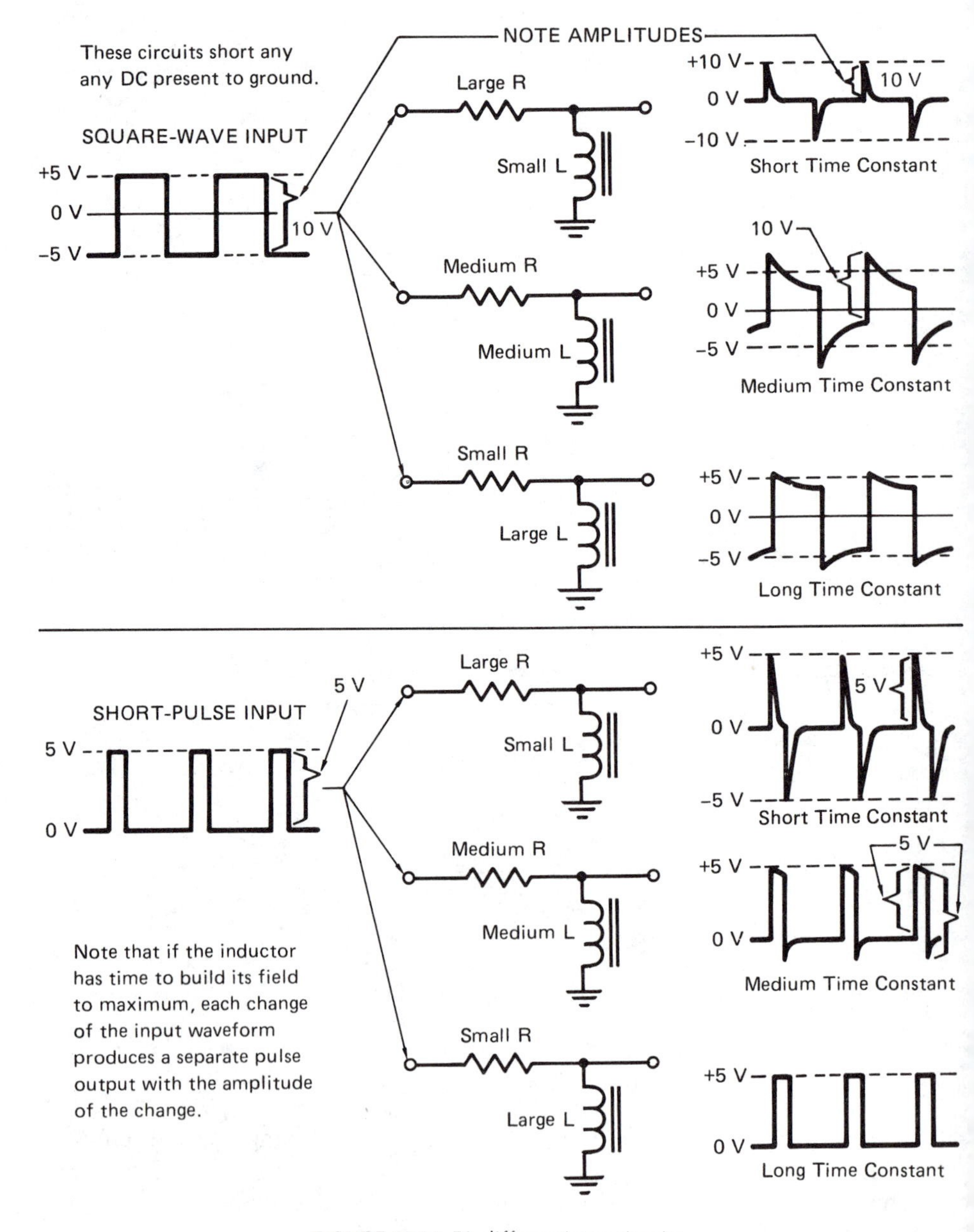

FIGURE 12-26 RL differentiator circuits.

to ground by the relatively low resistance of the inductor winding. As the time constant gets shorter, the inductor has time to reach full magnetic field strength. Because the inductor opposes any change in current flow through itself, the output from the differentiator circuit results in output voltage pulses that momentarily have up to *twice* the maximum input voltage.

SUMMARY

This chapter has covered some of the special techniques of tracing signals through pulse circuitry. When the reader has identified a defective stage in pulse circuitry, Chapter 16 will help find the defective component with passive tests.

REVIEW QUESTIONS

1. What is the duty cycle of a waveform that is positive for 2 μs and negative for 4 μs before repeating?
2. What does the small dot near a transformer winding on a schematic mean?
3. What is the prr of a 600-Hz waveform?
4. What should happen to the trace of a properly adjusted, triggered oscilloscope if the external trigger is disconnected?
5. How will an oscilloscope trace appear if triggered with a low-duty cycle signal and if the horizontal sweep rate is very fast?
6. If an event occurs only infrequently, what kind of instrument is used to observe such an event?
7. To what voltage value does the average VOM or DMM respond?
8. How can high-powered pulse circuits damage a voltmeter or an oscilloscope probe?
9. When a large amplitude signal is fed into a diode and capacitor, what is the logical result?
10. As the capacitor value increases, how is the charging time through a constant resistance affected?
11. What circuit is used to "average out" a waveform to its DC value?
12. What circuit is used to make sharp pulses from square waves?
13. How is the time constant affected if the resistance in an RC timing circuit is increased?
14. How is the time constant affected if the resistance in an RL circuit is increased?

chapter thirteen

Digital Troubleshooting Techniques

CHAPTER OVERVIEW

This chapter covers the principles of tracing signals through operating digital printed circuit boards. Binary circuits for the purposes of this chapter are those that use only two states: on/off, or high/low. These are shorthanded as "0" and "1" on most schematics. Chapter 14 will cover microprocessor circuits with bidirectional, three-state lines.

Good references to have on hand are Motorola's *Schottky TTL Data,* #DL121R1, *CMOS Data,* #DL105, and *MECL Device Data,* #DL122R1. See Appendix III for ordering information. For a very practical and easy-to-understand approach to ICs there are two books available, the *TTL Cookbook* (Howard W. Sams Publishing Co.) and the *CMOS Cookbook* (Tab Books). (See Appendix III.)

FIRST CHECKS

Many digital circuit problems can be repaired by a few initial checks. After opening the equipment or getting the card in question in hand, the board should first receive a very thorough visual inspection. The ICs should be carefully inspected for any signs of overheating, such as discoloration or a raised dimple in the center of the IC. If the reader has access to a solid-state tester, it may save much time and effort to make the checks suggested in Chapter 16 without power applied to the circuit. Most defects in digital circuits can be detected very efficiently with this instrument.

The next step would be to apply power to the board and check the voltage of all the DC power supplies on the board. A digital multimeter is desirable for this check. The 5-volt supply used by transistor-transistor logic (TTL) circuits is particularly critical and must be within the range of 4.75 V to 5.25 V. It should be adjusted for 5.00 V DC if an adjustment is provided. CMOS circuits will operate on supply voltage from 3 V to 18 V, and supply voltages are therefore not as critical. The output of all regulators should be checked to ensure that the proper voltages are present. While the circuit is in operation, check the ICs to see if any are overheating.

If the equipment is operating TTL circuits, it is a good idea to check with an oscilloscope to see if the supply line of 5.00 V is clean at various points around the board. The totem-pole output of the TTL IC is potentially responsible for causing very short transients on the supply bus. A very fast sweep must be used, as these transients are very short.

If the board uses ICs that must be clocked, there will be an oscillator somewhere on the board to produce the clocking signal. It will usually be of the quartz-crystal type for frequency stability. The output of this system clock should be among the first checks made. Make sure it is on the proper frequency. A few cycles off frequency can cause loss of synchronization with other systems. The symptoms of such a problem may look like a thermal intermittent because of the natural slow drift of an oscillator when reaching operating temperature. It is also possible for a crystal oscillator to "jump modes" and begin operating on twice or three times the frequency intended. This will be obvious if a frequency counter is used to check the clock frequency.

DIGITAL IC REFERENCES ARE ABSOLUTELY NECESSARY

When working with TTL circuitry (ICs that have identification numbers of 74XXX and 54XXX), you will find the *TTL Cookbook* an excellent book for gaining familiarity with this family of ICs. In order to have information available for all of the TTL family ICs and their pinouts, it will be necessary to have additional publications such as Fairchild's *TTL Data Book*. CMOS familiarization information is available in the *CMOS Cookbook* (CMOS chips have a 4XXXX or 14XXXX designation). Detailed pinouts and specifications are available in publications such as Motorola's *CMOS Data Book*. See Appendix III for ordering sources for these publications.

DIGITAL TROUBLESHOOTING BASICS

It is interesting to see a graph of the means by which ICs fail (Figure 13-1). Digital circuits must be turned on or off. Any voltage level between full on or full off is a bad level and points to a problem. These partial voltages and voltages that should be changing but aren't, are what the technician looks for in troubleshooting. Voltages that won't change are commonly referred to as being "stuck," and are shorted to either the ground or the supply bus. One exception is the "floating" or three-state lines of microprocessor circuits.

There are two special digital circuits that are exceptions to the rule that digital circuits require either high or low signals: the *Schmitt trigger* and the *delay multivibrator.* A digital circuit called a Schmitt trigger will accept any voltage input between 0 and the maximum voltage (Vcc) for the family and will give an output much like the mechanical action of a snap-action switch. Once the voltage reaches a critical level, the circuit snaps on very rapidly. The output will remain in this state until the input voltage falls to a lower level than that at which it turned on. At this lower level the circuit snaps to the off condition and

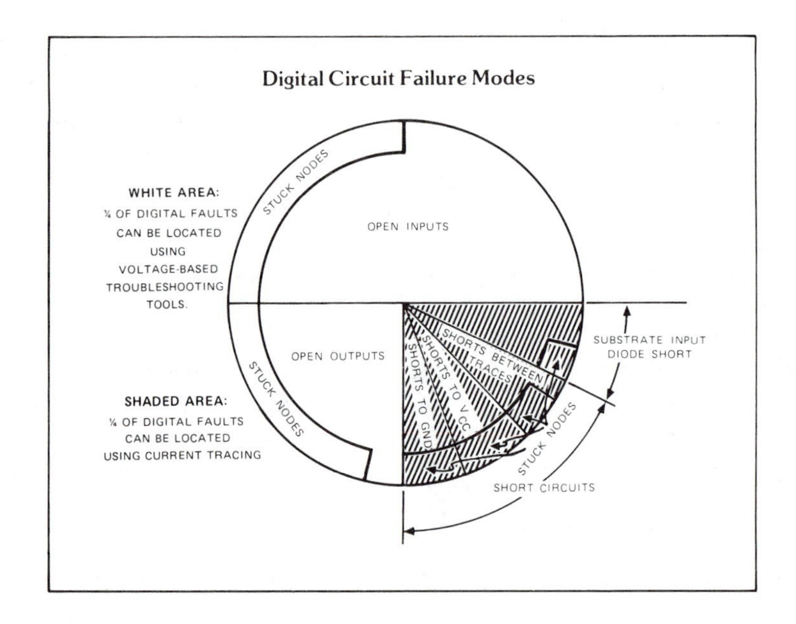

FIGURE 13-1 How digital ICs fail. (Photo courtesy of Hewlett-Packard Company)

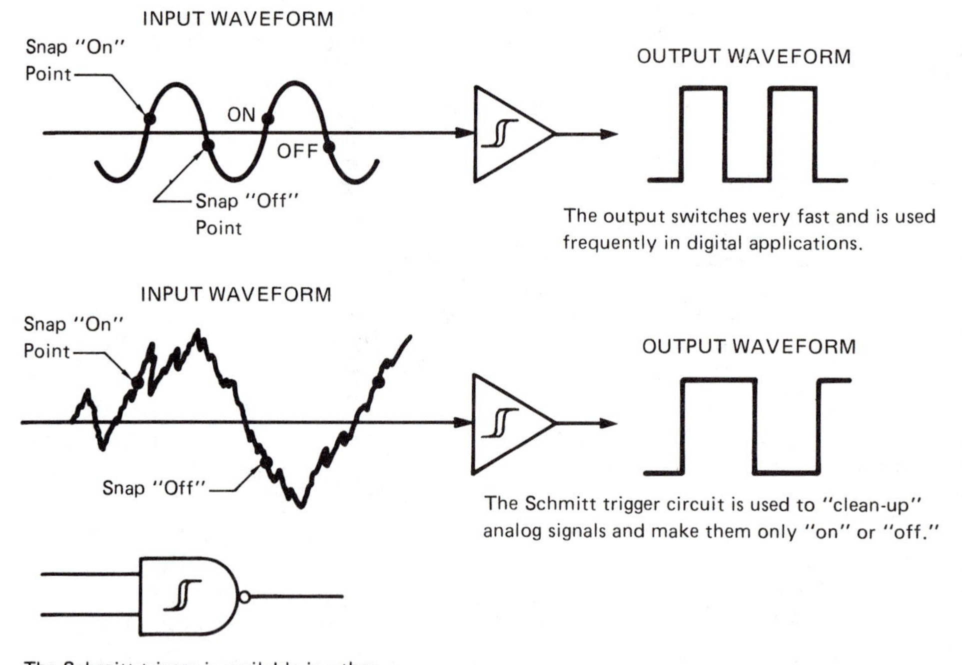

FIGURE 13-2 The Schmitt trigger action.

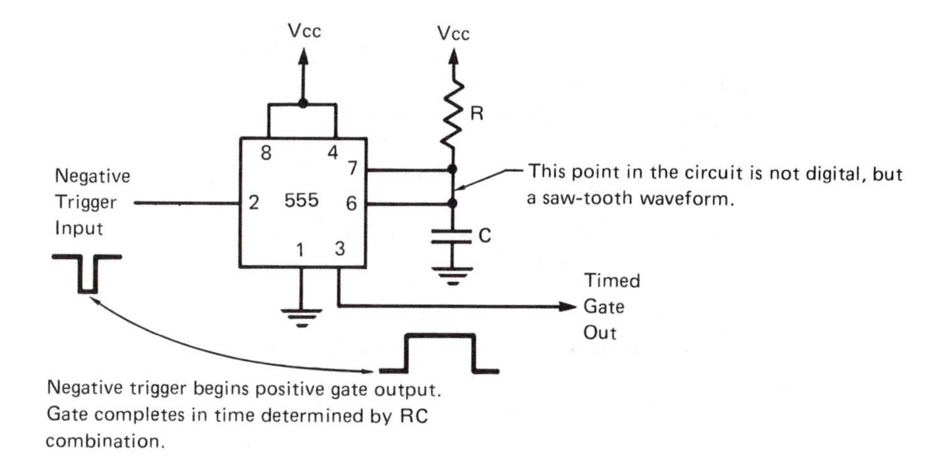

FIGURE 13-3 The delay multivibrator circuit.

remains there even if the input voltage fluctuates slightly positive again. The input must reach the higher critical "on" voltage before the circuit will snap on again. This chip is particularly useful for "cleaning up" analog input so that it is suitable for the pure on/off conditions required by digital logic. Figure 13-2 illustrates this snap action.

The second exception, the multivibrator chip, has a special pin that has a capacitor and a resistor connected to it. This pin is used to sense the timing of the RC circuit and produce a fixed delay based upon it (Figure 13-3.)

Of all the digital families available, three should be familiar to the servicing technician: the transistor-transistor logic (TTL), the complementary metal oxide semiconductor (CMOS), and the emitter-coupled logic (ECL) ICs. Only limited discussion will be given to ECL due to its less common use.

TTL Family Characteristics

TTL family characteristics dictate that the supply voltage (Vcc) for the entire family be very close to 5.0 V DC, between the limits of 4.75 V DC and 5.25 V DC. Voltages over 7 V DC will damage the ICs. A logic low signal for TTL is 0.8 volt or less. A logic high is 2.4 volts or more. Note that the logic levels are *not* as might be expected, 0 volts and 5.0 volts! (Figure 13-4).

TTL ICs are not as susceptible to static discharge as CMOS ICs, being by nature low-impedance devices. TTL requires a moderate amount of current to operate. A typical TTL chip may dissipate perhaps 60 milliwatts. There are several families within TTL that have characteristics tailored to special uses such as lower power requirement, higher current output, faster response, etc. Table 13-1 explains the characteristics in more detail.

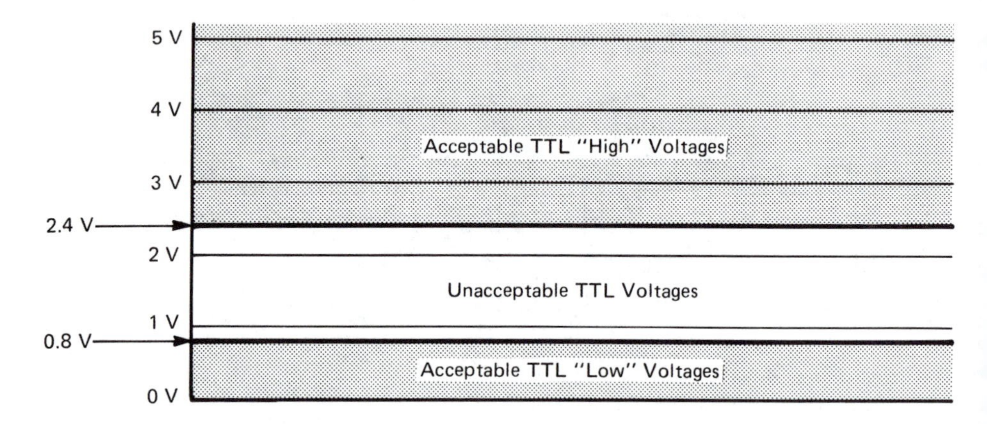

FIGURE 13-4 TTL logic levels defined.

TABLE 13-1 TTL Family Characteristics

FAMILY	SPEED	POWER/GATE	MAX. FREQ.
Regular TTL	10 ns	10 MW	35 MHz
High-Power TTL	6 ns	22 MW	50 MHz
Low-Power TTL	33 ns	1 MW	3 MHz
Schottky TTL	3 ns	19 MW	125 MHz
Low-Power Schottky TTL	10 ns	2 MW	45 MHz

The TTL family can be best identified by the numbering system that begins with either 74XXX or 54XXX. Details of the TTL numbering system and what the numbers signify are well covered in the *TTL Cookbook* (Howard W. Sams Publishing Co.). This is an excellent reference manual for technicians working with TTL circuits. The more common ICs are explained in detail, along with typical applications for them.

It is an interesting note that due to an internal resistor in many of the TTL ICs, the output may be shorted to ground, without disconnecting the IC from the circuit, to eliminate an input for testing purposes if the technician finds it an advantage. However, the presence of the resistor should be verified before grounding a TTL output (Figure 13-5). A characteristic of this IC family is that a "floating" TTL input pin will usually float high, producing an extraneous "high" input signal to the IC.

CMOS Family Characteristics

This family of ICs is identified by the 4XXX identification number series. Most CMOS ICs will operate with a supply voltage of from 3 volts up to 18 volts. This family is not critical as to power-supply requirements. Some CMOS chips

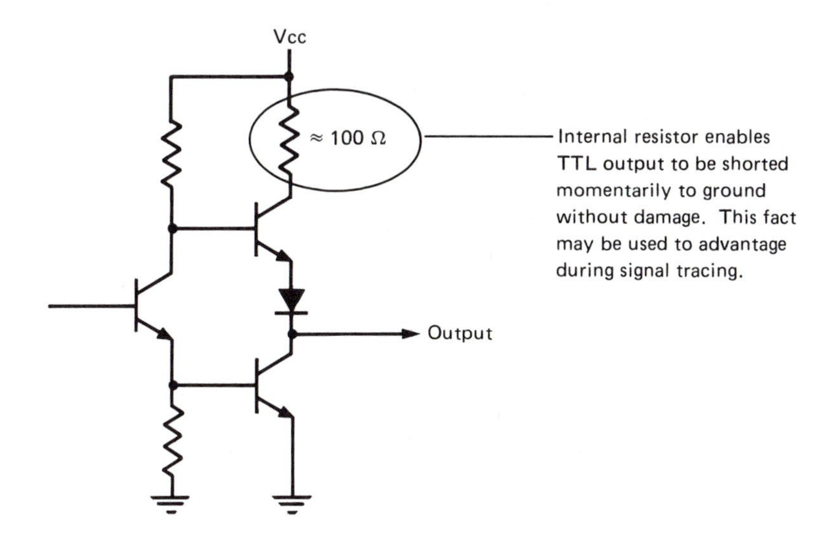

FIGURE 13-5 Typical TTL totem-pole output circuit.

when idling draw as little as 0.5 nanoampere (0.0000000005 ampere!). This is because of the open-circuit internal structure of the ICs. During operation some power is drawn, but it is far less than TTL circuitry where a single chip might draw anywhere from about 2 ma to 35 ma, or 0.002 to 0.035 ampere. The logic input levels to CMOS chips are defined in terms of the supply voltage (Figure 13-6).

A disconnected CMOS input will be erratic and may introduce some unexpected outputs from the chip. All unused inputs should be tied to either supply or to ground to prevent introduction of these signals. Circuit defects producing open inputs will produce unpredictable output levels.

For all of the power advantage of the CMOS line of ICs, they do have one drawback. CMOS ICs are very susceptible to damage by static electricity discharges. A good CMOS reference book will go into considerable detail about

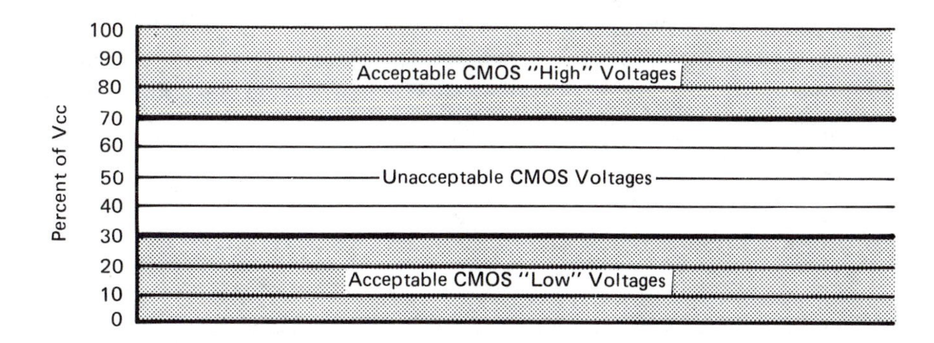

FIGURE 13-6 CMOS logic levels defined.

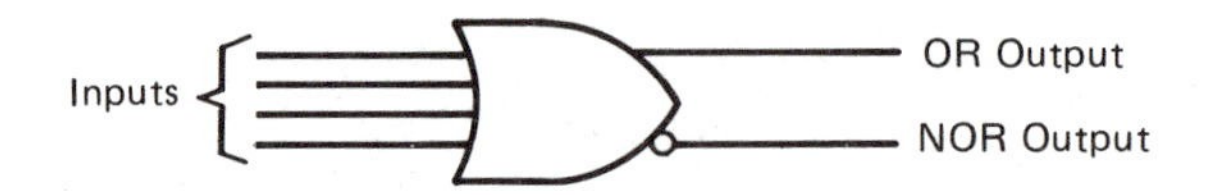

FIGURE 13-7 An ECL OR-NOR gate symbol.

a static-free place to work on these ICs. Abiding by these rules is something like using a seat belt. To be safe, it should be done. It is possible to get away with not taking the proper precautions for some time, but it is not recommended. Do the job right and avoid static damage by following the rules!

Briefly, CMOS ICs should be handled only in their original packaging, which is designed to be static free. Use a grounded workbench and touch the bench before picking up a CMOS device. Don't use static-generating materials, such as nylon or styrofoam, around CMOS. Insert or remove CMOS chips only when the power to the board is off.

ECL Family Characteristics

Emitter-coupled logic (ECL) is a family used for high-speed digital operations. One of the characteristics of ECL is that the gates have differential outputs (Figure 13-7.) The power supply for ECL is usually a negative 5.2 volts. This is sometimes labeled Vee. The positive supply leads of ECL are grounded and labeled Vcc. The "high" logic level is a negative 0.9 volt, and the "low" is a negative 1.75 volts (Figure 13-8).

There are several special tools that will be used frequently during troubleshooting digital ICs, if they are available. The job is easier if they are on hand.

1. Good test probes (see Figure 5-2)
2. The extender clip
3. An IC puller

See Figure 13-9 for an example of item 2.

FIGURE 13-8 ECL logic levels defined.

FIGURE 13-9 Extender clip. (Photo courtesy of AP Products Inc.)

THE DIGITAL SCHEMATIC DIAGRAM

Unlike the analog schematic diagram, which usually shows individual components, the digital schematic diagram is a combination of standard gate symbols and an interconnection diagram. Some of the blocks may not be la-

The functions of U_3, U_4, and U_5 are easy to understand and troubleshoot because of their shape, which defines their use.

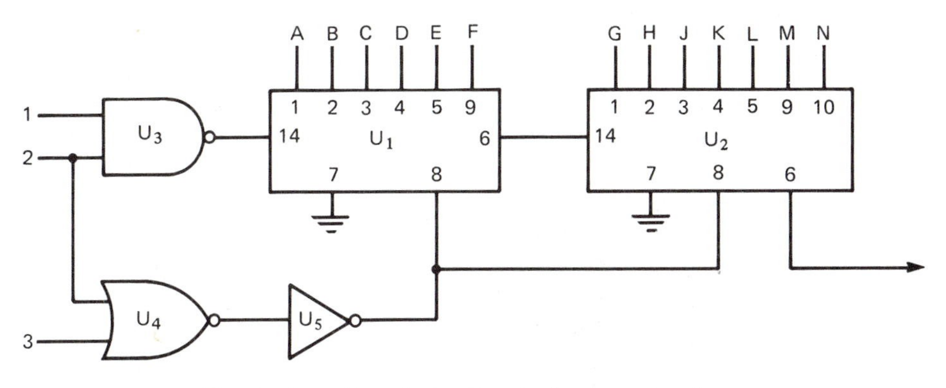

This circuit could not be intelligently signal traced until U_1 and U_2 were identified as to their use, their inputs, and their outputs.

The lines A through N could be either inputs or outputs. Only proper identification would determine their use.

Missing PIN numbers on a schematic may be unused or they may be very important. Look up *all* PIN numbers for troublesome circuits.

FIGURE 13-10 All digital schematic blocks must be identified to troubleshoot the circuitry.

beled such that the technician can tell at a glance what the functions of these blocks might be. See Figure 13-10 for an example of such a mixture.

Although a full schematic of a digital circuit might be rather intimidating, it is well to remember that each of the blocks can be tested for proper functioning, and that really isn't so hard. Don't let it overwhelm you. When using such a schematic, it is highly recommended that the servicing technician at least pencil in the important information right on the schematic. The use of the IC, such as "divide by ten," should be right there on the schematic. Usually all that has been provided is something like a circuit symbol "U3". This doesn't tell what the chip is doing in the circuit unless perhaps the parts list is consulted. The best reference is a data book for the IC family in use.

Another aid in troubleshooting that is particularly useful is to color-code the signal flow through the various ICs. Change color when the signal takes on new significance. Colors can be alternated and repeated further down the line.

When accounting for the pins of the IC, keep in mind that it is not unusual for the manufacturer of the equipment to connect unused input pins of the IC to either Vcc (the positive supply) or to ground without making any notation to that effect on the drawing. Often, even the pins supplying power to the IC and providing a ground are not shown on the schematic.

BASIC DIGITAL GATES

There are several digital ICs that are used very frequently as building blocks for large systems, and to tie larger scale ICs together with the proper signals.

The AND Gate

This gate requires *both* inputs to be "high" (positive) at the same time in order for the output to be positive. At all other times the output pin will be low (Figure 13-11).

This is a good time to explain the term *active* as it is used in digital electronics. The two inputs of this gate must both be positive for this circuit to function as an AND gate. These are *active high* inputs. By definition, the output of an AND gate goes high when the input conditions are met, therefore the output is an active high output, too.

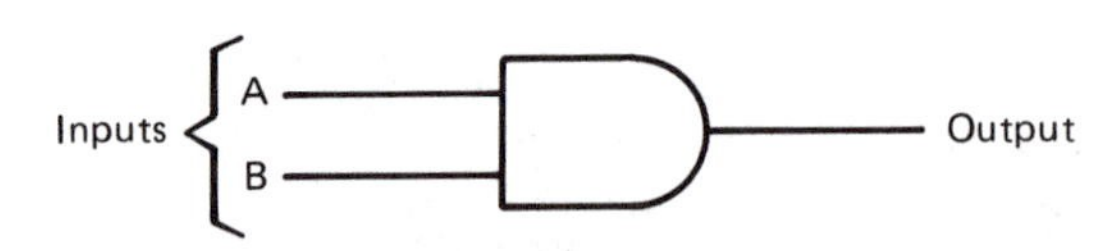

Both inputs must be high for the output to go high.

INPUTS A	INPUTS B	OUTPUT
0	0	0
0	1	0
1	0	0
1	1	1

FIGURE 13-11 The AND gate.

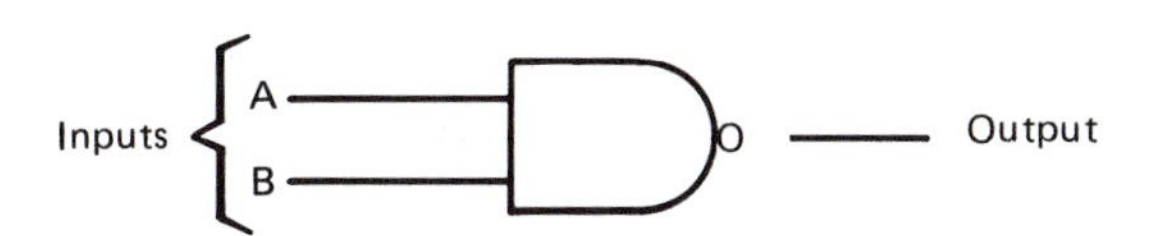

Both inputs must be high for the output to go low.

INPUTS		OUTPUT
A	B	
0	0	1
0	1	1
1	0	1
1	1	0

FIGURE 13-12 The NAND gate.

The NAND Gate

This gate is similar to the AND in that the inputs are still active high; both must go positive for the circuit to function as it should. In this case, however, the output goes "low" when the proper input conditions are met. The output is reversed and is referred to as being an *active low* output (Figure 13-12).

The OR Gate

This gate is shown in Figure 13-13. The OR gate will provide a "high" output when either the first *or* the second *or both* go high.

The NOR Gate

Like the OR gate, the NOR will have an active output if either input, or both, is active. The active output of the NOR gate, however, is *active low* (Figure 13-14).

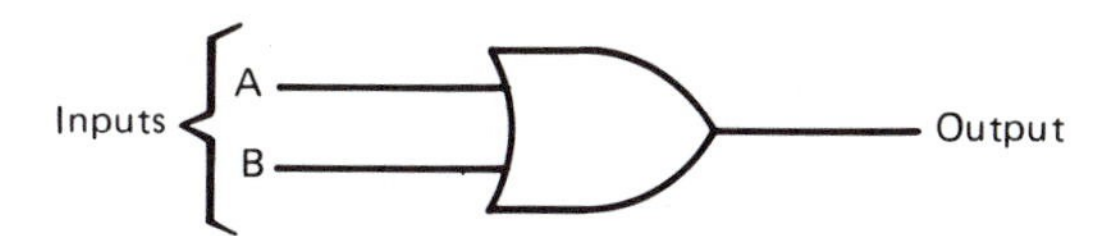

Either or both inputs must go high for the output to go high.

INPUTS		OUTPUT
A	B	
0	0	0
0	1	1
1	0	1
1	1	1

FIGURE 13-13 The OR gate.

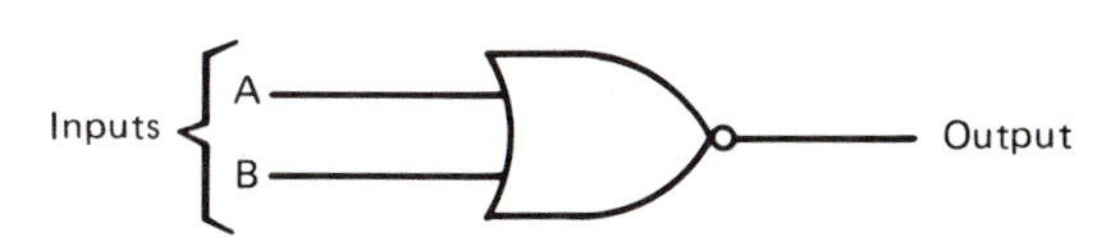

Either or both inputs must go high for the output to go low.

INPUTS		OUTPUT
A	B	
0	0	1
0	1	0
1	0	0
1	1	0

FIGURE 13-14 The NOR gate.

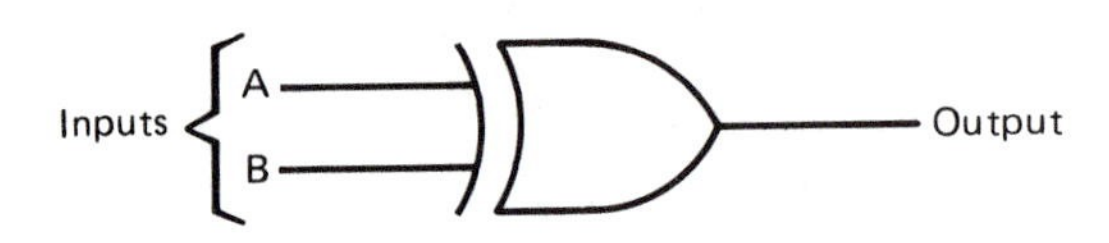

Either but not both inputs must go high for the output to go high.

INPUTS		OUTPUT
A	B	
0	0	0
0	1	1
1	0	1
1	1	0

FIGURE 13-15 The exclusive-OR (X-OR) gate.

The Exclusive-OR (X-OR) Gate

This gate is similar in appearance to the OR gate. The difference between the X-OR and OR is that the X-OR will *not* provide an active output if both inputs are high at the same time. In other words, the X-OR provides its normal high output only when *one* of the inputs is high (Figure 13-15).

The X-NOR Gate

This gate will provide an *active low* output only when *one* of the inputs is high (Figure 13-16).

The Inverter

This logic circuit performs the simple function of reversing a logic level from high to low, or vice versa (Figure 13-17).

The Driver

This circuit is similar to the inverter, but instead of inverting signals, provides more driving power to circuits on its output (Figure 13-18).

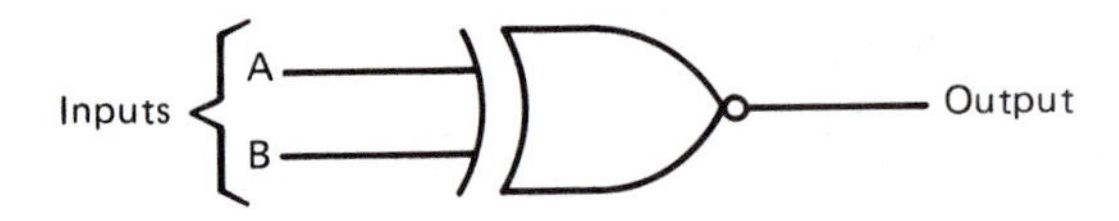

Either but not both inputs must go high for the output to go low.

INPUTS		OUTPUT
A	B	
0	0	1
0	1	0
1	0	0
1	1	1

FIGURE 13-16 The exclusive-NOR (X-NOR) gate.

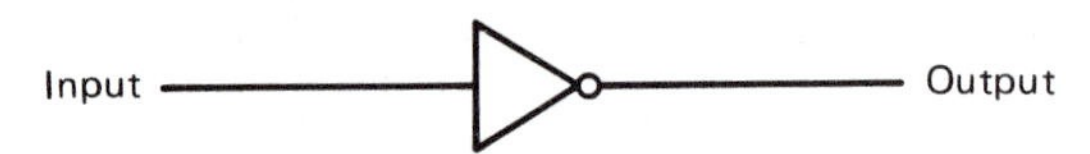

If the input is high, the output is low and vice versa.

INPUT	OUTPUT
0	1
1	0

FIGURE 13-17 The digital inverter.

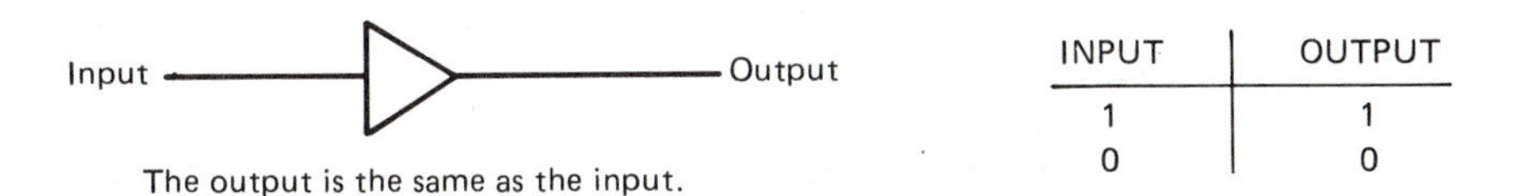

INPUT	OUTPUT
1	1
0	0

FIGURE 13-18 The digital driver.

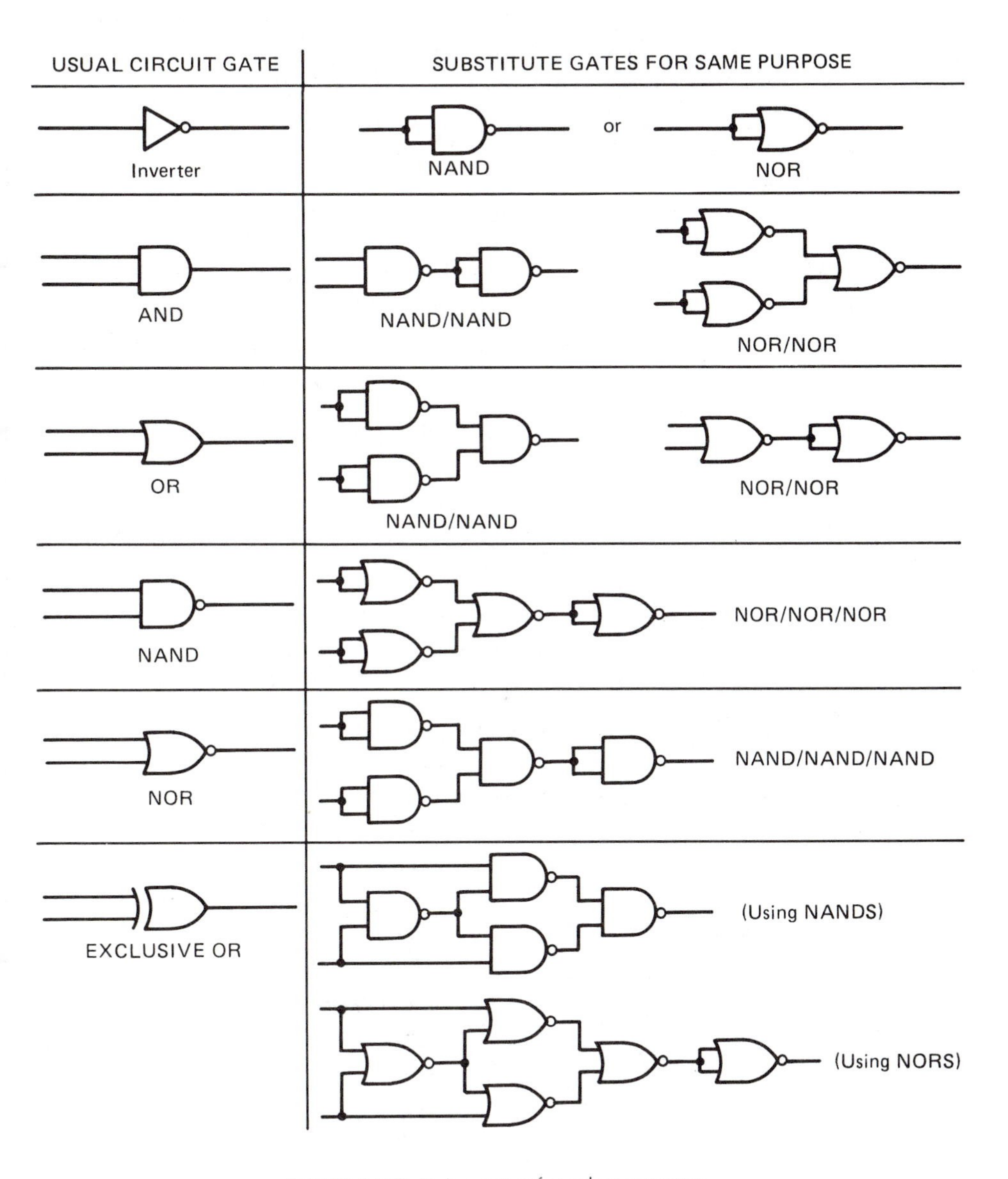

FIGURE 13-19 Using gates for other purposes.

It is common practice when designing digital equipment to use gates in combination to perform the functions of other chips that are not available on the board. When a combination such as this is encountered, it helps to trouble-shoot it as its equivalent rather than trying to interpret it all over again from scratch (Figure 13-19).

ICs other than the basic gates explained above include many different types: counters, converters, flip-flops, and microprocessor chips, to name a few. When a signal normally goes positive for an active or required input for a desired output from that stage, no special marking is used. If that pin must be negative, however, the input pin is designated with a small circle. In some circuits using letter designations for the pins, the letters have a bar over them, a practice common in microprocessor circuits. Inputs that are sensitive to the leading edge of a pulse rather than a set high or low level are shown with a small triangle. See Figure 13-20 for examples.

Other standardized inputs are as follows:

1. *Enabling* inputs must be satisfied in the proper active state, whether high (without a small circle) or low (with the circle) (Figure 13-21).
2. *Dual* inputs. See Figure 13-22 for an example of a dual-input condition that must be met before this stage within the chip will operate.
3. *Set* input makes the main output connection (designated Q) go "high" regardless of normal signal input. As long as this input is high, so is the output (Figure 13-23).

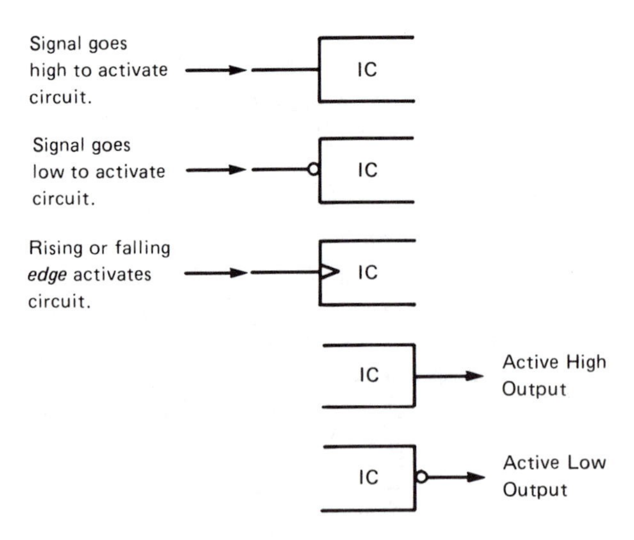

FIGURE 13-20 Digital signal coding.

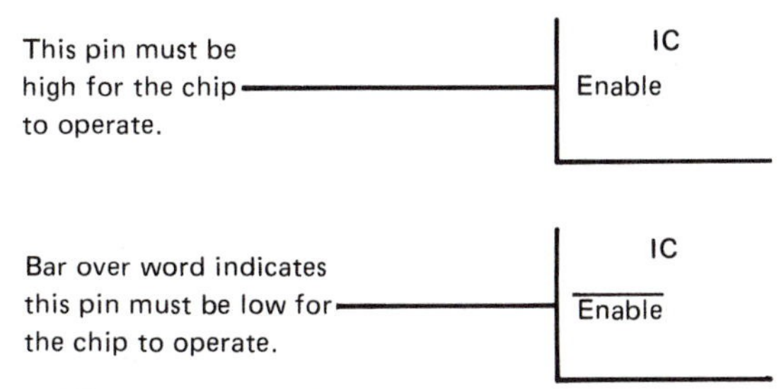

Lack of proper enabling signal may make a chip seem defective. Check all inputs before concluding a chip is bad.

FIGURE 13-21 The enabling pin.

ONE HALF OF A 74123 MONOSTABLE MULTIVIBRATOR CHIP

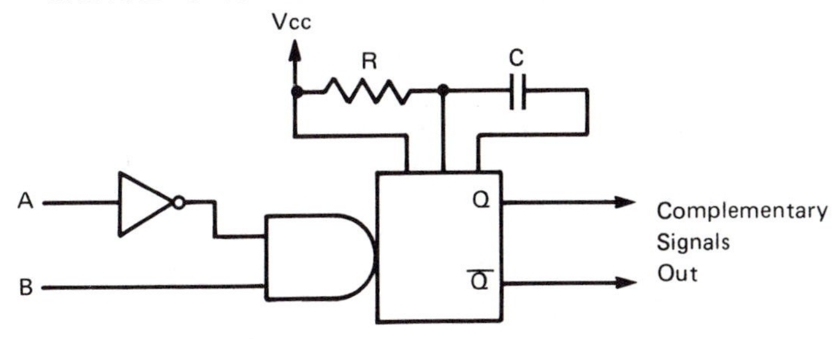

Holding input A low allows a positive trigger at B to activate the circuit.

or

Holding input B high allows a negative trigger at A to activate the circuit.

or

These two inputs can be used together to gate-trigger pulses into the circuit, triggers on one, a gating signal on the other.

FIGURE 13-22 Dual inputs must both be satisfied to activate circuit.

TYPICAL FLIP-FLOP CIRCUIT

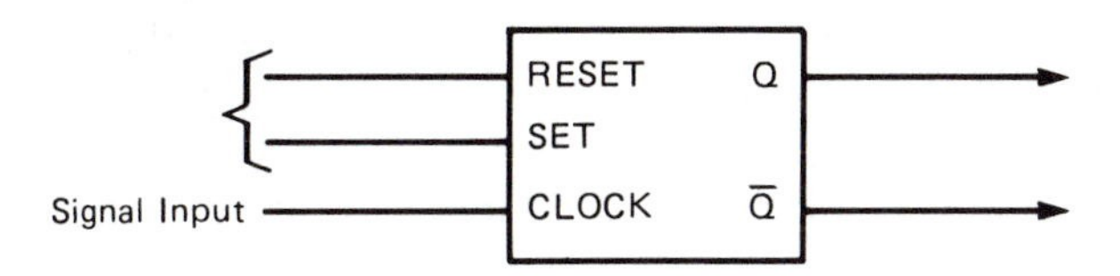

If the set input is held high, the circuit locks Q high and Q low. No incoming signals will have effect. If reset is held high, the circuit locks Q low and Q high. No incoming signals will have effect.

The set and reset inputs may cause a chip to appear defective. Check *all* inputs before assuming the chip is bad.

FIGURE 13-23 The set and reset inputs.

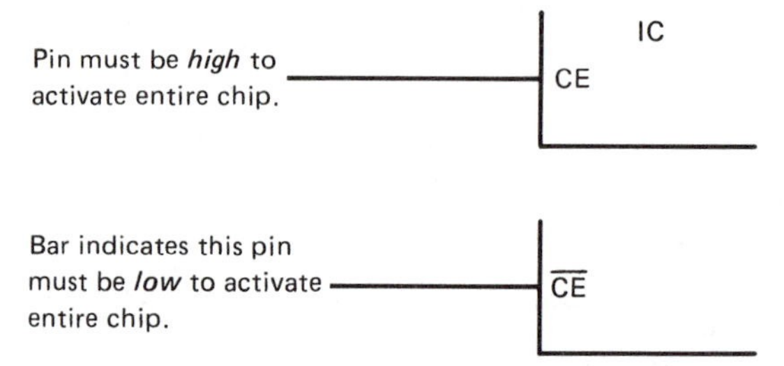

The chip enable pin must have proper signal applied for the chip to be active. In three-state ("floating") applications, absence of chip enabling signal opens circuit output lines from the chip.

FIGURE 13-24 The chip enable pin.

4. *Reset* input makes the main output connection (designated Q) go "low" regardless of other inputs, and stay there as long as the reset line is high (Figure 13-23).
5. *Chip enable* is similar to enable, above, but must be satisfied with its *active* state or the entire chip is essentially floating. This designation is used with microprocessor chips such as ROM, RAM, and input-output chips. This pin is usually marked with an overscore bar if it is active *low* (Figure 13-24).

INSTRUMENTS FOR DIGITAL CIRCUIT TROUBLESHOOTING

Source Instruments

The signal-tracing inputs to digital logic circuits are usually the waveforms normally in the circuit. however, there is one instrument available that can be used to force a short pulse into logic circuitry for troubleshooting. It is designed to pulse signals into circuit traces without disconnecting any of the ICs. The output stage is forced high when low, and forced low when high. Due to the very short duration of the pulse, the ICs are not harmed by this instrument. It is called the *logic pulser,* and one model is made by Hewlett-Packard (Figure 13-25).

This instrument can be used on TTL or CMOS circuitry. It is connected to the circuit's supply bus for power. Due to the ingenious circuitry within, it will pulse a high circuit low or vice versa without the attention of the operator. It will also provide the proper pulse width for the circuit under test, giving a shorter pulse for TTL than for CMOS. This shorter TTL pulse is required to overcome a saturated transistor without providing enough energy to damage the transistor. Since it is driven from the supply voltage for the IC, it will have the proper amplitude of pulses—higher for most CMOS circuits than for TTL, for example. Pulsers made by other companies than Hewlett-Packard may not have sufficient drive to overcome a saturated transistor; the specifications sheet for the instrument should be consulted to verify if this is the case.

FIGURE 13-25 Logic pulser. (Photo courtesy of Hewlett-Packard Company)

This instrument is very useful when used with the current tracer to track down shorts on PC boards. It is used with the logic probe to exercise ICs and signal trace through them. It can also be used for clocking ICs.

Load Instruments

The oscilloscope is often used to trace signals through logic circuits. This has been the traditional approach because the oscilloscope will show the pulses quite well and is often readily available for the technician to use.

Oscilloscope Probe Circuit Loading. The perfect attenuator probe does not exist. The common 10X probe has a circuit-loading capacity of about 10 or 11 pF. This capacity will be placed in parallel with any inherent capacity of the logic circuit. In so doing, the rise time of the voltage observed will be longer than it is when the probe is not attached. There will be more effect in high-impedance circuits. Although this loading effect may lengthen the rise time of a logic signal, it should not present any particular problem from a troubleshooting standpoint. If, however, a circuit behaves differently if the probe is applied, this effect should be kept in mind before immediately assuming that there is a circuit malfunction. Remember that defective, open inputs will be affected by the probe or any other changes in circuit loading. This is particularly true of the ICs of the sensitive CMOS family.

Quick Signal-Tracing Scope Settings. Use of the oscilloscope on digital circuits requires more skill in the use of the instrument than that needed for routine low-frequency or RF troubleshooting. The technician must be very familiar with the triggering controls, particularly the level and mode controls. Improper triggering of the oscilloscope can make pulses appear when they should not and disappear from the screen when they should be there. Although the "automatic triggering" mode may be used for simple troubleshooting where either the presence or absence of a signal is being observed, sophisticated use of the oscilloscope requires skilled use of the triggering controls.

Digital Timing Analysis. When analyzing digital signals, the oscilloscope's "triggered" function should be used in preference to the automatic triggering mode. This will ensure that the pulses displayed will have the proper timing relationships and will not drift across the screen in a random manner when their repetition rate is low. To test whether the triggering circuits are properly set for digital troubleshooting, temporarily remove the trigger. In the absence of a triggering signal, there should be no trace. If there is a trace, the triggering level control or other adjustment is not correct.

To set the triggering circuit properly, proceed as follows:

1. Set the triggering source switch to the appropriate setting. If a separate signal will be used to trigger the oscilloscope, use the "external" position of the trigger source switch. Provide the trigger via a proper oscilloscope probe to the "external trigger" connector. If, however, the signal will be provided by one of the signals to be observed on the screen, set the trigger source appropriately to select "internal" triggering source and set appropriate switches to select the proper oscilloscope input channel for the trigger.
2. Switch the *trigger mode* to DC if the pulses have a low-duty cycle. AC triggering may be used otherwise.
3. Attach the triggering source probe, either the external probe or the vertical channel probe for internal triggering, to the point in the circuit where the trigger will be sampled. Since digital signals are all pretty much the same, either high or low in voltage, you can later change the circuit source of the trigger using most any other digital signal.
4. Next set the *trigger polarity.* For most troubleshooting work this will be the (+) or "positive" position. If the technician desires, this control can be changed to (−) or "negative" later on to take advantage of the scope's ability to trigger on negative-going waveforms.
5. The last adjustment is to set the *triggering level.* This control is turned from the left or CCW end to the right until the scope produces a trace. Note the point at which the control is set when the sweep begins to trigger reliably. Then continue turning to the right until the sweep again disappears. Note the position of the control again and reset the control to a point midway between the two. This level should trigger the oscilloscope sweep when the triggering source is approximately halfway between its low and high logic levels. As the triggering source probe is then moved throughout the circuitry, the scope should produce a sweep whenever there is a good logic signal at the triggering probe and no sweep when the probe is disconnected or grounded.

The triggering signal chosen from within the circuit is of considerable importance. The trigger must occur at the same time or slightly before the event to be observed. The triggering source and sweep speed are adjusted until the desired presentation appears on the screen. Thus, a thorough familiarity with the oscilloscope's controls is required to set them up quickly for the best picture. If some difficulty is experienced in obtaining a stable presentation on complex waveforms, review the use of the "holdoff" control, explained in Chapter 12.

Using the vertical channel-switching options of "alternate" or "chopped" sweeps on a dual-trace oscilloscope requires understanding use of the instrument for signal tracing and timing checks on digital signals. Whereas on analog circuits it doesn't make much difference which mode of vertical switching is used because the signals are always there on both channels, on digital circuits the signals are often *not* there *at the same instant of time* on each trace. The use of the "alternate" trace option makes the oscilloscope beam sweep across the full width of the screen while displaying the signal of one channel. When that sweep is concluded, the sweep begins again, displaying the signal of the second channel. This results in missing some of the events occurring on one channel while the other channel is being observed.

A better choice of sweep mode for digital troubleshooting would be "Channel 1" or "Channel 2" input only, when using a single probe for digital signal tracing, or the "chopped" sweep function when using two vertical channel inputs at the same time. The use of the "chopped" vertical channel display provides essentially simultaneous sampling of both the input channels. The rate of sampling between the two channels is fast enough that the two signals appear on the screen apparently simultaneously. The sampling rate between the two signals is not synchronized with either signal and therefore will drift, causing the missed portions of a waveform due to the switching during one sweep to be filled in during succeeding sweeps.

Oscilloscopes with more than two vertical channels may be available for troubleshooting, but they may not be worth the effort in setting them up. The technician can use them as single or dual-trace instruments for simple signal tracing.

A Third Input for a Dual-Trace Oscilloscope. The "external" triggering input may be used as an extra input to the oscilloscope. With a dual-trace scope this would give three inputs, enough to test many of the commonly used gates. See Figure 13-26 for an example of how to use the scope to test the two inputs and the output of a NAND gate.

To obtain timing relationships between two or more signals, *at least two* inputs to the oscilloscope are required. If a single signal is observed without an external triggering source, it is shown at the left of the screen, since it is the triggering signal. If the probe is then shifted to another signal, that signal will also be shown at the left of the screen. This can lead the technician to believe that the two signals occur at the same time, unless time is taken to figure out what is happening within the triggering circuits and the timing involved. To observe time differences a minimum of two probes is necessary. One of the

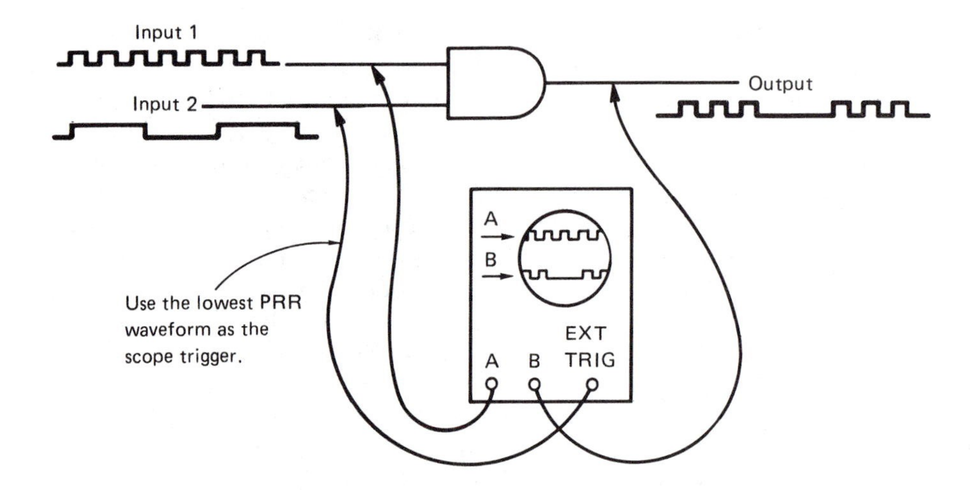

FIGURE 13-26 Using the external trigger as a third input to an oscilloscope.

probes will be connected to each of the signals in question, and the oscilloscope must be set to trigger on one of those signals.

There are other instruments available to trace most signals through digital circuits with more speed and efficiency than an oscilloscope.

The Logic Clip

This small instrument (Figure 13-27) is attached to the IC in question and presents the high or low status of all 14 or 16 of the pins simultaneously. The

FIGURE 13-27 Logic clip. (Photo courtesy of Hewlett-Packard Company)

logic clip of Figure 13-27 may be attached to either TTL or CMOS chips. It finds the supply and ground leads without any effort on the part of the operator—all automatic within the clip. The status of all 14 or 16 of the remaining pins are shown on the LEDs on the top. A corresponding pin that is high will have the LED on, and vice versa. An LED that is partially lit indicates that the IC pin associated with that LED has a signal in an operating circuit. The amount of the LED brilliance compared to the LED on the supply pin of the IC will give a very rough indication of the duty cycle of the signal. A very short duty-cycle signal will not be seen, as the logic clip has no pulse stretcher to make the signal detectable.

The logic clip may be used with TTL and CMOS up to 30 V DC. It may be used to check truth tables if the incoming pulses can be halted and the IC held in a static condition. The state of each of the pertinent pins is noted and the IC allowed to advance a count. This is best done using the logic pulser, discussed earlier in the chapter. The new state of the pins can then be observed for proper indications. During static testing such as this, the presence of any half-lit LED would indicate a bad logic level to be further investigated.

The Logic Probe

This instrument is so handy that it has been called "the digital screwdriver." It is a small probe the technician uses to make go-no-go decisions throughout the circuitry. It will detect the presence or absence of logic signals and will stretch very short pulses out so that the technician is aware they were there. It will also define bad levels (between high and low), or three-state buses. Figure 13-28 shows the logic probe.

The logic probe has a selector switch for choosing between the TTL and CMOS families. Its only other control is a switch to reset a memory, which can be used to detect the arrival of a single pulse. When the switch is pressed, the tip light goes out and remains so until the receipt of at least one pulse, whereupon it lights and remains lit. The logic probe can be used with existing circuit signals or with the logic pulser as a source of pulses.

The Current Tracer

The current tracer (Figure 13-29) is usually the final instrument to be used of the instruments so far discussed. This instrument will trace the flow of current through a circuit trace. It is a valuable tool in finding shorted traces due to

FIGURE 13-28 Logic probe (the screwdriver of logic tools). (Photo courtesy of Hewlett-Packard Company)

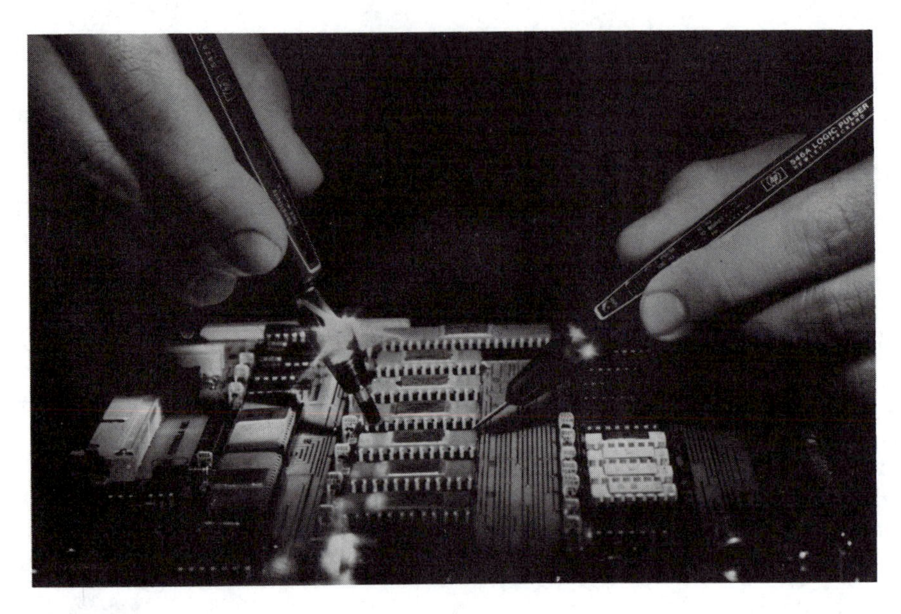

FIGURE 13-29 Using a logic pulser with a current tracer to find shorts in printed circuit boards. (Photo courtesy of Hewlett-Packard Company)

solder bridges or other board faults, or shorted ICs. The ICs can even be shorted to ground or the supply, and will be found with the current tracer. The reason it is valuable is that the conventional voltage-measuring instruments cannot find the specific location of a short when there are many possible paths. The voltage instruments are useful for finding which trace is stuck, then the current tracer is used to find the component responsible for the short.

Normal circuit currents are usually quite low. If a short occurs, the normal circuitry will try to drive considerably more current than normal into the short, whether high or low. The current tracer can detect these increased currents in operating circuitry using existing signals.

Another way to use the current tracer is to use the logic pulser as a source and follow a current path with the tracer to find the short. The current tracer also has a pulse stretcher so that the operator can see the relatively short pulses that the tracer is following.

The Logic Comparator

This instrument allows the technician to verify if an operating circuit is occasionally missing pulses or is erratic in such a way that other instruments miss the cause of the problem. Its principle of operation is quite simple. A duplicate IC of the one under test is inserted into the comparator and by means of a multipin clip is compared parallel to the one in question. The IC used as a good comparison is driven by the same input signals as the IC under test. If the

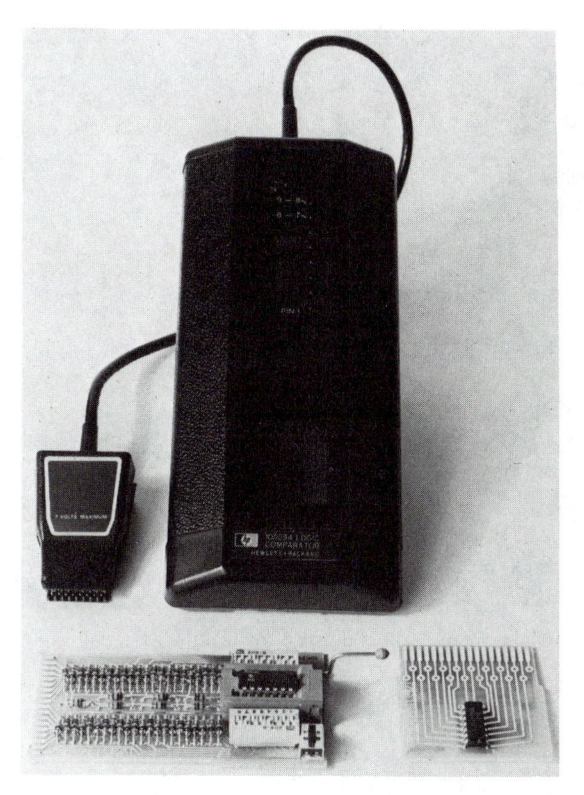

FIGURE 13-30 Logic comparator. (Photo courtesy of Hewlett-Packard Company)

IC under test varies at all in output from the good one, the comparator lights a light to indicate a discrepancy (Figure 13-30). The logic comparator is particularly good for finding missing pulses, glitches, and other intermittent problems caused by defective ICs. A good stock of duplicate ICs known to be good is necessary to use as "standards."

The Logic Analyzer

The logic analyzer (Figure 13-31) is used principally with microprocessor systems during manufacturing research and development. It is not yet in common use for routine troubleshooting and will not be discussed in any detail in this practical guide.

The Signature Analyzer

The signature analyzer for troubleshooting microprocessor circuits is an innovation of the Hewlett-Packard company (Figure 13-32). Its principle of operation is similar to that used in an analog system where there is a library of

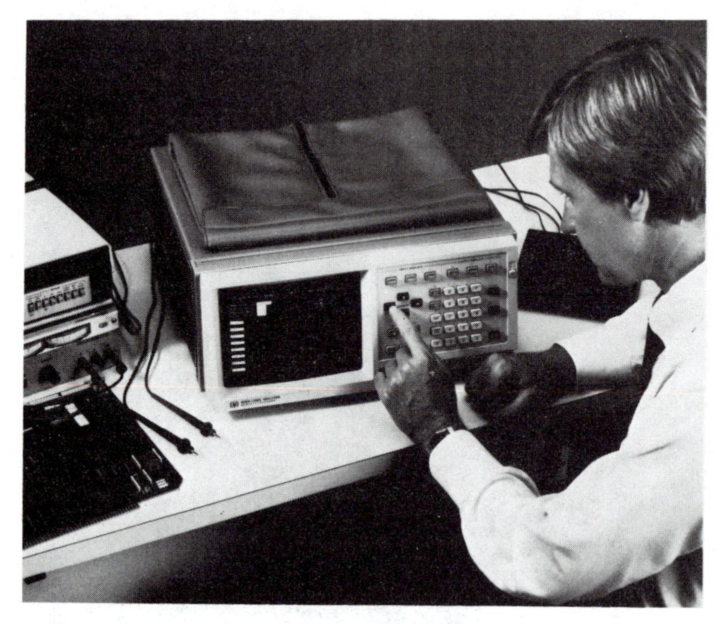

FIGURE 13-31 Logic analyzer. (Photo courtesy of Hewlett-Packard Company)

precise oscilloscope waveforms available for standard signals within the equipment under test. The signature analyzer uses a standard input to the microprocessor system and instead of presenting an analog output, it gives an alphanumerical display to be compared to those values in separate documentation.

An advantage of signature analysis is that the troubleshooter, with proper software and documentation, can tell quickly which circuit is malfunctioning.

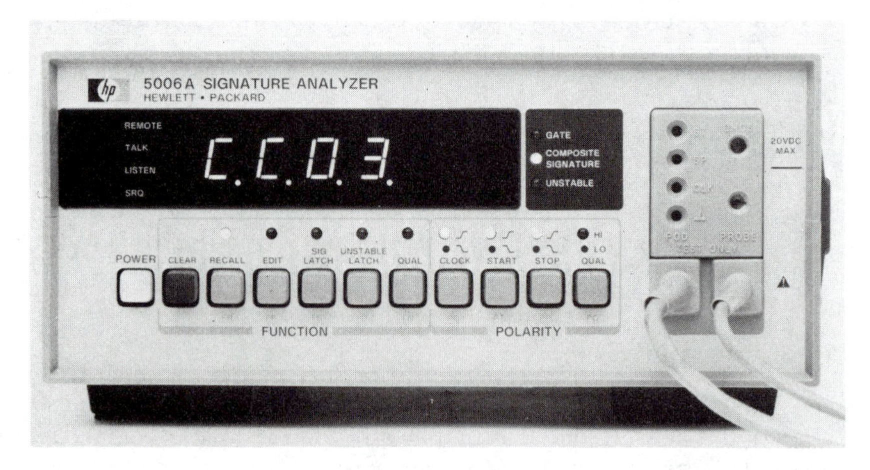

FIGURE 13-32 Signature analyzer. (Photo courtesy of Hewlett-Packard Company)

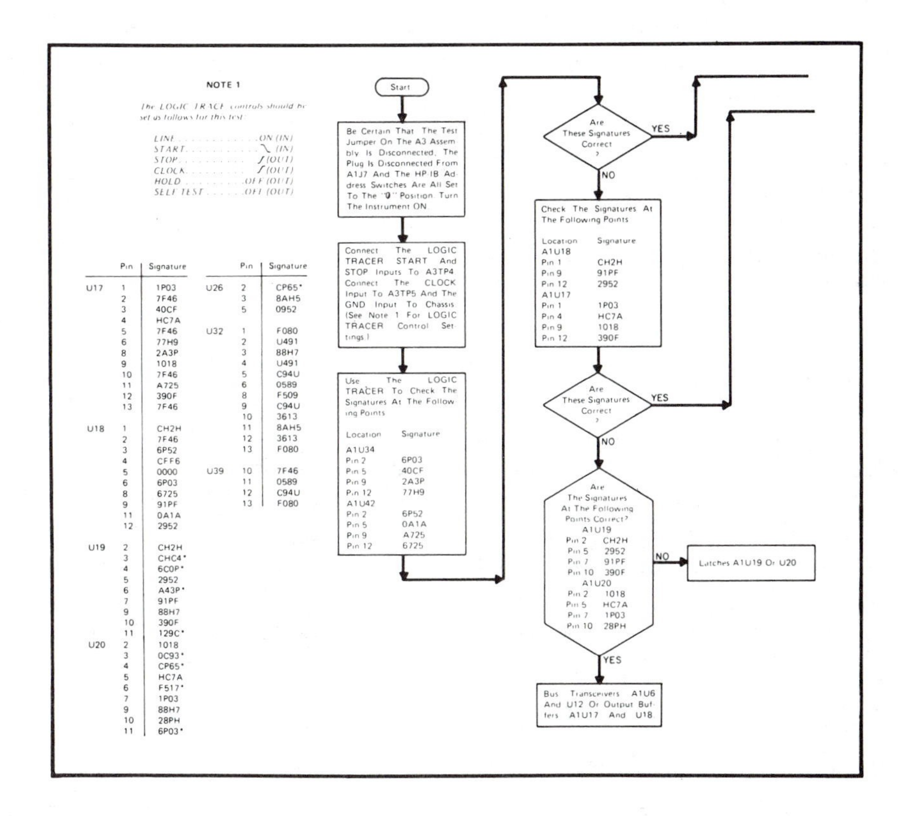

FIGURE 13-33 An example of a signature analysis troubleshooting tree. (Photo courtesy of Hewlett-Packard Company)

The biggest disadvantages are that the technician must have documentation with which to compare readings, and the equipment under test must be operated with specific software for the run. Those few technicians fortunate enough to have this test equipment available, will, however, find it a very fast and efficient way to troubleshoot compatible microprocessor equipment.

Troubleshooting using signature analysis test equipment consists of following a troubleshooting "tree" to find the defective part while monitoring special test signals designed to exercise the circuits. Figure 13-33 is an example of such a "tree."

TRACING SIGNALS THROUGH DIGITAL CIRCUITS

Servicing Diagrams Are Indispensable

When tracing signals through a digital circuit, the instruction book is an absolute must. Without proper information, only the dead circuit checks of

Chapter 16 have any chance of success. Proper information should include the schematic diagram and an explanation of the circuit, the theory-of-operation section. Shotgunning a digital problem by replacing parts at random in the hope of accidently finding the problem is seldom worthwhile unless the circuit is very simple, with few parts.

The instruction book should first be looked over carefully. Get the general layout of the board and what it does. If there are several boards involved, their use may be generalized according to function by the use of series of part numbers. For instance, the power-supply components might all begin with a "100" series number like R123 or C104. The input signal board might have a series of "200" numbers, and so forth.

Clocked and Unclocked Logic

There are two general kinds of logic circuits. The first are those that pass signals right through them immediately upon receipt, like the basic gates discussed earlier, "D" flip-flop circuits, and others. The other kind of circuit is called clocked logic. These chips require the timing pulses of a system clock to step signals through, one operation per clock pulse. Input signals occurring at other times are ignored. These circuits were designed to prevent runaway rippling of signals through the circuits. These circuits are particularly common in microprocessor applications.

The Object: Look for Signal Activity First

Circuits have been designed to take into account the delay of signals as they progress through the ICs of a digital circuit. This is called *propagation delay.* The rise and fall time of circuits has also been considered. The troubleshooting technician should not ordinarily be expected to troubleshoot such problems on established circuits. If the work is in research and development of circuits, this may not be the case. The technician may have to find these difficult problems in the normal course of work. A solid foundation in normal troubleshooting will be necessary.

The pins of the ICs are test points. They are accessible from both sides of the board, a handy feature in itself. It takes a bit of practice to remember to count *clockwise* on the *bottom* of the board and *counterclockwise* on the *top.* Chips are usually laid out on the board so that all of the #1 pins of the chips are toward the same edge of the board—but check the ICs to be sure! It will really confuse the issue to encounter a chip that is upside down from the rest, and not to have noted this difference.

When practical, all signal readings should be taken at the IC pin on the top of the board. The reason this is recommended is that whether or not there is a socket, one leg of an IC can be bent under, not making contact with the board. This is a common problem with automated insertion machines at the factory.

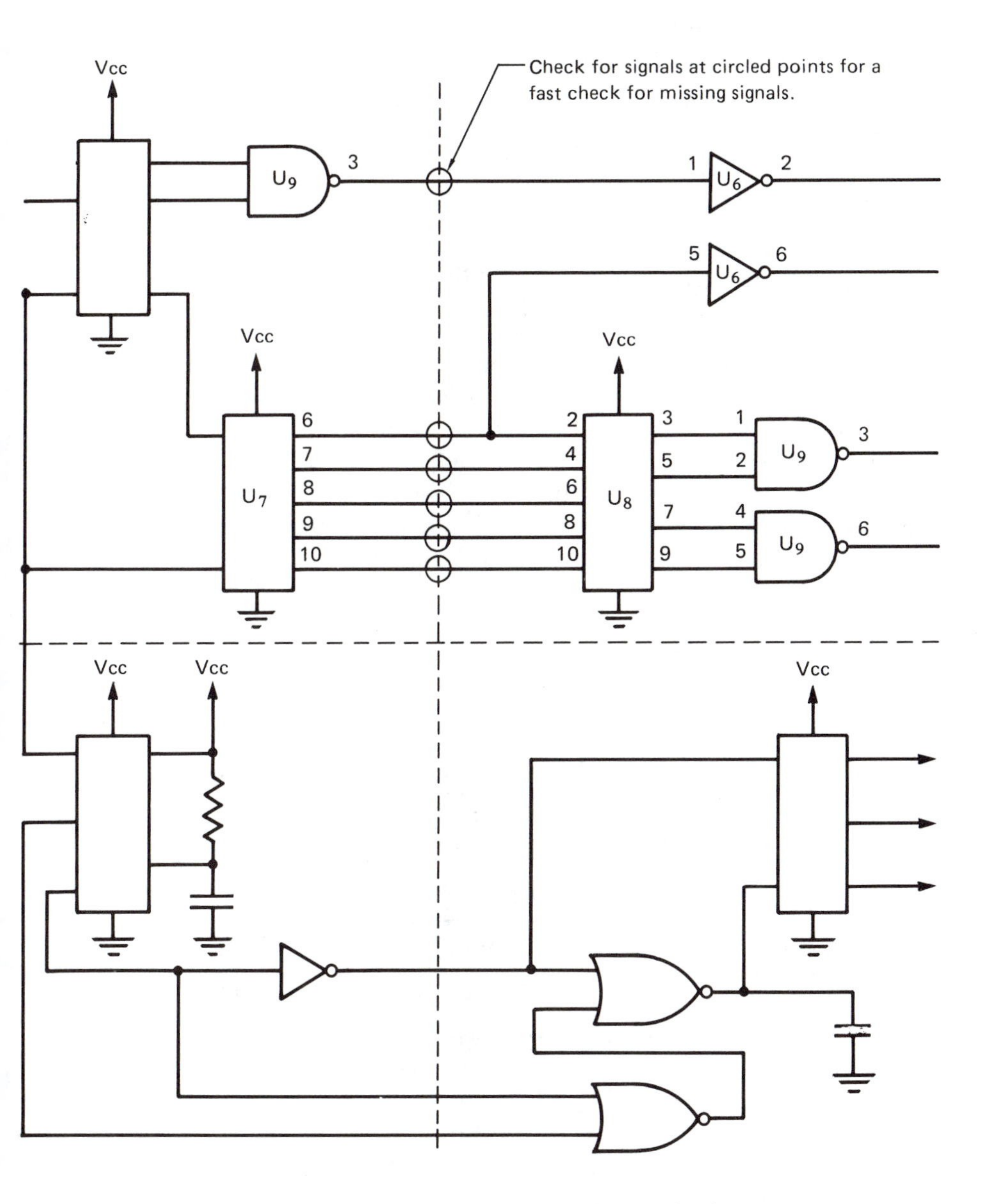

FIGURE 13-34 Identify each chip and output lead.

Sockets can also develop bad connections to produce the same symptoms. Reading at the pin of the IC will make detection of these problems easier to recognize.

With the schematic at hand, draw a light line down the approximate center of the schematic, between any ICs. This will be the beginning of the half-splitting method of signal tracing. Identify each of the ICs on either side of this line as to their use and location on the board. With the aid of any references available, identify all of the output pins of the ICs producing signals crossing this line, along with the termination pin numbers of the load ICs (Figure 13-34). With this schematic as a guide, first determine if signals are present where they should be. Depending upon whether or not the signals are all where they should be, it will be necessary to half-split again, in one direction or the other.

Setting Up a Scope for Signal Tracing

The vertical channel of an oscilloscope should be set up for DC input signals, and the zero-reference line should be at the center of the scope. Set the sensitivity of the oscilloscope so that the maximum voltage (Vcc) is at the top line of the scope (ECL circuits would set 5.2 V DC at the bottom). It is a big help to put a mark on the oscilloscope face to remind the technician to look for levels that are not acceptable. For TTL circuits, this means a mark on the scope face at 2.4 V and at 0.8 V. Any voltage falling *between* these lines is a bad level and should be regarded as a definite indication that the problem is near at hand. (Figure 13-35). CMOS circuits should be set up at the 30-percent and 70-percent values of the Vcc supply. ECL would be set up for −0.9 V and −1.75 V.

The marking on the CRT can be a small piece of very narrow black tape such as is used to lay out printed circuit traces. An alternative is to use any one

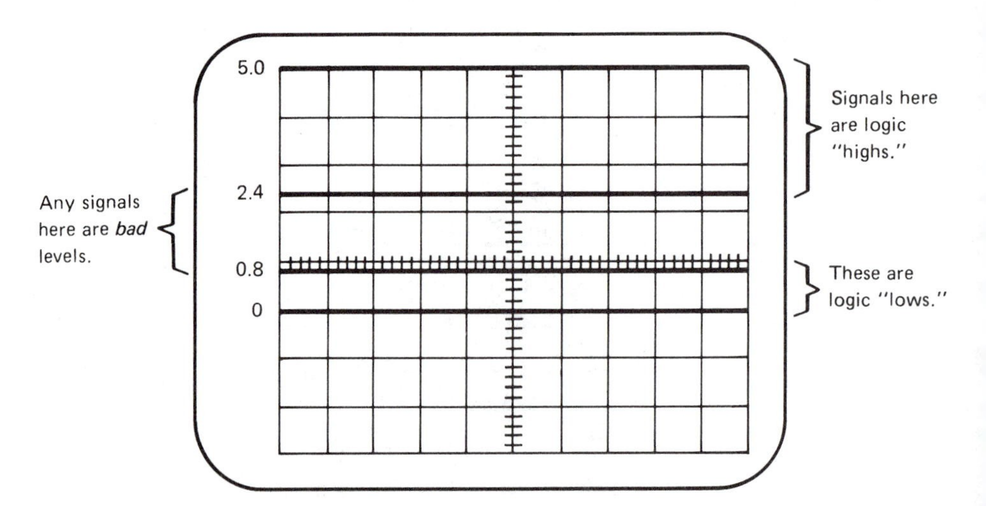

FIGURE 13-35 Marking an oscilloscope face with TTL logic levels for rapid detection of bad signals.

of the special pens for marking on glass and plastic. If the oscilloscope face is plastic, this marking must be done with care to prevent scratches.

When looking for missing signals, the "internal" triggering circuits and "auto" triggering may be used.

Full-Speed Testing

Full-speed testing is simply using the existing signals of the circuit, running at their normal speeds, to look for missing signals. Look with either an oscilloscope or a digital logic probe at each of the lines that cross the center line of the schematic, drawn as described a few paragraphs ago. Look for normal signal activity at each of these lines.

If You Find a Missing Signal. When you find a pin that apparently has a signal missing, the circuit diagram should be consulted to verify that the signal should indeed be there. In microprocessor applications, it is possible that that particular chip is not being "called upon" by the microprocessor chip, and certain signals should not be there. Be sure to check the special pins on the chip that sources the signal. Be sure that it is not being disabled by another input such as Set, Reset, Enable/Disable, Chip Select, or in the case of the microprocessor itself, an Interrupt line. For chips having them, all of these inputs must be correct for the chips to work.

If a Pin Is Stuck High or Low. The next items to check for proper signals are the inputs. If the input signals are *not* okay, you will have to continue tracing backward to find out why. If the inputs are good and the output of the chip is bad, several possible reasons for this exist:

1. There may be an internal short in an IC connected to the bad pin.
2. There may be an internal open inside the source IC.
3. There may be a short on the PC board, external to any IC.
4. The proper input conditions of a gate may not be present to produce any output. This should be verified, using an oscilloscope or a comparator.

The next step is to turn off the power and measure the pin that has no signal with an ohmmeter. Measure from the pin to the supply voltage bus and then to ground. If the lowest resistance is less than 1 ohm, look carefully for a short on the circuit-board trace. If the lowest resistance is a few ohms, either the source or one of the load ICs is shorted. Either of these shorts may be traced efficiently with the pulser/tracer combination of instruments (Figure 13-36). If there is no short circuit according to the ohmmeter, it is most likely that the source IC is open internally.

If a Pin Is Stuck at About 1.5 V (TTL Only). This is a possibility when a TTL output open-circuits. The input circuits following will pull themselves to this approximate level, which is easily detectable with the oscilloscope. It should fall right in the middle of the "no-man's land," between the acceptable logic limits.

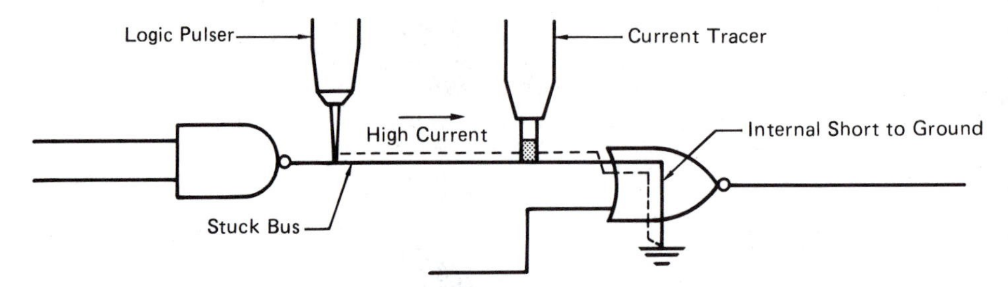

FIGURE 13-36 Using a logic pulser and a current tracer to identify defective component.

Watch for a CMOS Open Circuit. If a CMOS input is open-circuited, it will be extremely sensitive. The circuit output may or may not change state with the slightest input circuit disturbance, such as a finger touching the board or the application of a probe.

If the Signal Level Is Bad Only Part of the Time. A short between inputs will tie two other circuit outputs together. When two outputs are tied together, they will "fight" each other *sometimes,* when they disagree. If they agree, the output level will be acceptable (Figure 13-37). Look for the chip or circuit-board short that is tying two inputs together. You should be able to identify another output signal that has a similar problem, and then be able to identify the chip that has an input from each of the two signals.

If all of the signals appear normal down the center of the schematic,

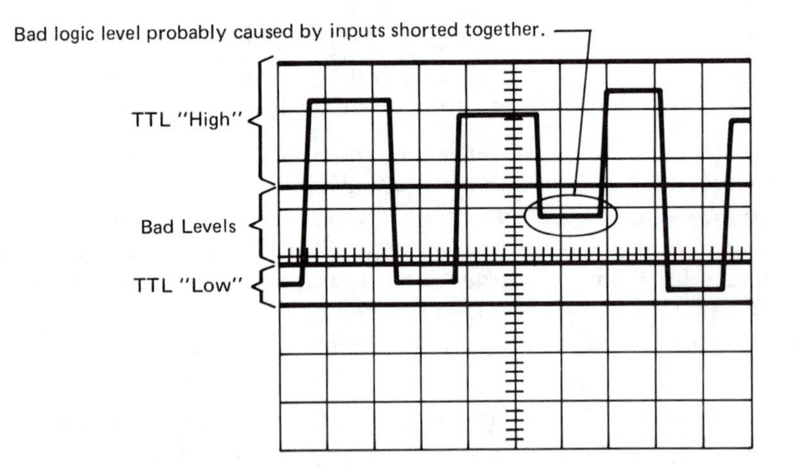

FIGURE 13-37 Bad logic level can be seen on an oscilloscope.

move to the right and draw another line similar to the one you drew down the middle. If you find many signals missing, trace them back to their source. Continue this half-splitting as necessary to find the single common problem causing the whole mess.

If You Don't Find a Missing Signal on the Board. If all of the signals on the board appear normal from one end of the circuit to the other, there is one problem that may be causing these symptoms: An AND, NAND, X-OR, or X-NOR gate with an internal open input can mask a problem. The output will be ignoring one of the inputs. The reason this can happen is that with some internal failures, one of the inputs will pull high inside the chip. There will be no response to the outside signal being normally applied. The chip will assume that that input is always high. Thus, the OR and NOR gates cannot cause this problem—they will "lock up" and won't pass signals if one of the inputs goes high.

Other, more complicated chips can cause similar masked problems. A counter, for instance, could have shorted or open circuitry and still be producing an output, although the output would not be valid. This possibility could be checked by testing each of the chips except those mentioned above for proper truth-table output. Although this wouldn't be too difficult to do with the simple AND, NAND, X-OR, and X-NOR gates, it could get a bit tedious with the counters and other chips when operating at full speed. Using a comparator at this point would be a very good idea—all of the outputs and inputs of a good and a suspect IC are compared in full operation, at full speed.

Keep in mind the possibility of two inputs of a single chip shorting together, causing a poor level part of the time as the two inputs conflict. An oscilloscope can help verify this as a possible problem.

Single-Stepping Logic Circuits

In order to detect counting problems at a rate that will allow the technician to analyze actual counts, it may be necessary to stop the normal counting of a circuit and single-pulse check the circuit for proper counting sequence. This can be done easily, without breaking the PC trace or removing a clock circuit component, if a circuit card extender is used. Some extenders have switches installed to break selected traces for troubleshooting purposes. If the extender board in use does not have such a feature, a bit of Scotch tape on the appropriate pin of the board may accomplish the same purpose. (Don't forget to remove the tape at the end of the job!)

The logic pulser can be used to single-pulse circuits for this purpose. This is also a time when the technician can make good use of the multipin monitoring capability of the logic clip (Figure 13-38). The logic probe may also be used to monitor counter pins one at a time.

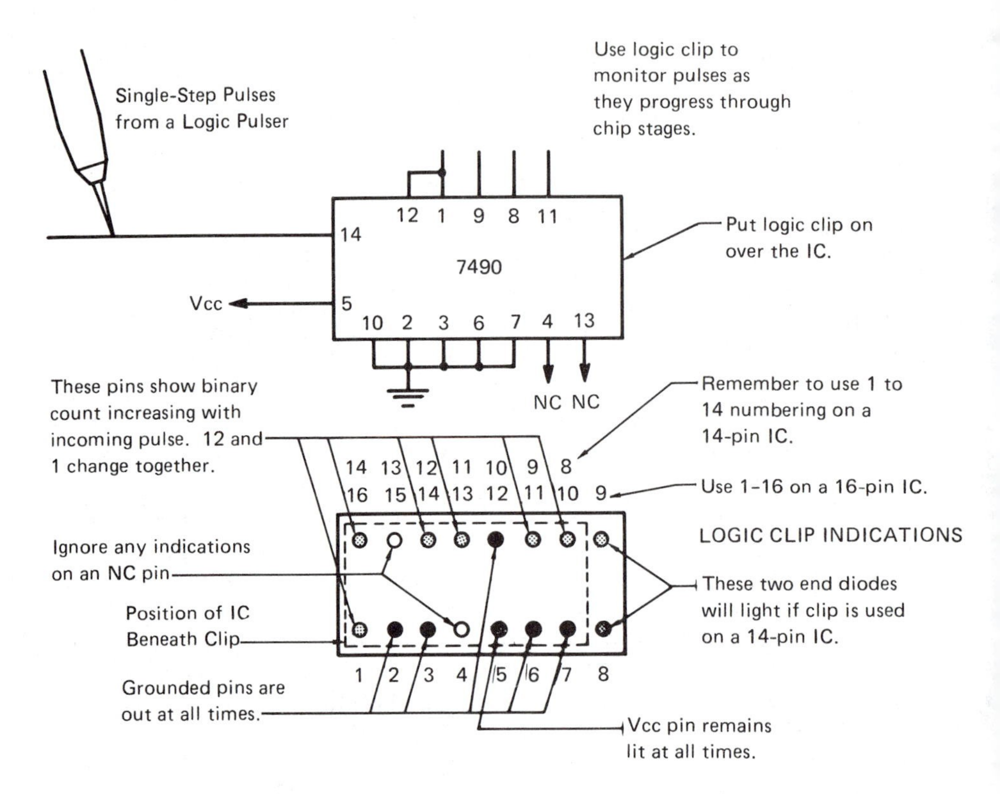

FIGURE 13-38 Using the logic clip with a logic pulser to single-step logic digital ICs.

SUMMARY

This chapter has covered, step by step, the procedures to follow in troubleshooting digital circuit boards. The technician should be able to repair 95 percent of digital cards with these procedures if the proper schematics, test instruments, and tools are available. The remaining five percent or so of digital problems involve obscure problems such as timing errors and circuit glitches and may be very difficult to find. Such circuits are probably best returned to the manufacturer to repair.

REVIEW QUESTIONS

1. What voltage does TTL operate from, and what tolerance is acceptable?
2. Name two instances of digital ICs that will not have digital waveforms at one of their pins.
3. If a CMOS circuit is operated from 10 volts, what is the least voltage that can be accepted as a "high" level? The most voltage that can be considered a valid "low"?

4. The AND gate: ______ input(s) must be ______ for the output to go ______.
5. The NAND gate: ______ input(s) must be ______ for the output to go ______.
6. The OR gate: What combinations of inputs go "high" for the output to go "high"?
7. The NOR gate: What combinations of inputs go "high" for the output to go "low"?
8. How is the X-OR gate different from the OR gate?
9. How is the X-NOR gate different from the X-OR gate?
10. What does a bar over pair of letters ,such as $\overline{CE}$, mean?
11. What does a small triangle on a schematic at the input to a chip mean?
12. What is the quick check that can be made to see if an oscilloscope is properly set for external triggering?
13. Why should the "alternate" sweep mode *not* be used to see timing differences between two vertical channel signals?
14. Name four uniquely digital, small troubleshooting instruments.
15. Why is it a good idea to use the IC pins on the top of a board for tracing digital signals, rather than the circuit traces on the bottom of the board?
16. What are two reasons that a TTL level might be at approximately 1.5 V?

chapter fourteen

Troubleshooting Computer Circuitry

CHAPTER OVERVIEW

This chapter will present the basics of troubleshooting microprocessor circuits. These are the circuits used in today's modern personal computers, from the most basic 8-bit machine to the most sophisticated 32-bit machines that are now coming on the market.

ADDITIONAL KNOWLEDGE REQUIRED

Microprocessors, in particular those used in personal computers, require troubleshooting knowledge and techniques in addition to the principles covered so far in this book. To understand some of the technical instructions in the various technical manuals and for computers in general, the technician must have a thorough understanding of binary, hexadecimal, and BCD (binary-coded decimal) numbering systems. The internal workings of the computer must be at least basically understood. How a microprocessor gets its instructions and data bytes and how it generates memory addresses should be understood thoroughly.

Depending upon the depth to which the technician must repair computers, additional detailed knowledge should be acquired regarding special computer chips such as the disk drive controllers and parallel and serial input/output (I/O) chips, to name a few.

Why Computer Circuits Are Different

The circuits used in microprocessor applications differ from those discussed in Chapter 13 in that there is a third state of operation besides either "high" or "low." This third state is the "open circuit." The digital signals of the previous chapter originated in a single chip, and the signals went from there to one or more receiving chips. In computers, however, things get a bit more complicated. Signals are passed back and forth between many chips, all connected to common lines. While many chips can receive signals from a single line at the same time, only one of the chips can send information. If two were to try to send at the same time, there would be a "fight" over who had control. This is called "bus contention." The possibility of bus contention is avoided by making unneeded chips an open circuit. The open circuit is the third possible condition used in computer circuits.

TYPES OF COMPUTER PROBLEMS

Computer problems come in three broad types:

1. hardware problems (power-supply and circuit-board problems; disc-drive problems)
2. software problems
3. setup problems (internal setup-switch problems)

Computer hardware problems are covered in this chapter, except for disk-drive problems, which will be covered in Chapter 15. Software and setup problems will be discussed later in this chapter.

HARDWARE PROBLEMS

Power-supply and circuit-board problems can occur. Troubleshooting of power supplies was covered in Chapter 6, and that chapter is applicable to working on computer power supplies. Computer power supplies may have up to 220 V AC inside them, even when operating on 115-V AC lines, so special caution must be exercised.

The Fastest Hardware Repair Method of All

The first step in troubleshooting a computer is to eliminate external equipment as possibly being involved in the failure. Printers, monitors, and keyboards can be easily disconnected, or different ones substituted, to see if the problem remains.

If a hardware problem is suspected, the most efficient way to prove that it is a hardware problem and where it is within the computer is to borrow known good cards from another computer and try them. When changing cards, be careful to *turn off the power* before removing or installing them. *This is very important!* Also, be sure to observe all of the normal precautions to prevent static-discharge damage. Substitution of cards will require that new cards have exactly the same jumpers and switch settings as the originals. Changing cards between earlier and later models has its own hazard, in the form of subtle changes that can produce weird results like inoperation, strange outputs, or even circuit-board damage from unintentional cross-connection of critical lines. Be very sure that the cards are indeed interchangeable before plugging them in and applying power.

Substitute the good cards one at a time, and test after each substitution for the presence of the malfunction. If a card substitution cures the problem, it is reasonable to assume that the last card replaced has a problem. At this point, it may be economical simply to return the defective card to the manufacturer for repair, whether or not under warranty.

Finding a hardware problem within a computer using traditional signal-tracing methods is possible but seldom worthwhile.

LET THE COMPUTER TEST ITSELF

Powerup Diagnostic Routines

Some of the more sophisticated computers have programs in ROM (read-only memory) that test for the proper connection of a keyboard and test the system's RAM (random-access memory), the heart of the computer. These programs in ROM are called *powerup diagnostics*. These programs are an "extra," not necessarily supplied on all computers.

Generally speaking, the powerup diagnostics cause a delay of a few seconds upon applying power to the computer. If there is a problem with the keyboard connection or an error detected in the checking of the RAM, an error code will show on the screen. If a single bit in a memory of up to perhaps 704,000 kilobytes is found defective, the diagnostic routine will signal the operator that something is amiss. A code is flashed on the screen that will guide the technician into the proper area for troubleshooting. The program may even pinpoint the specific chip to be replaced. This method of testing RAM chips is far superior to any other, particularly in view of the fact that most RAM chips these days must be refreshed many times a second to retain their memory anyway. Memory chips that must be refreshed like this are called *dynamic* RAM chips.

In the absence of an error in the powerup routine, the computer may signal the operator that all is well with a short "beep" after the diagnostic program is completed. All of the memory is automatically checked by the program in less than a minute! That troubleshooting speed can't be beat with a manual system.

Disk Diagnostic Routines

A computer in at least a minimum working condition, sufficient to read a program from disc or tape and execute it, can often troubleshoot itself to the board level and sometimes to the specific circuit, even to the defective chip in the case of memory chips. This takes a special troubleshooting program called a *diagnostic program*. The program is constructed so that various portions of the computer are tested, repeatedly if the technician desires, to detect improper operation. Error indications are given on the screen to show the technician when there is a problem and where it is.

Diagnostic programs usually come on a diskette and are loaded into the computer's RAM for execution, just like any other applications program. A diagnostic program can test many functions of the computer, such as:

1. the system motherboard
2. the system RAM
3. the keyboard
4. a color or monochrome monitor
5. the installed character ROM
6. the parallel port
7. the serial port
8. the disk drives
9. the game port

Diagnostic programs can allow the operator to put the computer into a loop where, if desired, the computer will spend "forever" simply going over and over and over any test desired. For example, if a single bit in ¾-million bytes is suspected of occasionally failing, a technician can set up the computer to continually test memory and if a failure is detected, the defective memory bank or chip is noted. If desired, the printer can be used to log this information while the computer continues testing. Now, *that* is convenient troubleshooting! An instruction manual for the diagnostic program may be necessary to interpret codes shown on the screen in the event of a failure.

Computers that can run diagnostics are the better choices for critical applications, such as for business purposes. The availability of a diagnostic program with understandable documentation for it is a principal consideration for applications that require a minimum downtime for repairs.

ADDITIONAL CHECKS TO MAKE

Check the power-supply voltages, preferably with a digital voltmeter. Remember that TTL supply voltages are critical. If adjustable, TTL supplies should be set to 5.00 V DC. ECL should be set to −5.2 V DC. CMOS is not as critical, so the manufacturer will specify any adjustments to be made.

If the computer is not operating well enough to load a program, it is quite possible that there is a problem in the system RAM, preventing the program from loading and executing. If this is the case, it may be possible to use the mother-board system-memory DIP switches to disable a few banks of memory and thus verify, if the computer suddenly begins working, that it is indeed a memory problem. If the memory banks are progressively switched out, a defective bank can be identified. This method will not directly identify a specific bad chip, but may indicate that it is in a particular bank of eight or nine chips. If the computer seems completely dead, check to be sure that the system clock is working and at the proper frequency. A frequency counter or a communications receiver is excellent for use in checking the clock frequency.

The next quick-and-dirty suggestion would be to *turn off the power* and check the resistance of each of the address and data lines to ground and/or Vcc. They should all be high resistances, on the order of a megohm or more. Any different or low reading should be investigated. This simple test will check many chips at once, since the pins of many chips can be connected to each of the address and data lines.

As a final possibility, call the manufacturer or a local representative and discuss your problem. If you have a common problem, you may get a very quick and effective repair over the telephone.

STATIC STIMULUS TESTING

While some of the digital circuits of a computer can be tested using traditional means, the problem of dealing with high-speed, three-state operation can greatly complicate troubleshooting of the address and data buses of a computer. These have been among the most difficult computer problems to troubleshoot.

An innovative solution to this problem is offered by Creative Microprocessor Systems for 8-bit computers. The concept is very simple: Remove the microprocessor and plug into its place a system of switches. With these switches any desired address may be set up and data either passed to or obtained from that address. This company offers a small Static Stimulus Tester (Figure 14-1), which replaces any of four microprocessors, the Z80, 8080, 8085, or 6800. Using this tester, problems in computer address decoders, control bus lines, and data lines can be isolated with relative ease. The voltage levels on the microprocessor lines are held steady for as long as the technician desires, making them suitable for signal tracing with conventional digital means, even a voltmeter.

FIGURE 14-1 Static Stimulus Tester. This unit replaces the microprocessor and allows troubleshooting with a voltmeter in digital circuits. (Photo courtesy of Creative Microprocessor Systems, P.O. Box 1538, Los Gatos, CA 95030)

The proper use of the Static Stimulus Tester is well detailed in the instruction manual that comes with the unit. Step-by-step instructions are provided for performing four basic tests:

1. Reading a data byte from a memory location.
2. Writing a byte to a memory location.
3. Reading data from an input port.
4. Writing data to an output port.

Every bit of each of these bytes can be individually traced from origin to destination. Each of the above tests involves the address decoding, I/O port decoding, and support chips involved in these functions. Thus, most of the chips on the mother board become available to traditional means of testing.

Because of the grass-roots approach to troubleshooting that the SST provides, it is an excellent tool to use on computers that are unable to run diagnostics because of a drastic problem. The SST fills a basic "analysis gap" that sophisticated equipment such as signature and logic analyzers cannot fill. These instruments require a running computer to be of any value at all. The Static Stimulus Tester is also an *excellent* tool for learning microprocessor theory.

SOFTWARE PROBLEMS

Software problems can cause a computer to "crash." A crash occurs when the computer does not respond to the keyboard, stops working, or perhaps writes nonsense characters on the screen. Crashes occur when programs for different computers or even different versions of the same program are loaded and run. The technical reasons why this can happen are beyond the scope of this book, but the repair technician soon becomes acutely aware of the fact that software problems can and do occur frequently.

Problem 1: Programs that have run well in the past can suddenly become indigestible to a computer if the disk has been damaged by mechanical means, such as disk surface contamination or dimpling the disk surface while writing on it with a ball-point pen. Disks should be marked only with a soft, felt-tipped pen pressed very lightly. The magnetic records on the diskette are easily ruined by using paper clips to hold temporary notes to them or worse, placing them near a magnet! It is all too easy to erase valuable data by incorrect manipulation of files, as when copying disk or files from one drive to another. Accidental formatting of the wrong disk is another good way to ruin all of the data previously recorded on it.

Because of possible erasure of disks, it is always a good idea to make an extra copy of important information and programs on a separate disk in case the original becomes unusable for any reason. This copy of the original is called a *backup* disk, and the original is called the *working* disk. If the backup is kept in a separate location, a disaster befalling the working disk will not also affect the backup.

Problem 2: Software problems are indicated if the original disk suddenly becomes unusable, but the backup disk works okay, or if new, untried software won't work, but old programs will. The obvious solution to software problems is to obtain software that will work with the computer in question. Making a new copy of the backup disk is the solution to problem 1, above. Problem 2 is not as easy to solve. Some assistance in this matter can be obtained from the store at which the computer was purchased, from the distributor of the new software, or from the computer manufacturer's local representative, if any.

FIGURE 14-2 The system board setup switches, which must be set correctly for a computer to operate properly.

OPTIONS AND HARDWARE SETUP PROBLEMS

When optional cards are added to or deleted from a computer, certain settings must be made to the computer and the option card. This is generally done with the use of small switches, commonly eight of them to a bank, called DIP switches (dual in-line plastic) or with jumper arrangements (Figure 14-2). The proper settings must be made by referring to the instruction manuals, both for the computer and for the option card. Setting these switches to the wrong settings can produce error indications when attempting to use newly added memory or other options. Check *all* of the switch settings on the option card and the mother board of the computer when making an option change on a computer, or if the computer has "never worked right" after changing an option card (Figure 14-3).

FIGURE 14-3 Selector switch on option board.

SUMMARY

This chapter has presented some of the major troubleshooting tips for repairing computers and microprocessor circuits. By no means comprehensive, this chapter should nonetheless enable a technician to tackle repair problems on most computers at least with a comfortable measure of confidence. This chapter did not address problems frequently encountered with the mechanical aspect of computer operation, the disk drives. This subject is covered in the next chapter.

REVIEW QUESTIONS

1. Besides high and low levels, what third state is used in microprocessor circuits?
2. What is the fastest method to prove a hardware problem in, for instance, a plug-in circuit card?
3. What is a program used for testing a computer called?
4. Why do some computers take considerable time, up to 45 seconds or more, to "come to life" after power is first applied?
5. What is removed to use a Static Stimulus Tester?
6. If a software problem is suspected, what test can be made to verify if this is the cause? Assume that the diskette in question may have been damaged.
7. If a new option card is installed in a computer and then the computer crashes, what should be checked first, and very thoroughly?

chapter fifteen

Troubleshooting and Repairing Floppy Diskette Drives

CHAPTER OVERVIEW

Although not strictly electronic items, personal computers and the diskette drives closely associated with them are popular enough to justify a chapter devoted to the servicing and repair of the most popular floppy diskette size in use, the 5¼". The terms used in servicing the diskette drives (radial, azimuth, and skew) are explained. Special alignment diskettes available from Dysan are also explained, along with a presentation of their specialized diskette drive alignment tester, which may be available to the technician.

The 5¼" floppy diskette drive is widely used in personal computers. Other sizes in limited use include the 8", the 3¼" and the 3½". Since these sizes are not used as frequently, they will not be covered. The methods of alignment are basically the same, however, with appropriate diskettes to test them.

CAUTIONS

Before getting into the mechanics of the drives, some cautions are in order to help the technician prevent further damage to the drives and the diskettes used in them:

1. Never close the door on a dual-sided diskette drive without a diskette in the drive. To do so puts the glass-like heads on the top and bottom sides of the disk in direct contact with each other. A glass-to-glass contact like this invites chipping the heads. A chipped head can instantly ruin any diskette inserted into the drive. Transporting a drive or computer with drives installed without the protection of a diskette or a cardboard protector is especially dangerous.
2. When inserting diskettes into the drives, insert the diskette slowly and carefully, all the way in. Close the drive latch or door carefully too. If possible, insert the diskette and clamp it when the drive motor is running, as this will assist greatly in properly seating the spindle into the hole with less stress on the edges of the diskette hole.

The drive mechanisms for floppy diskettes are mechanical, and therefore they are subject to wear and failure. The mechanical tolerances of a floppy diskette drive are very tight, and slight dimensional changes in either the drive (due to a "bump") or in the diskette itself (warping, shrinkage, etc.) can produce misalignment and read/write errors.

FLOPPY DISKETTE GEOMETRY

This chapter will consider the geometry of the most popular data recording diskette, the soft-sectored 40-track 5¼" diskette. The soft-sectored diskette has a single index hole, where the hard-sectored has multiple index holes. The index hole is visible through the diskette jacket, near the center (Figure 15-1). The signal from the index hole is also used to indicate to the computer that there is a diskette in the drive and that the drive is turning. The index hole is about 4000 microseconds wide at the proper rotational speed of exactly 300 rpm. The exact width of the index pulse is not critical; from 2000 μs to 5000 μs is often acceptable.

Radial Alignment

The tracks on a diskette, usually 40 of them, are concentric. These tracks begin at the *outside* of the diskette with track #0 and number upward toward #39, near the center hole (Figure 15-2). Diskettes with 35 tracks are now obsolete. Using 35-track diskettes in a 40-track drive may cause the head to run into the end of the shorter access slot. This can cause alignment problems of the drive.

The stepper motor of the drive moves a precise amount at each phase increase of the electronic circuits driving it. From the home position, track 0, the

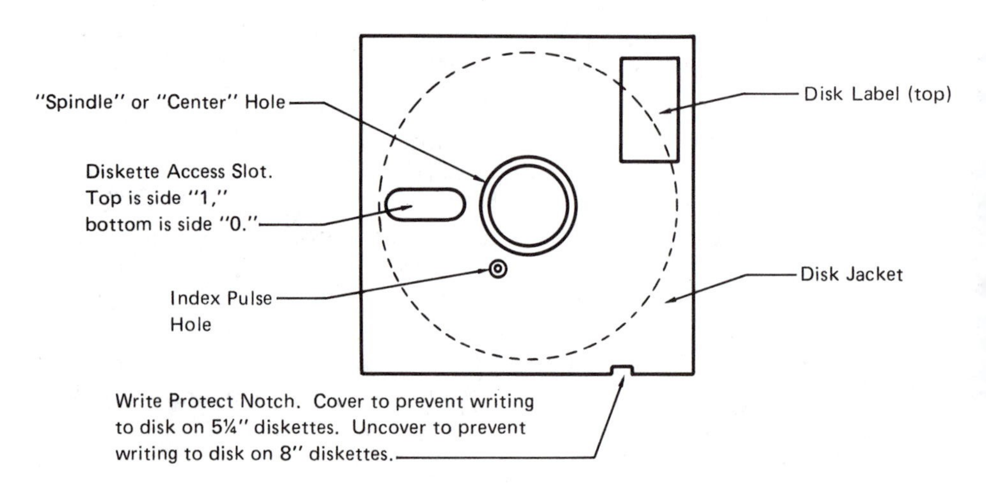

FIGURE 15-1 Parts of a floppy diskette.

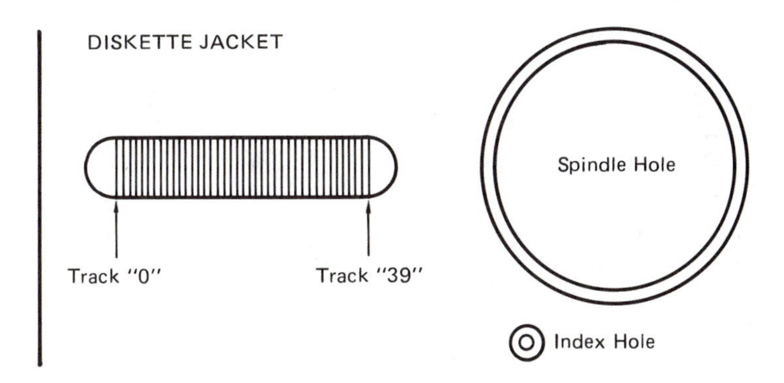

FIGURE 15-2 Track layouts of a 5¼-inch diskette drive.

stepper moves a precise amount to each given track. There is no means on the standard floppy drive to "lock onto" a track. The stepper motor is "blind" and goes a certain distance. If the track is where it is supposed to be on the diskette and the head moves the right amount, the two shall meet and everything works just fine. But it often happens that the stepper motor becomes dislodged slightly from the centerline of a track. Once out of true alignment position on a track, it has no way of finding it again. This sometimes happens if the head mechanism is

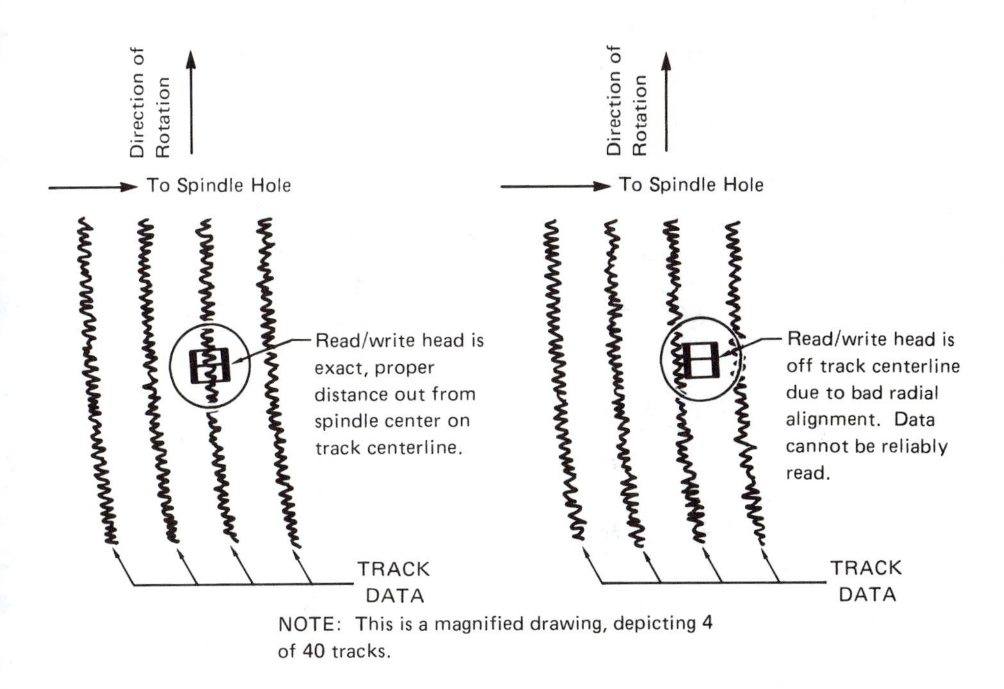

FIGURE 15-3 Exaggerated drawing showing radial alignment and misalignment.

merely friction clamped to the stepper motor shaft by a coupling that can slip (Figure 15-3).

This misalignment can be corrected by loosening the coupling and putting the head back on the proper track, then retightening the clamp securely. An alternative way to realign slight misalignment problems is to move the entire head and stepper motor assembly slightly with a cam-head screw. The method used depends upon the drive concerned. Recurring problems of slippage and the resulting misalignment can sometimes be cured by drilling and pinning the stepper motor shaft to the coupling, thereby permanently putting the drive into alignment.

Radial alignment is performed using data read from an alignment diskette by the *bottom* or "0" head. Radial alignment of the top "1" head is performed only after the bottom head is accurately placed. Adjustment of the top head is a factory adjustment. It is necessary to adjust it for radial, azimuth, and index-to-date specifications simultaneously. Adjusting for two of these specifications is difficult by itself, but getting simultaneous adjustment of all three is very time consuming and frustrating.

Azimuth Alignment

The heads of a diskette drive must meet the data at precisely a 90-degree angle. Looking directly down on a drive, any rotation of the head about its center is a misalignment of that head's azimuth (Figure 15-4). The only head of a single-sided drive and the bottom "0" head of a double-sided drive cannot be changed in azimuth because these heads are usually set into epoxy and are therefore not adjustable.

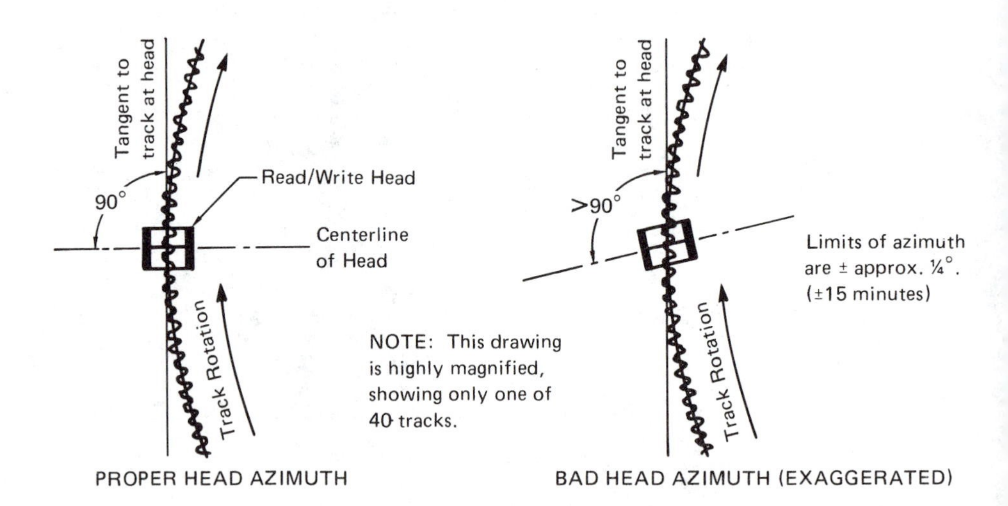

FIGURE 15-4 Exaggerated drawing showing head azimuth and misalignment.

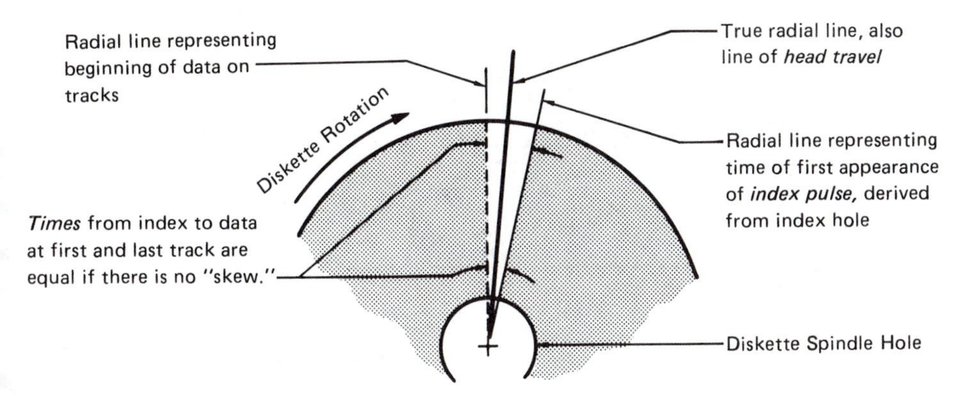

REPRESENTATION OF A GOOD, "NO-SKEW" CONDITION

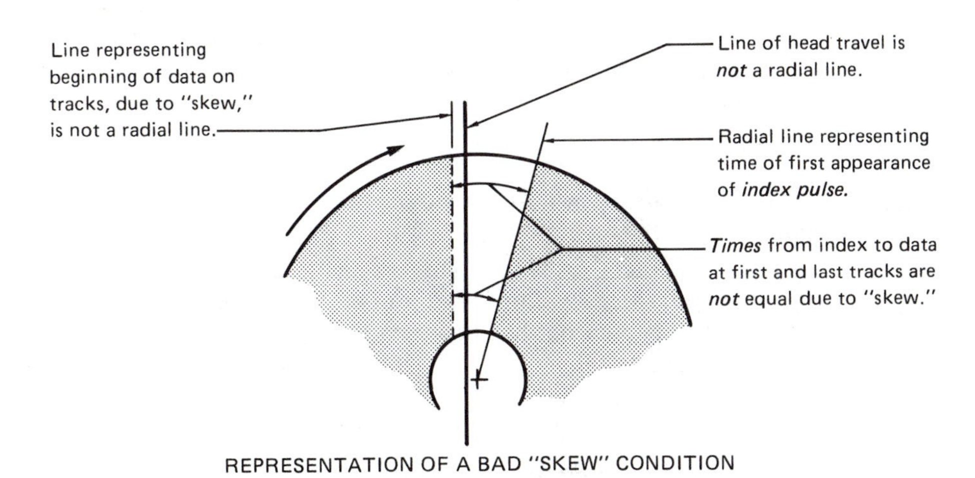

REPRESENTATION OF A BAD "SKEW" CONDITION

FIGURE 15-5 Explanation of "skew" of a diskette drive read/write head.

Checking for Skew

Skew is a measure of the amount by which the head, in moving from the outside to the inside of the diskette, departs from a true radial course straight to the center of the diskette (Figure 15-5). Skew can be measured by noting the time between the beginning of the index pulse and the first reading of data on an oscilloscope. If there is no skew, the time between these two events will remain the same whether the drive is reading track 0 or track 34 (Figure 15-6).

TESTING AND ALIGNING DISKETTE DRIVES

A diskette drive that is out of alignment may very well read a diskette that it has written upon. The written path and the reading path are the same, so the signal is retrieved without problems. Such a drive, however, could be totally

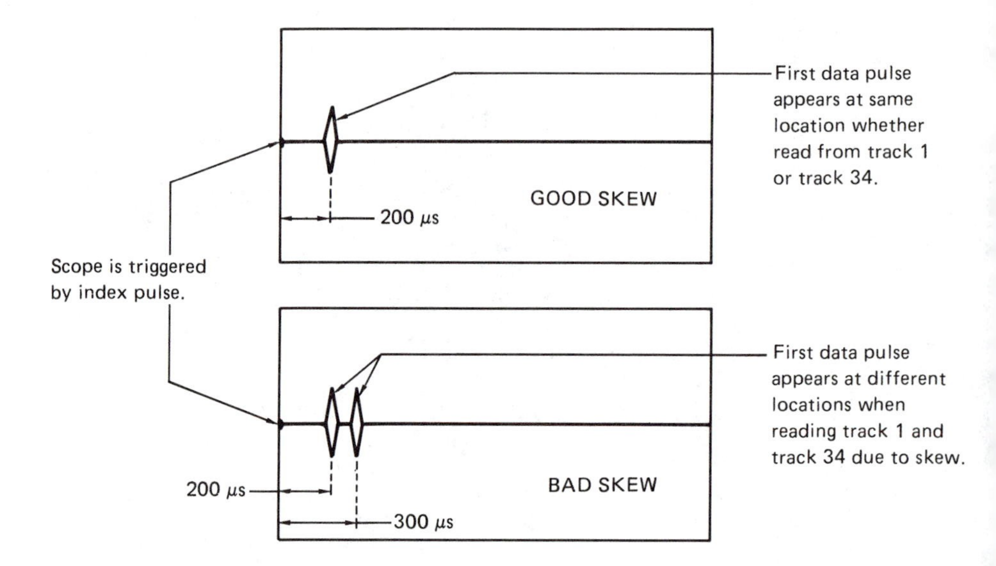

FIGURE 15-6 Oscilloscope presentations of good and bad skew conditions.

unable to read discs written on by other drives that are in proper alignment. Thus, the lack of interchangeability of discs from one drive to another is an indication of head alignment problems. This problem can be localized to one drive if a known good diskette will not read properly on one of the two drives. A known good diskette is one that has been written by a drive known to be in good condition or by one of the alignment diskettes to be discussed, the Dysan DDD diskette.

There are several ways to approach the analysis of a diskette drive:

1. Simply obtain information about the present state of the drive, using a computer to operate the drive.
2. Realign or otherwise repair a known defective drive, using a computer to operate the drive.
3. Analyze problems of a drive and correct them, using specialized test equipment.

Getting the Present State of a Drive Using a Computer

A diskette drive can be checked for alignment without taking the cover off a computer. A special diskette is available from Dysan that is recorded in such a way that the reading head in a drive under test reveals drive problems by the way in which the drive reads the data on the test diskette. This diskette is called a digital diagnostic diskette, or DDD. It is used with another diskette, the software necessary to read the DDD in a second drive (Figure 15-7).

The Dysan DDD diskette, number 508-400, is used with the special software to examine double-sided, double-density drives. Test diskettes for single-sided

FIGURE 15-7 Digital diagnostic diskette and drive diagnostic program. (Photo courtesy of Dysan)

and single-density drives are also available. The following important tests are among those available:

1. radial alignment
2. azimuth alignment
3. skew
4. index pulse width
5. index-to-data timing

In order to use the DDD diskette, the computer must be able to load and execute a program with at least one drive. This method of testing drives provides an excellent way to monitor gradual changes in drive efficiency, thus helping to prevent sudden failures of the drive from interfering with the use of the computer during critical tasks.

Realigning a Drive Using a Computer

Dysan's analog alignment diskette (AAD), number 224/2A, is precisely recorded to test double-sided diskette drives and to put them back into alignment. Diskette number 224A is used for single-sided drives. In order to use an AAD, it is necessary to apply power to the drive under test and supply the proper signals to place the head electronically at specific tracks. An oscilloscope is also required to monitor the signals produced by the read operation. The step-by-step details of the alignment procedure are covered in the AAD accompanying instruction booklet and will not be repeated here. The definitions of radial, azimuth, and skew at the beginning of this chapter are necessary to understand the instruction book, however.

Using a computer to provide the command signals for alignment with an

AAD requires that signals be brought out of a computer to the drive under test. One way of doing this is to remove or disconnect the computer's "B" drive, and bring out both the power signals and the command signals with homemade cable extensions. This method will allow the technician to work on the drive conveniently and have access to either side of the drive while it is operating.

The drive head will need to be positioned on the test diskette tracks for the alignment procedures. To use a computer, the BASIC program of Appendix IV will work on many computers, particularly those commonly referred to as "the compatibles." It may be modified as necessary to fit other computers. To use the program, load BASIC, all three programs, run "menu.bas", and follow the prompts.

Repairing Drive Problems with Specialized Equipment

Technicians who work on many drives may consider using Dysan's special Performance and Alignment Tester. This small instrument takes the place of a

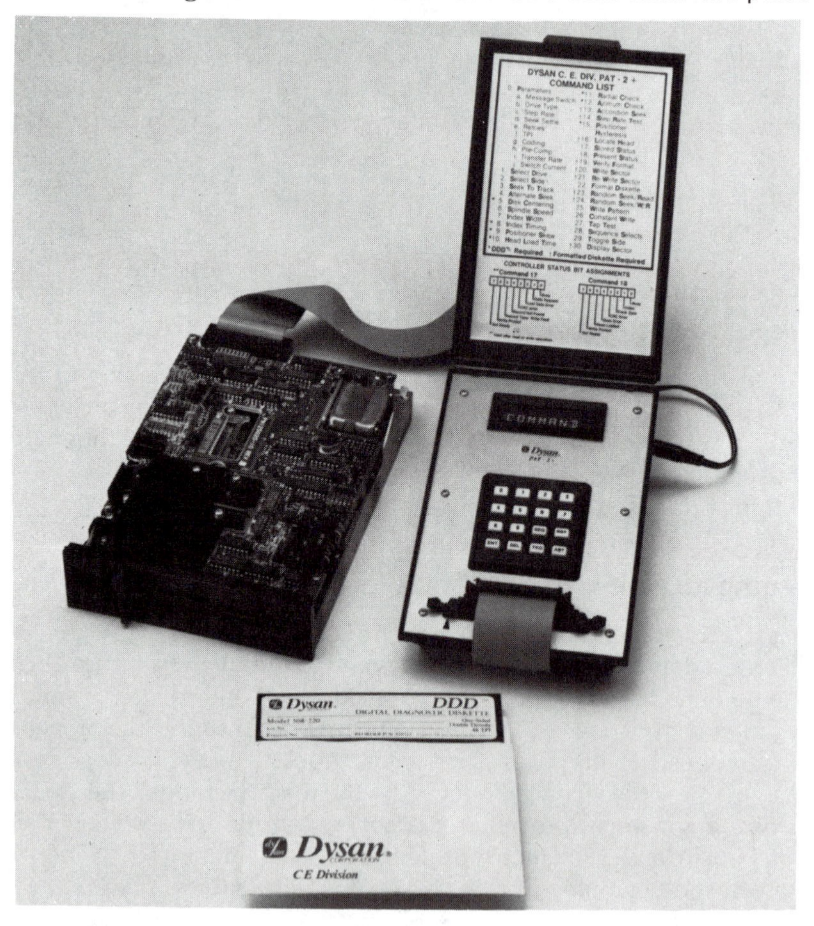

FIGURE 15-8 A disk drive exerciser and alignment tester. (Photo courtesy of Dysan)

computer in testing drives. It not only provides the proper command signals to control the head in the drive for aligning a drive with the analog alignment diskette, it also interprets the digital signals produced by the digital diagnostic diskette and gives a readout of the alignment status of the drive.

The only equipment needed for aligning drives with the Performance Alignment Tester is the power supply for the tester that comes with it, the DDD and/or AAD diskettes, and a power supply providing +5 and +12 volts for the drive under test (Figure 15-8). The documentation provided with the 508-400 DDD refers to the radial alignment tracks as "progressive offset" tracks. Tests for skew use tracks 0 and 34, called "index format" tracks. Tests for azimuth also use track 34. These differences between track names may not be immediately evident.

SUMMARY

This chapter covered the equipment and procedures necessary to align the most popular diskette drive used in personal computers. The mechanical aspects of the drive were explained sufficiently for the technician to understand the procedures presented with the test diskette and alignment tester.

REVIEW QUESTIONS

1. How can a computer know that the disk drive is turning?
2. Is track 0 of a diskette at the inside or outside of the diskette?
3. Which head is put into radial alignment first, the "0" or the "1" head?
4. Which head, the "0" or the "1" head, is on the bottom?
5. What would you need to perform fast checks of your disk drives with no test equipment and without dismantling your computer?
6. What minimum equipment would you need to align a defective drive?
7. What, basically speaking, does the Performance Alignment Tester do?

chapter sixteen

Dead-Circuit Testing

CHAPTER OVERVIEW

There are two occasions for dead-circuit testing. One is after the technician has signal traced a problem down to a stage or several possible components. The object is to verify that the suspected component is really bad, before removing it from the circuit if at all possible. The other occasion occurs when a technician is handed a circuit and asked to repair it with no other information. No schematic, no cables to connect it to power, no history of likely failures—in fact, the technician may not even know what the board is used for! With only an ohmmeter to test such circuitry, there is little chance for successful repair unless by good fortune one of the easily tested, high-failure items, such as SCRs or filter caps, tests bad. More subtle failures, such as a short in one of many ICs, would take a great deal of time to find with an ohmmeter and would probably not be profitable in terms of technician time versus the cost of a new board. The solid-state tester, however, has been designed to find faults under these difficult conditions.

DON'T RUIN YOUR OHMMETER!

The common way of damaging an ohmmeter is to apply voltage to the test leads. Don't use an ohmmeter of any kind on a circuit that has power of any kind applied. On 115-V AC-operated devices, pull the plug. This avoids accidently contacting the 115 V AC that is still present on the back of the power switch and at the fuse clips. Battery-powered equipment should have the batteries removed for similar reasons.

Electrolytic capacitors in the power supply have sufficient storage capability to damage an ohmmeter if they are not discharged by a resistor for that purpose, called a bleeder resistor. Vacuum-tube circuits need bleeder resistors for personal safety reasons, but lower voltage supplies may not always have them because the voltages do not pose a safety problem. The problem is still there for an ohmmeter, however. Discharge the filter capacitors of the power supply before taking resistance readings with any meter, but particularly with the VOM. Some DMMs can take considerable abuse on the ohm scale, but the VOM is very easily ruined.

A negative ohmmeter reading means danger, that there is a source of voltage in the circuit that should not be there when taking resistance readings. Occasionally, when testing electrolytic capacitors, an ohmmeter will charge up a capacitor, which will hold a charge for a period of time. Later tests of the capacitor with reversed leads may momentarily show a negative resistance reading. This should not be harmful to the meter, as the voltage used is very low.

USING THE VOM ON THE RESISTANCE FUNCTION

Using the VOM to test semiconductors can result in ruined components. The VOM can deliver up to 125 milliamperes on the RX1 scale into a short circuit. Currents far below this value can burn out small transistor junctions instantly. On the higher ranges, a VOM may place as high as 30 volts across an open circuit. This, too, can ruin semiconductors. Whenever possible, the VOM should be used to check semiconductors on the RX100 or RX1000 ranges only, and then with due care for the fragility of the semiconductor under test.

Before using the VOM, it is a good idea to be sure that the instrument reads to the full right-hand "zero ohms" mark when the test leads are shorted together. Using the control for this purpose, set the needle at the "zero ohms" mark. If the ohmmeter battery is weak, the meter needle will not reach the "zero ohms" mark at the far end of the scale with any setting of the zero adjustment. The following hint can save some frustration while using the lowest (RX1) scale. Many ohmmeters will not hold a consistent zero on this scale and are sensitive to mild thumping, which makes the needle change its zero setting. This is often because of the construction of the flashlight cell that powers the ohmmeter circuit.

Figure 16-1 shows the discs used on the top and bottom of some cells to

FIGURE 16-1 Dry-cell battery with end disks. The end disks in some cells can cause an erratic reading on low ohms' scale of some ohmmeters.

make them look more attractive. These end discs depend upon a rubbing contact with the ends of the cell for the cell to work. A very satisfactory way around this is to remove the end disks and solder two short leads directly to the ends of the cell. These leads are then soldered into the ends of the battery clips in the ohmmeter. This makes small jumpers directly from the cell into the circuitry, avoiding four potential problems at mechanical contact points. See Figure 16-2 for the finished wiring job.

Reading the resistance scales of a VOM takes a bit of practice. The VOM has reversed scales. (The VTVM and other meters may not.) The resistance is read from the meter scale, and then the multiplier indicated by the range switch is affixed to obtain the actual value of the resistance. Be sure to keep your fingers off the probe tips when using the resistance ranges. This avoids putting your body resistance in parallel with the circuit, thereby giving erroneous readings.

In testing semiconductor junctions with the VOM ohmmeter function, the polarity of the leads is important if you wish to identify the cathode of a diode, or to determine whether a transistor is an npn or a pnp type. It is a good idea to check your ohmmeter's lead polarity before such a need arises.

FIGURE 16-2 Installing a jumper, soldered at both ends from the cell to the circuitry, will improve ohmmeter zero-set.

Checking Ohmmeter Lead Polarity

With the test leads in their proper jacks in the VOM (red positive, black negative), attach the test leads to a diode that already has the cathode end plainly marked. The cathode of a diode is usually marked with a black band around it. Using the RX100 or RX1000 scales, take a resistance reading across the diode. Note the approximate resistance (an open may be indicated). Then reverse the leads and take another reading. Of the two readings one should be very low (indicating to the right on the scale). Put the probes on the diode so that this lowest reading is present on the meter.

If the *black* test lead is now on the *cathode* of the diode while reading the lowest of the two readings, there is no reversal of the test lead polarity when switching to the ohms function. But, if the *red* lead is on the cathode, the meter could confuse you when using it later. A small label saying something such as "Reverse leads on ohms" should be attached to the meter front in a convenient location as a reminder. Then, when junctions are tested, the technician will be reminded to reverse the leads at the meter (red in the negative jack, black in positive) so that the colors of the leads show the correct polarity, thus avoiding confusion.

To tell if a transistor is an npn or a pnp, place the red lead of the ohmmeter on the base of the transistor, and the black lead on the emitter or the collector. If the ohmmeter reads a low resistance, indicating forward conduction of the junction, the transistor is an npn. If the ohmmeter shows an open, the transistor is a pnp.

USING THE DMM ON THE RESISTANCE FUNCTION

The digital multimeter (DMM) has several advantages over the VOM. With its higher accuracy, digital readout, and repeatability of readings, little is left to operator interpretation.

The DMM uses very little current to test resistances. In most cases this is good, but not in the case of testing semiconductor junctions. A minimum test of a junction requires that it pass a reasonable current. Many DMMs cannot do this. Junctions are better tested with the VOM, so the junction has a higher current passed through it. A DMM with a special range for testing semiconductor junctions can be used for this purpose. Fluke meters have such a range, marked with a diode schematic symbol. These meters pass about 1 milliampere through junctions to test them. Besides passing a decent current through the junction, the meter reads the *voltage* across the junction rather than the junction resistance. This voltage reading also enables the technician to distinguish germanium (0.3 V) from silicon (0.7 V) junctions.

Autoranging of the DMM resistance scales is an attractive feature to use in most cases. The technician is able to keep both hands busy with the probes without having to pause to change meter ranges to get a good reading. When testing semiconductors, however, the autoranging feature can cause the meter to search from one range to another, endlessly trying to find the resistance of a junction, which is changing resistance on each resistance range.

Another handy feature available on some DMMs is the "beeper" or continuity tester. This allows the technician to follow wiring through cables, traces around circuit boards, etc., without having to look at the meter. When using the "beeper" circuit, keep in mind that the beep will indicate continuity even if there is considerable resistance in the circuit. This resistance may be as much as several hundred ohms. Be familiar with your meter, and know this cutoff value. A typical cutoff resistance for the beeper is 175 ohms.

When using the DMM on the resistance scales, keep your fingers off the probes to avoid introducing error due to putting the resistance of your body in parallel with the component under test.

USING A SOLID-STATE TESTER

The Huntron Tracker™ (solid-state tester) shown in Figure 16-3 is a very valuable addition to the test bench and as a field troubleshooting tool. Like an ohmmeter, this solid-state tester provides its own testing voltage at the probe tips. The Huntron, however, does not use DC. It uses a sine wave voltage, which tests components for conduction in both forward and reverse directions. Patented circuitry limits the current to the circuit under test. Three ranges are provided for low, medium, and high impedance. The highest range can even be used to test the breakdown voltage of zener diodes to about 45 volts.

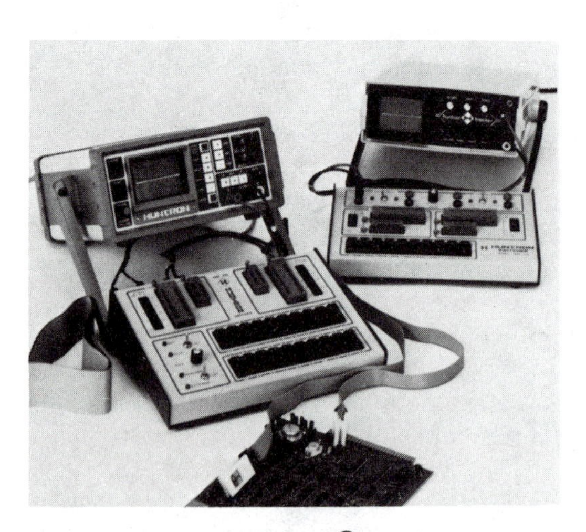

FIGURE 16-3 Huntron Trackers™, new and old models, and accessory Switcher™. (Photo courtesy of Huntron Instruments Inc.)

OPEN

SHORT

DIODE

CAPACITOR

RESISTANCE

DIODE AND
PARALLELED
RESISTANCE

FIGURE 16-4 Basic solid-state tester waveforms.

The Tracker is simple to use. Select the proper impedance range, apply the probes, and look at the screen. Figure 16-4 gives a few basic examples of what the Tracker might show. An excellent publication is supplied with the instrument and gives detailed operating instructions and sample presentations for good and bad components.

This instrument may be used on any circuitry without danger of harming components thanks to the special circuitry it employs. Only one component should not be tested with this instrument (or with DMM or VOM, for that matter): the microwave diode. This tiny diode is rated at such a low current that it requires special test techniques.

The solid-state tester is particularly handy for testing components while they are still in the circuit. You can even troubleshoot a circuit board without knowing what the board is used for. There are four ways to use this instrument:

1. The individual display interpretation.
2. Comparison with recorded patterns.
3. Comparison with known good circuitry.
4. Using it with a Huntron Switcher(TM).

The first method means simply applying the probes to the circuit, then noting and interpreting the presentations. This method requires familiarity with the basic presentations in Figure 16-4 plus a few more. Presentations of components in parallel, particularly, require some interpretation. This is easily learned with a few hours' practice.

The second method requires less familiarity with the scope's presentations because pictures of correct presentations for each test point are made available to the technician by the employer. The correct presentations may be in the form of sketches or photographs. This method is useful where a large number of identical cards are serviced and building the necessary library of presentations is a cost-effective method.

The third method compares scope presentations between a known good board and a defective one. Readings are taken at identical points on each board. The Tracker has special provisions for using this method. It contains circuitry to switch between two "hot" leads, one for each of the two circuits. The instrument then switches from one circuit to the other, making an overlay of the two presentations on the Tracker scope. Any difference results in the scope presentation appearing to jump back and forth (Figure 16-5).

The fourth method is an automated extension of the third. Instead of a technician placing the probes on two boards manually, an accessory unit called the Huntron Switcher(TM) will compare all the leads of one IC to another, one pin at a time, by automatically sequencing through the pins (Figure 16-6). This is a good method for large-scale testing of boards or for defect detection prior to boards being placed into service.

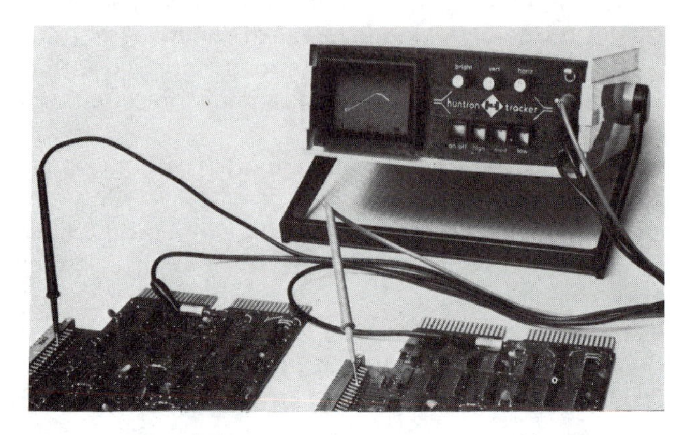

FIGURE 16-5 Using the switching feature to compare two waveforms automatically. (Photo courtesy of Huntron Instruments Inc.)

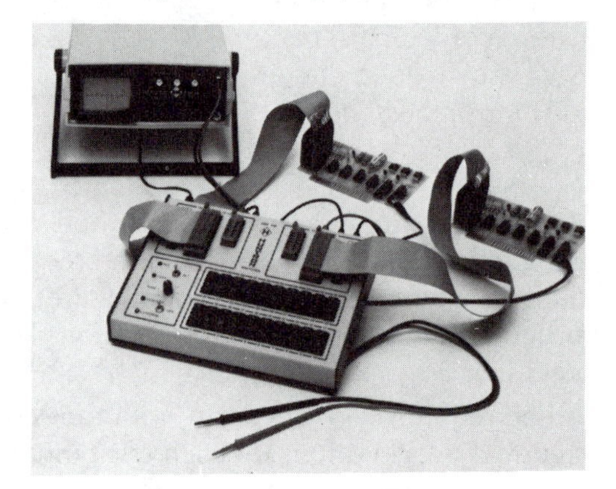

FIGURE 16-6 Using Huntron's Switcher™ to compare ICs on two boards automatically. (Photo courtesy of Huntron Instruments Inc.)

MAKING THE CHECKS

The first step in dead-circuit testing depends on how much is known about the failure, whether or not schematics are available, and what test equipment is available. If a defective stage has already been isolated by signal tracing or visual checks, a few additional checks in that stage with either a solid-state tester or an ohmmeter will probably isolate the bad component. At the other extreme, if nothing is known for sure, perhaps not even the use of the circuit board, then troubleshooting should probably take the following course:

1. Check the large electrolytic capacitors for signs of leakage and for possible internal shorts with an ohmmeter.
2. Check the larger semiconductors such as SCRs, power transistors, and diodes for shorts.
3. Check for a short circuit from the supply lead, if identifiable, to ground.

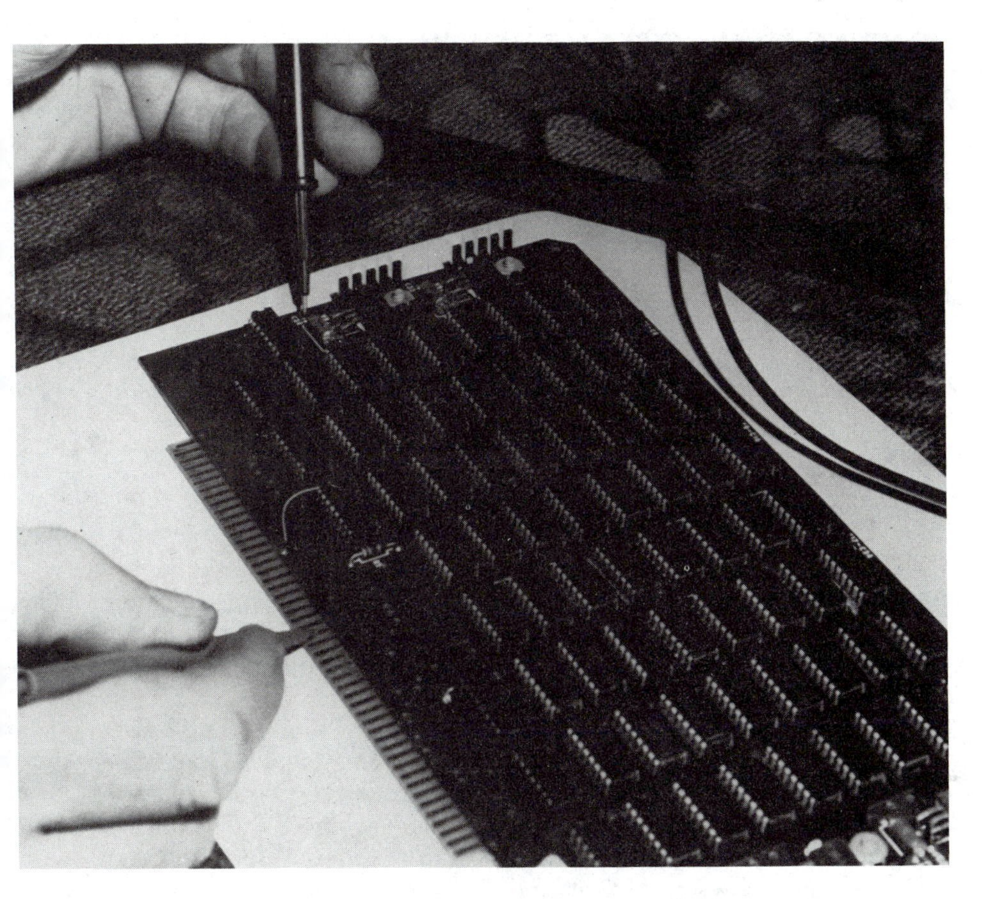

FIGURE 16-7 Checking the edge connector with a solid-state tester. This is a good first attempt at finding a problem on a circuit board.

If no problem is found after these attempts, further troubleshooting may not be cost effective if a new board can be obtained at reasonable cost.

If a Tracker is available, much more can be done to find the board defect. The first step in using it to test an entire large circuit board is to check the edge connector of the card, if any (Figure 16-7). Unusual waveforms should be traced to the defective component and corrected before proceeding. After finding and correcting any defects discovered at the edge connector, continue testing and clearing defects at the rest of the connector contacts.

Having cleared all problems at the edge connectors, mentally divide the board into four sectors and test each IC pin or component lead for possible problems that may not have shown at the board connector. Finish one sector before moving on to the next. If a defect is determined to be a short and if there are many paths involved, consider using the logic pulser and the current tracer to find the defective component. More about this later in this chapter.

FINDING SHORT CIRCUITS ON PRINTED CIRCUIT BOARDS

An Ingenious Tip

There are two distinct kinds of shorts that occur on printed circuit boards: the "hardwire short" of a solder bridge or a shorted capacitor (usually a supply-line bypass capacitor) and the "semiconductor short," a short within an IC. Once a short is determined to be the cause of a failure, measure its exact resistance with an ohmmeter. A resistance of a few ohms or more is almost always a shorted IC, where as a value of a few tenths of an ohm will be a solder bridge or shorted capacitor. This single trick can be a valuable time-saver. A short of a few ohms will be called a *semiconductor short* and a short of a few tenths of an ohm a *hardwire short* in the remainder of this book. Credit for this tip goes to Mr. Bill Hunt of Huntron Instruments Inc.

Finding a Short Circuit

There are several traditional methods of finding a short circuit on a printed-circuit-board trace:

1. Cut the circuit-board traces wherever necessary to bracket the problem. This method may require many breaks in the traces to finally find the defective component. This method is time consuming and damaging to the board.
2. Apply a large current through the trace and "smoke out" the short. This approach can be even more devastating than method 1. Some means must be taken to limit the current supplied to the trace. Any discoloration of the trace, particularly in any narrow portions where the resistance is higher, means that damage has already occurred. Turn off the current flow immediately and resort to some more civilized method. Once in a while this method will cause the defective component to get hot enough to identify it before causing any board damage. As long as the current flow is not excessive, it is a method that can be used. It works reasonably well for a semiconductor short, but will invariably damage a board if pursued too vigorously to find a hardwire short.
3. Take out the component or lift appropriate legs of ICs from the board until the defective component is found. This is time consuming but less damaging to the board than the two methods above.
4. Selective shotgunning is making a guess as to which of the components may be defective. The technician knows that there are only X number of possibilities and begins removing parts until the short clears. When there are only a few components, this is not a bad approach.
5. If the short is a semiconductor short, you may suspect an IC as the cause. Monitor the short circuit resistance with a DMM, which reads low values of resistance better than a VOM in many cases. Use a circuit board cooling mist such as "Freeze Spray," or use heat from the tip of a small iron, and proceed to change temperatures down the rows of ICs until one is found that seems temperature sensitive. This is probably the shorted IC.

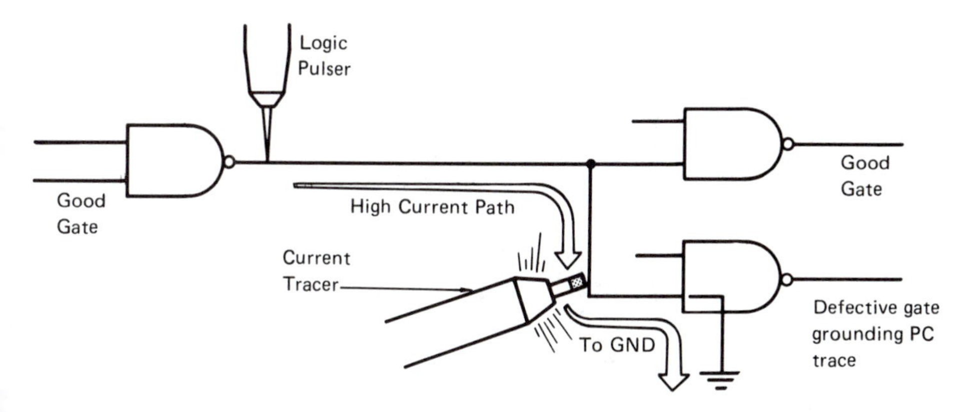

FIGURE 16-8 Using the logic pulser and current tracer together to find a shorted circuit-board trace.

Another method is available that allows the technician a fast, sure way of pointing directly to the bad component or finding where the board is shorted from one trace to another. Figure 16-8 shows how to introduce a source signal from the logic pulser and pick up the high current resulting from a short with the current tracer. Using this combination of instruments, short circuits of both the semiconductor and hardwire varieties are easily found. The shorts can be to ground, to supply (Vcc), or to other traces on the board. Shorts on multilevel (sandwiched) boards can also be found, since the current tracer does not have to physically contact a trace to detect the current flowing through it.

TESTING COMPONENT RESISTANCE IN-CIRCUIT

Whenever a resistance reading is taken on a circuit board, the effects of other components on the board play a major part in interpreting the reading of the ohmmeter. If the component under test had no other components connected to it, the interpretation of the resistance reading would be rather simple. Resistors should read the value marked on them, capacitors should be open, and inductors should have a low-resistance reading.

Interpreting ohmmeter readings across components still installed in a circuit requires considerable thought. Suppose you wish to test the resistor for proper value in the circuits of Figure 16-9. A resistor paralleling the resistor under test should cause the resistance reading of the combination to be lower than you would expect for one resistor alone. If the ohmmeter reads out the value of one of the resistors, it is reasonable to assume that the other resistor is open. If the ohmmeter reads open, both resistors are open.

A capacitor paralleling a resistor should have no effect on the ohmmeter reading other than an initial charging effect if the capacitor is large enough. A shorted reading across this combination probably means that the capacitor is shorted and that the resistor is all right. An open reading will mean that the resistor is open.

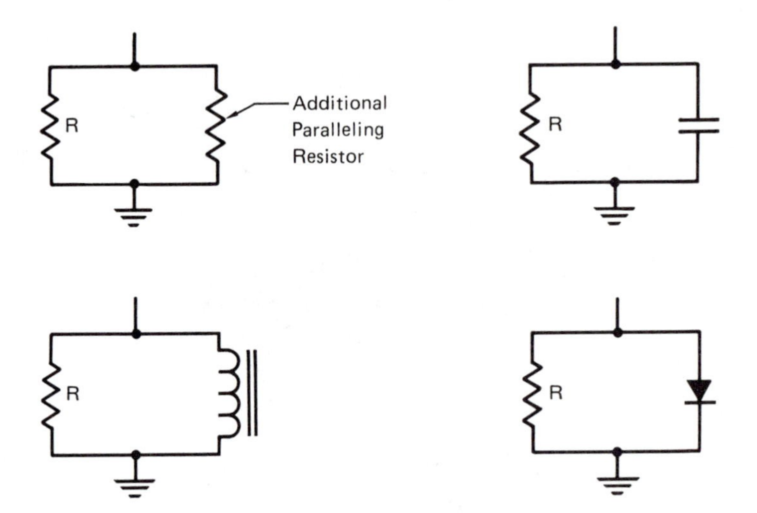

FIGURE 16-9 Perils of testing a resistor in-circuit.

An inductor paralleling a resistor will cause a very low resistance reading, since the inductor winding has a low resistance compared to most resistors. The exact effect of this combination will depend on the ratio of the inductor-winding resistance to the value of the resistor. If the combination reads open, both components are open. If the combination reads the value of the resistor alone, the inductor is open.

A semiconductor junction across a resistor may read much lower than the value of the resistor alone, depending upon the polarity of the ohmmeter leads. This fact gives the technician the chance to test the junction in the circuit, keeping in mind that the junction will not test "open" in one direction as it would out of circuit, due to the resistor. Lack of a low resistance in one direction *may* indicate an open junction. Some digital multimeters test components with such a low voltage that a junction may not have sufficient voltage to turn on. There is also some effect of the resistor in reducing the amount of voltage present in the circuit: the lower the value of the resistor, the lower the circuit voltage, and the less the chance of the junction being turned on.

Don't Forget the Power Supply Is There

Resistance readings taken with the components in the circuit are subject to every parallel path that the current can take. This includes the power supply. This fact may not be obvious on the schematic, where the power supply is not shown connected to every point that it actually is in the equipment. Therefore, a reading across a filter capacitor, for instance, may at first indicate a leaky capacitor. Further inspection of the power-supply circuit will probably reveal one or more paths through it to ground, often in the form of a resistor to discharge the filter capacitors or paths to ground in the regulator stage.

The variety of component combinations is almost endless, so it is not practical to attempt to analyze all combinations and what the problems might be in this book. However, if the technician understands the foregoing principles thoroughly, each resistance situation can be thought through *before* the resistance reading is taken. Then, when the reading is taken, the technician can be confident of the action to take. Remember that cold solder joints can make a component seem open when all that is needed is some heat and a bit of solder to cure the problem. Consider this possibility and look carefully for breaks in the circuits and cold solder joints before declaring components bad.

The Schematic, Again

Most of the above discussion assumes that the schematic is available. Ohmmeter readings that don't seem right can be very frustrating without circuit information. If no schematic is available and the circuit-board traces are complex, only limited information can be deduced:

1. When testing resistors, an open means the resistor is open. A short indication probably means some other component is in parallel with the resistor, and the reading is useless. A proper reading probably means the resistor is all right.
2. When testing capacitors, an open reading means the capacitor is not shorted—but it may be *open.* (A Tracker can tell if the capacitor is okay.) An intermediate value is useless, for the capacitor may be leaky or it may be paralleled by good components causing the reading. A very low reading *may* indicate a shorted capacitor. The lower the reading, the better the chances that the capacitor is indeed defective.
3. In testing inductors, an open reading indicates a bad, open inductor. An intermediate reading probably indicates the same, depending upon the resistance of the inductor winding. Some small inductors using tiny wire and thousands of turns have thousands of ohms of resistance, whereas large inductors have fractional ohm readings. A short or near short probably means the inductor is all right, but the proper resistance of the inductor winding should be known to interpret the reading. A reading appreciably lower than the proper resistance reading for the inductor points to possible shorted turns in it.
4. When testing diodes or bipolar transistor junctions, the polarity of the ohmmeter leads must be considered. One direction should show a substantially lower resistance than the other. Remember that some digital multimeters may not turn on a junction, as discussed earlier. If used, the DMM should be used on the special diode range, where the reading may be in volts, rather than ohms. A lower voltage shown with the probes in one direction across the junction indicates forward conduction. If the indicated resistance (or voltage) is very low or zero, the junction *may* be shorted. It may also be parallel with an inductor or very-low-value resistor, or it may have a

shorted capacitor across it. This last possibility can be eliminated by taking the reading of the junction in the opposite direction. A high reading here would eliminate the possibility of a paralleling shorted component and would also indicate that the junction is good. A similar low reading would confirm either a paralleling component causing the low reading or a shorted junction. A very-high-resistance reading in both directions indicates an open junction, assuming the DMM is used on the diode range.

In the absence of schematics, the solid-state tester assumes even more importance because it will read capacitors, inductors, and junctions in circuit, easily identifying them.

SUMMARY

This chapter has covered what to do if handed an unfamiliar circuit to test without benefit of schematic, mockup, or information on the circuit. It also provides some guidelines in performing in-circuit resistance checks to identify the one of several possibly defective components.

REVIEW QUESTIONS

1. What is the most common way of damaging an ohmmeter?
2. What does a negative indication on an ohmmeter mean?
3. If a low resistance is read when the black lead of an ohmmeter is on the cathode and the red lead is on the anode, are the ohmmeter leads reversed in polarity?
4. Why is the average digital ohmmeter not satisfactory for testing diodes and transistor junctions?
5. What is the typical cutoff value that defines continuity versus an open circuit, for a digital multimeter's continuity function?
6. Can a solid-state tester such as a Huntron Tracker be used on energized circuits?
7. What would a vertical line on a solid-state tester indicate?
8. What circuitry is incorporated within a solid-state tester to compare two separate circuit boards?
9. Name two of the five methods available to find a short circuit on a printed circuit board.
10. You are measuring the resistance of a resistor in a circuit. The reading is much lower than it should be. What is the probable cause?
11. You measure a resistor in a circuit and find the reading much higher than it should be. What does this tell you?
12. You measure the resistance of a resistor in a circuit and find that you get very different readings when you reverse the ohmmeter leads on the resistor. What is causing the difference in the readings?

chapter seventeen

Replacing the Part and Analyzing the Failure

CHAPTER OVERVIEW

The workmanship of a repair separates the professional from the amateur technician. It reflects upon the individual's competence, and so it is important to do a professional job.

SOME MECHANICAL TIPS AND CONSIDERATIONS

Always turn off all power and incoming signals from test instruments when removing components. This is particularly important when working with CMOS ICs and FET circuits.

Components such as resistors and capacitors can be damaged if their wire leads are bent too closely to the body of the part. Transistors have glass seals that can also be damaged by bending too closely. A broken seal will allow the transistor to absorb moisture and may eventually cause a failure.

Be careful not to drop parts on the floor. The shock of hitting concrete can cause failures. If you drop a part on a hard floor, consider discarding it and using a new part.

Using a poor pair of cutters can make a sharp "snapping" action that can damage components. Use sharp cutters that cut clean, and use the jaws as deeply as you can, close to the pivot point. Cutting at the tip of the jaws encourages the "snapping."

Some components like metal-film resistors and small capacitors can have their values changed if they are chipped by the careless use of tools such as long-nose pliers. Grip the component by one of its leads rather than by the body. It is a good idea to use a pair of long-nose pliers or hemostats as a *heat sink* to help conduct the heat of soldering away from the component. Place the tool next to the body of the component, gripping the incoming lead tightly. Leave the tool connected until several seconds after the soldering operation. A typical soldering operation should take five seconds or less.

When removing some components, the technician may find that a rivet must be removed. Simply drill them out with a drill slightly smaller than the diameter of the rivet shank.

When replacing power transistors and diodes, be sure to use a special heat-conducting grease such as *silicon grease* to help conduct away the normal operating heat of the component to the chassis or heat sink. This conducting grease has the consistency of a light lubricating grease. It does not conduct electricity. It is sold in small tubes at electronics supply houses as transistor heat-sink compound. A *small* amount of the grease spread over the mounting surface of a large transistor will greatly increase the life of the component because the component will not get as hot as it would without the compound.

Operating areas that subject equipment to a lot of vibration, such as aircraft and mobile installations, require special precautions against the equipment falling apart. Every nut and bolt should be provided with lockwashers or locknuts to help prevent loosening. Stranded wire should also be used in these areas or where the wiring will be moved frequently. Solid wire tends to break more easily than stranded wire under such conditions.

POINT-TO-POINT WIRING REPAIRS (NOT PRINTED CIRCUITS)

When removing components from wired circuits, take particular note of the connections of polarity-sensitive components. When removing transformers, take note of the wire color coding so that the replacement may be installed in exactly the same manner as the original. This is especially important in applications that are phase sensitive, such as transformers with windings in series and transformers used in oscillator circuits. Such transformers may have lead identification dots on the transformer and on the schematic to help install them properly.

Lead dress (the exact position of wiring) may be important in high-gain or high-frequency circuits. Make repairs so that the finished job looks exactly as it did originally.

If it is necessary to unlace wiring to make a repair, be sure to relace the wiring when the job is done. A quick way of doing this is to use small nylon Ty-Raps(TM).

PRINTED-CIRCUIT-BOARD REPAIRS

Work on printed boards requires a few special tools to make the job easier. A pair of small wire-cutters that have accurate mating of the jaws right to the very tip are important. These can be used to cut off component leads close to the printed circuit board. If possible they should have one side without a sharpening bevel. This will enable more accurate placement of the cut for precise trimming of component leads and a flat cut rather than leaving a tapered, sharp component lead (Figure 17-1).

Medium and small screwdrivers of both the (+) (Phillips) and (−) (straight) blade types are needed. A two-in-one screwdriver is a good idea, especially for a technician's small field toolbox.

For bench work a good magnifying glass is a useful item. The projection lens from an old 35-mm projector is ideal because it does not distort toward the

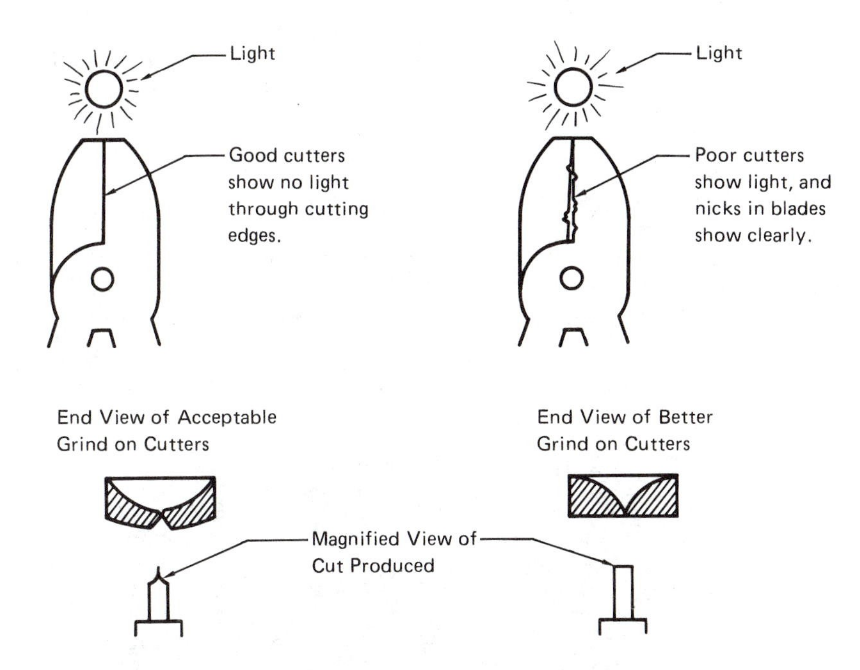

FIGURE 17-1 Selecting a pair of wire cutters.

edges of the field of vision. A magnifier is used to closely inspect PC boards and for reading the ultrasmall lettering on some components.

If you are on good terms with your dentist, you may be able to bargain for some very handy tools. One is the explorer, a pointed instrument good for scratching and probing about on PC boards. The most useful but least likely item you might get from him is a pair of small hemostats. They are extremely useful

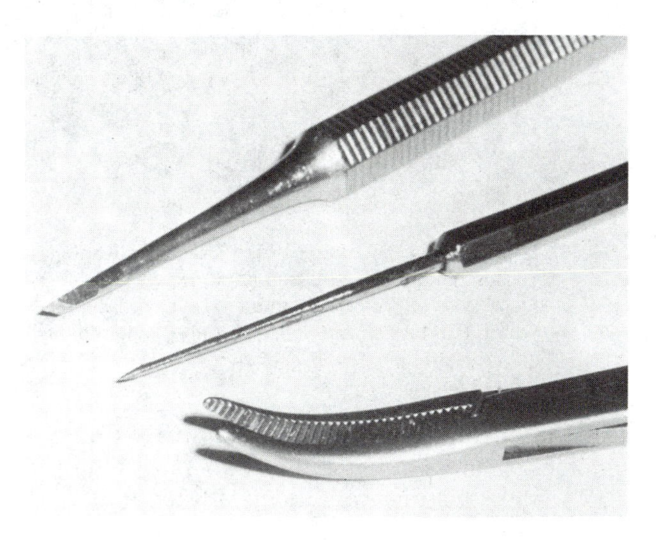

FIGURE 17-2 Useful dental tools for electronics work.

when working with the tiny components of the PC board. Another useful tool is the miniature chisel, used to make temporary breaks in PC board traces (Figure 17-2).

SOLDERING ON PRINTED CIRCUIT BOARDS

Soldering on printed circuit boards must be done with a temperature-controlled iron in order to minimize damage to the board. One such iron is shown in Figure 17-3. Never solder on a board with power applied to the board! The tips of many irons, such as that pictured in Figure 17-3, are directly connected to the power-line ground. Working on a "hot" board could cause extensive damage due to accidental grounding of the circuit.

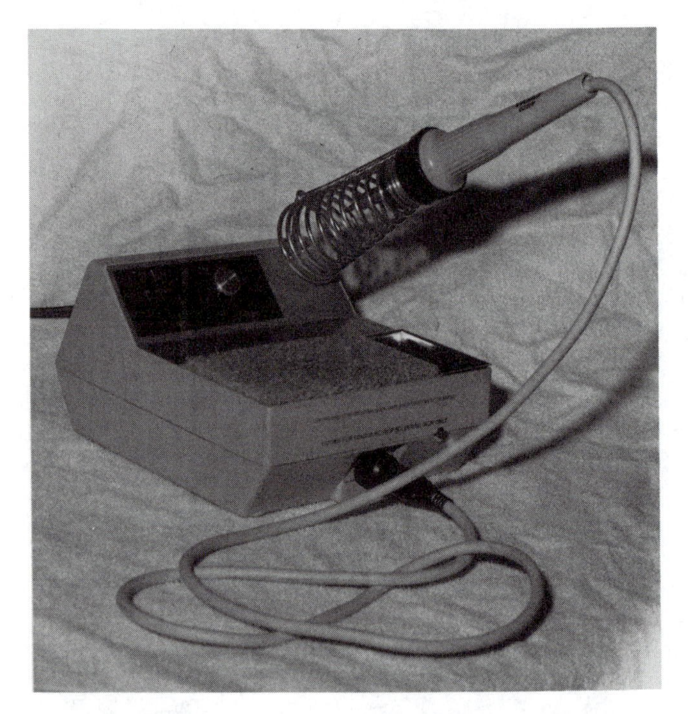

FIGURE 17-3 This Weller soldering iron has several advantages. It comes up to temperature in half a minute; it is small and very easy to handle; and it is completely safe to use on even sensitive CMOS circuits. (Photo courtesy of The Cooper Group—Weller Plant)

WHERE TO GET PARTS

The professional technician and the hobbyist will find quality parts from hundreds of reputable manufacturers at electronics distributors across the country (Figure 17-4). In the local directory they are listed under "Electronic Equipment and Supplies." These parts houses often give discounts for purchasing in quantities and will ship C.O.D. to established customers.

FIGURE 17-4 A typical distributor of electronic parts. (Photo courtesy of Radar Electric Co., Seattle, Washington)

Parts distributors differ from company-owned, quantity-sales chains that carry only their own brands of merchandise, equipment, and parts. Such chain stores cannot offer either the quantity or quality of parts to do a professional repair job.

REPLACING PARTS

Components with many leads must be removed by taking away as much of the solder as possible before wiggling the leads loose one at a time. This is also the preferred method to use with *all* components. Two methods are popular for removing solder, the wick and the vacuum. See Figure 17-5 for illustrations of each.

After the solder is removed, each component lead should be pressed from side to side in the board hole until it is free of solder. In many cases this is accompanied by a small "click" as the lead is broken free of the last bit of solder. Scratching over the pins of an IC that have broken free of solder produces a distinctly different sound from scratching one that is still being held by solder on the other side of the board.

Components with only two or three leads may often be removed by heating the connection on the bottom of the board and gently rocking and pulling the component from the top. As each lead is heated in turn, it is pulled more from

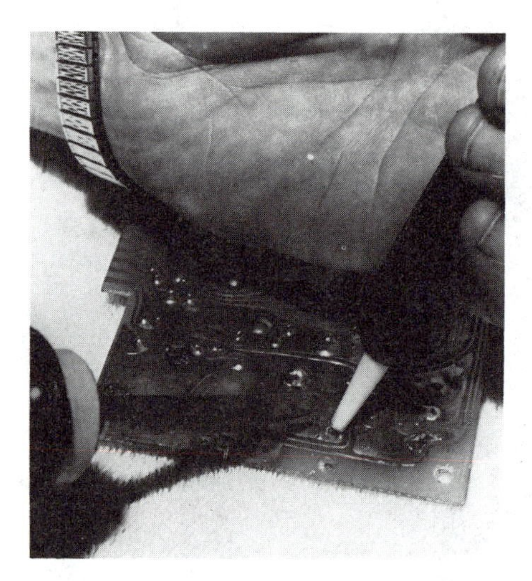

FIGURE 17-5 Two methods of solder removal: vacuum (left) and wick (right).

the board with each rock. This method is used most often when a vacuum or wick is unavailable. This method doesn't work with multilegged components.

Sometimes when access to the bottom of a printed circuit board is restricted, it is expedient to smash a suspected component and fasten the new component to the old leads (Figure 17-6). Still another way to remove ICs and other multi-pinned components is to clip off the leads at the component body, then remove each of the "stubs" left in the board one at a time with a pair of hemostats and the iron (Figure 17-7). Be careful not to pull on the tips remaining in the board

FIGURE 17-6 Replacing a component when the underside of the board is not accessible.

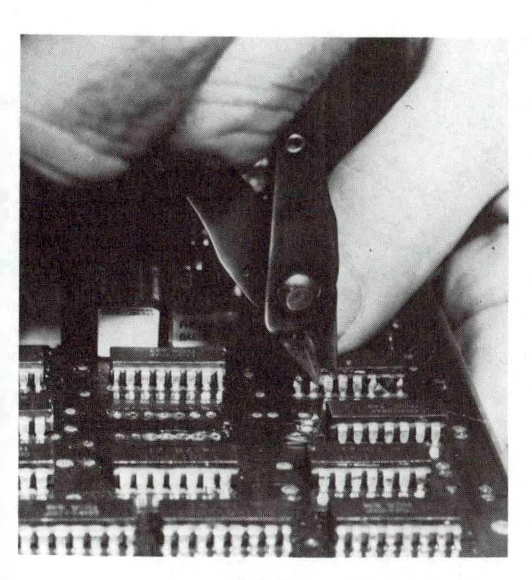
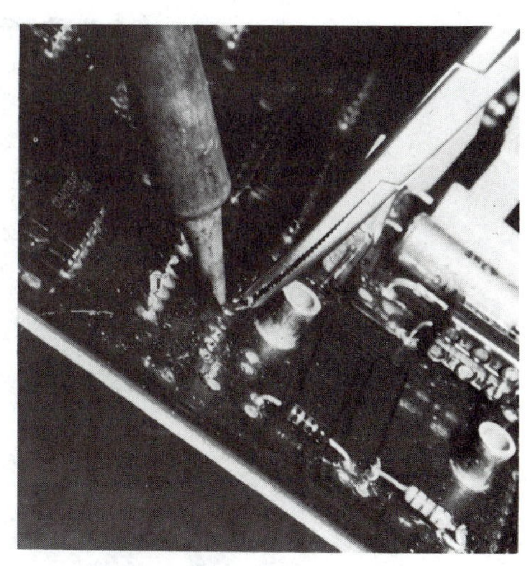

FIGURE 17-7 Removing an IC by clipping from board and removing stubs.

until the solder has melted completely through the board (in the case of plated-through holes). It may be very easy to pull a trace right off the board if too much pull is applied.

CLEANING AND REPAIRING DAMAGED BOARDS

When the old component has been removed, the board should be cleaned carefully with alcohol so the connections can be given a very close inspection. Look for lifted traces and damage caused by previous work on the board or the desoldering operation. Figure 17-8 shows how lifted traces and open circuits may be repaired with short lengths of bare wire. Such damage can be minimized by using a temperature-controlled iron and making connections and desoldering operations as quickly as possible, consistent with quality work.

CHECKING THE BAD COMPONENT

The component removed from the board should be checked to verify that it is indeed defective. If the troubleshooting was done correctly, the removed part should test "bad." If it does not, an error was made in troubleshooting, and the part may or may not be reinstalled. If the removed component is okay, do not necessarily reinstall it right away. It may be a big advantage in further troubleshooting to have the component out of the circuit. This can leave other components "open circuited" in the area in question, possibly making it easier to find the problem remaining, using the dead-circuit testing techniques of Chapter 16. If the removed part does not test bad, troubleshooting must continue until the defective part is found. Whether the old, good part will be reinstalled rather than replaced with a new part is a matter of judgment. The availability of the

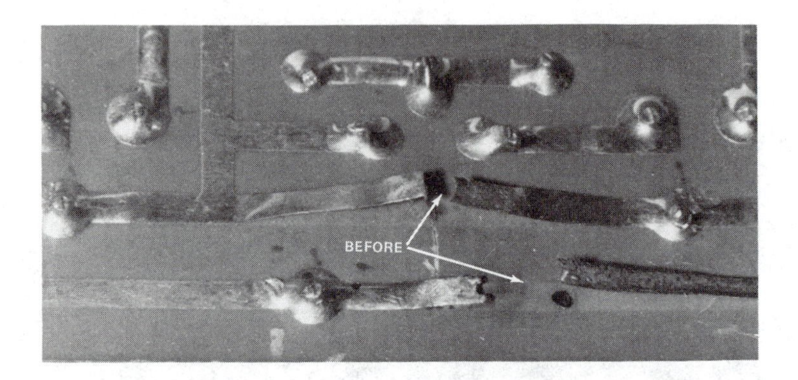

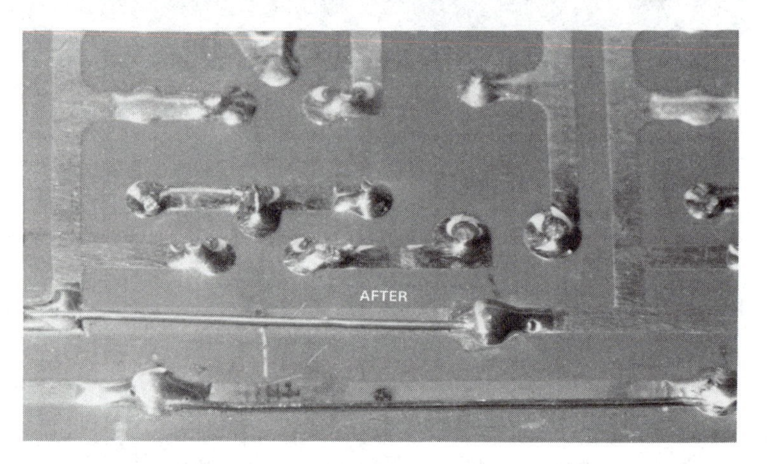

FIGURE 17-8 Repairing burned or lifted circuit-board traces.

part, its cost, and how much heat had to be used to remove it are all factors to consider.

Most components can be tested either with an ohmmeter or a solid-state tester. The ohmmeter is used to compare readings on a known good component compared with those on the suspect one, while the solid-state tester tests the components by exercising them. This gives a much better analysis of the failure than the ohmmeter.

The final authority as to whether a component is good or bad is to substitute a known good component in its place. If the new one works and the old one didn't, the evidence is pretty conclusive that the old one was defective. Components with two or three leads can be held in place fairly well for a quick power-on test in most cases. For ICs it will be necessary to ensure good contact of each of the pins before applying power. This can be done by "toothpicking." See Figure 17-9 for an example of how this is done. Get the new IC into the circuit the right way! Beware of getting it turned around and likely ruining it by forcing the supply current through it the wrong way. The IC is placed down against the board and the toothpicks are gently inserted from the opposite side of the board,

FIGURE 17-9 Temporary installation of an IC by using tooth-picks to assure contact.

into the holes alongside the pins, thereby forcing them into good contact with the sides of the holes. This method will work well only with PC boards that have through-plated holes, however. With single-sided boards the IC will probably have to be soldered in for a valid test.

Other instruments, such as capacitor analyzers and reactance bridges, are available for testing individual components, but they are seldom worth the time and effort taken to operate them. A quick substitution is generally preferred over the use of these instruments. They are used for research and development, but are seldom used by the repair technician.

The following sections describe out-of-circuit tests that can be made on individual components to determine if they are defective.

Testing Diodes

Solid-State Tester. Check according to instructions provided with the instrument. (See Figure 17-10 for a typical presentation.) Use the "low" range.

GOOD DIODE

BAD DIODE
(CURVED PORTION)

GOOD DIODE WITH LOW
RESISTANCE IN PARALLEL

FIGURE 17-10 Testing diodes with the solid-state tester.

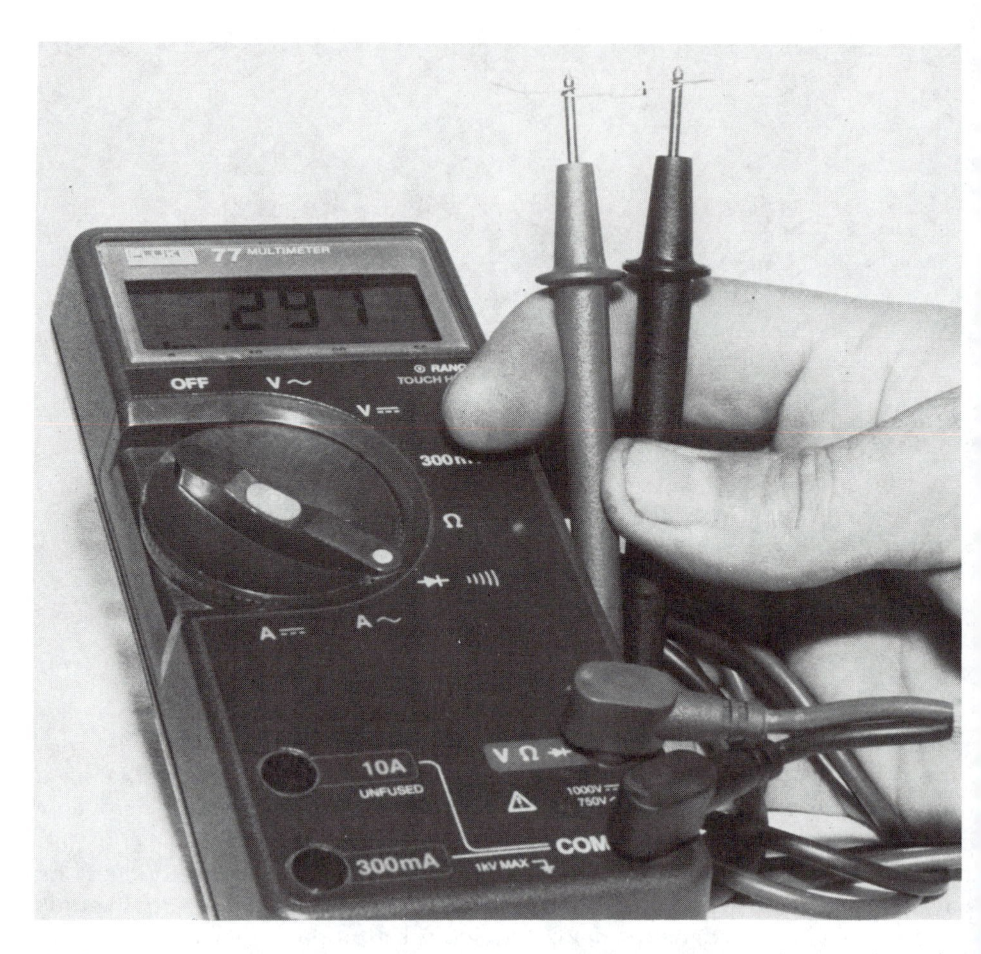

FIGURE 17-11 Testing a diode using the special diode function of a digital multimeter. This scale reads out the voltage drop across the diode rather than its resistance. Note the reading of nearly 0.3 volts, indicating a germanium diode. (Reproduced with permission of the John Fluke Mfg. Co., Inc.)

DMM. This instrument may not have sufficient test current capability for reliable junction testing. Consult your DMM instruction book. If there is no special provision for junction testing, the DMM is not recommended for this use. Some of the better DMMs provide a special scale for this purpose, which produces up to about a milliampere of test current for properly checking semiconductor junctions (Figure 17-11).

VOM. A VOM will generally give better readings on semiconductor junctions than many DMMs because the VOM produces more current on the ohms scales. Check the front-to-back resistance using a single scale such as the RX100 for high-current (greater than an amp) diodes, and the RX1000 scale for the smaller, signal diodes. One direction should be low in resistance, the other direction an open.

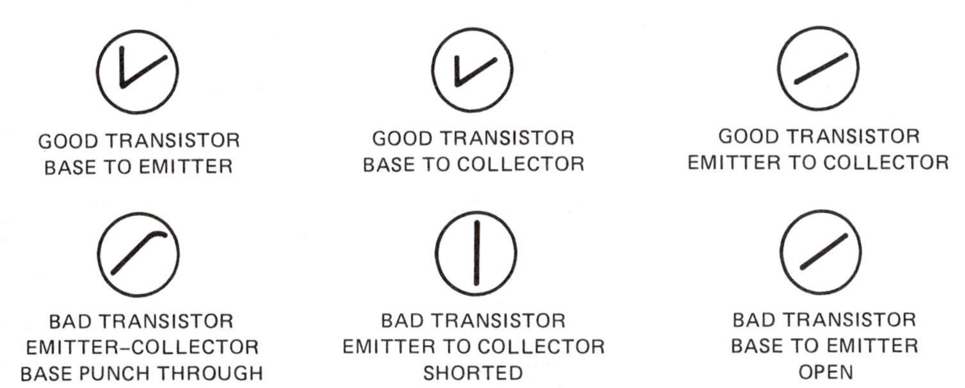

FIGURE 17-12 Testing bipolar transistors with the solid-state tester.

Testing Bipolar Transistors

Solid-State Tester. Check according to instructions provided with the instrument. (Figure 17-12) gives basic presentations of good and bad junction transistor presentations.) Use the "low" range.

DMM. This instrument may not have sufficient test current for reliable junction testing. Consult your DMM instruction book. If there is no special provision for junction testing, the DMM is not recommended for this use. If there is provision for junction testing, the DMM should show, for the six readings obtained, two

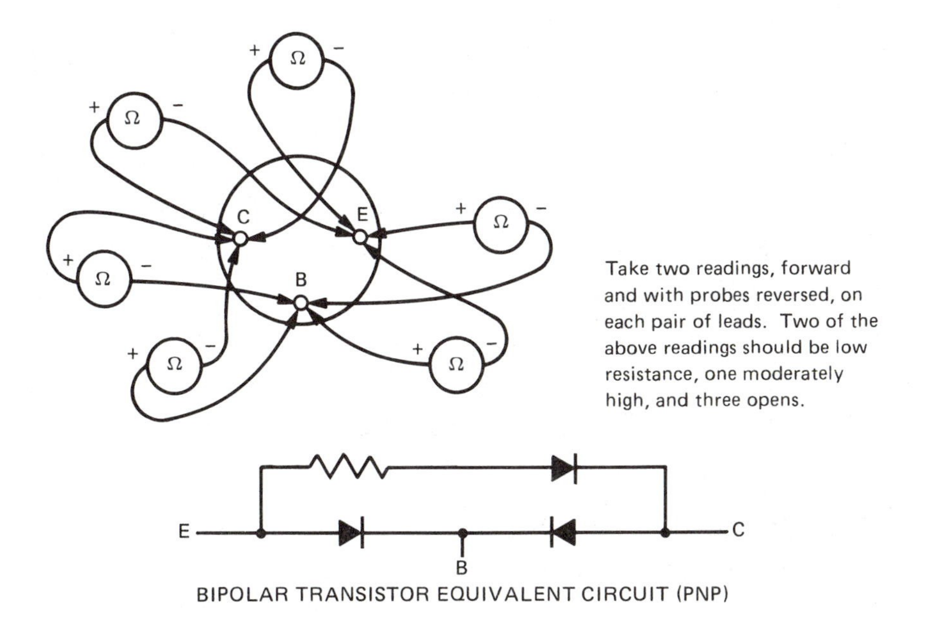

FIGURE 17-13 Taking six readings with an ohmmeter to determine if a bipolar transistor is defective.

TABLE 17-1 Sample Ohmmeter Readings on Several Bipolar Transistors

OHMMETER LEAD POLARITY		BIPOLAR TRANSISTORS					
+	−	**npn Power 2N1486**	**npn Signal 2N388**	**npn Signal 2N377**	**pnp Power 2N1164**	**pnp Signal HEP636**	**pnp Signal 2N2907**
Base	Emitter	2000 Ω (X1000) 33 Ω (X10)	600 Ω (X1000)	800 (X1000)	∞	∞	∞
Emitter	Base	∞	∞	∞	10 Ω (X10)	800 (X1000)	3 K (X1000)
Base	Collector	2000 Ω (X1000) 40 Ω (X10)	700 (X1000)	700 (X1000)	∞	∞	∞
Collector	Base	∞	∞	∞	10 Ω (X10)	600 (X1000)	3 K (X1000)
Emitter	Collector	∞	∞	∞	200 (X10)	∞	200 K (X1000)
Collector	Emitter	∞ (X1000) 500 K (X100,000)	50 K (X1000)	75 K (X1000)	∞	∞	∞

Figures in parentheses are range used.
∞ = open

diodes forward biased and four open readings. See the following paragraphs on how to obtain the six readings.

VOM. This test consists of making six readings on the three leads (Figure 17-13). A VOM will generally give better readings on semiconductor junctions than most DMMs because the VOM produces more current on the ohms scales. (Some of the better DMMs provide a special scale for this purpose which produces up to about a milliampere of test current for properly checking semiconductor junctions.)

Using a single scale of RX10 or RX100 for power transistors and RX1000 for smaller transistors, take the six readings. Two of them should be low resistance, one medium resistance, and three open readings. Table 17-1 shows examples of readings obtained with a Triplett 630 and a few miscellaneous transistors.

A Simple Transistor Checker. If many transistors are tested routinely, it might be a good investment in time to build the simple transistor checker of Figure 17-14.

Testing Darlington Transistors

A Darlington transistor is actually two transistors with their collectors connected together. Test it as a bipolar transistor, but allow for two junctions in series rather than one between the emitter and base leads. Figure 17-15 shows the internal connections of a Darlington device.

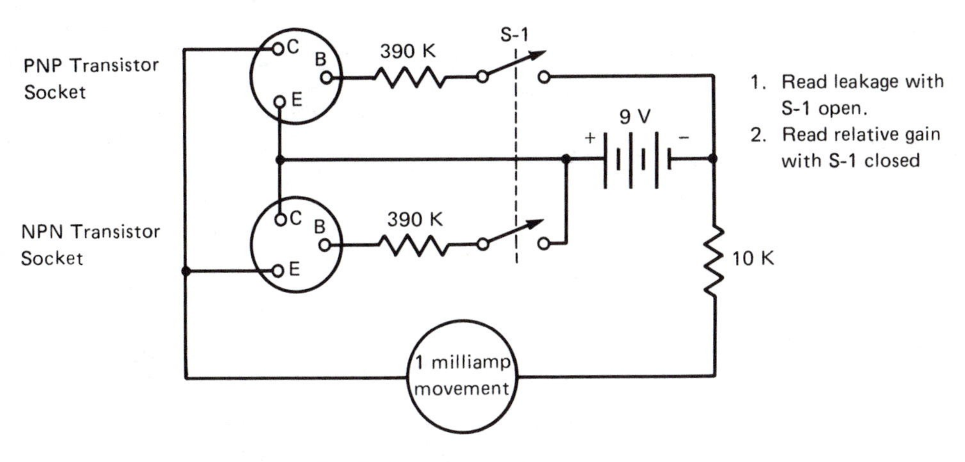

FIGURE 17-14 Simple bipolar transistor tester.

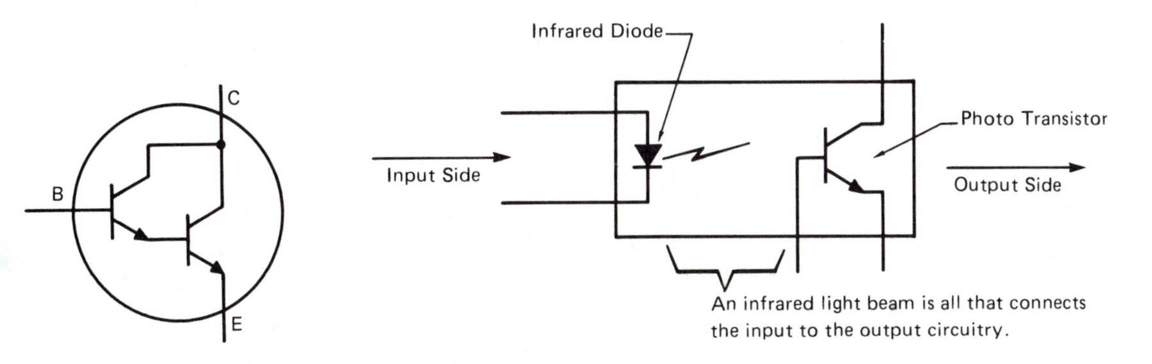

FIGURE 17-15 Darlington-connected transistors.

FIGURE 17-16 Internal construction of an optical isolator.

Testing Optical Isolators (Opto-Isolators)

The optical isolator is an LED and a transistor (sometimes a Darlington pair) within a single IC package (Figure 17-16). Testing the opto-isolator is done in two steps: testing the diode section as an LED and testing the transistor section as a bipolar transistor.

Testing FETs

Solid-State Tester. Check according to the special instructions provided with the instrument. Figure 17-17 shows the correct and incorrect presentations of the different types of JFETs and MOSFETs.

VOM/DMM. Due to the very high impedance of these semiconductor devices, substitution of a known good FET of the same type in the circuit is recommended.

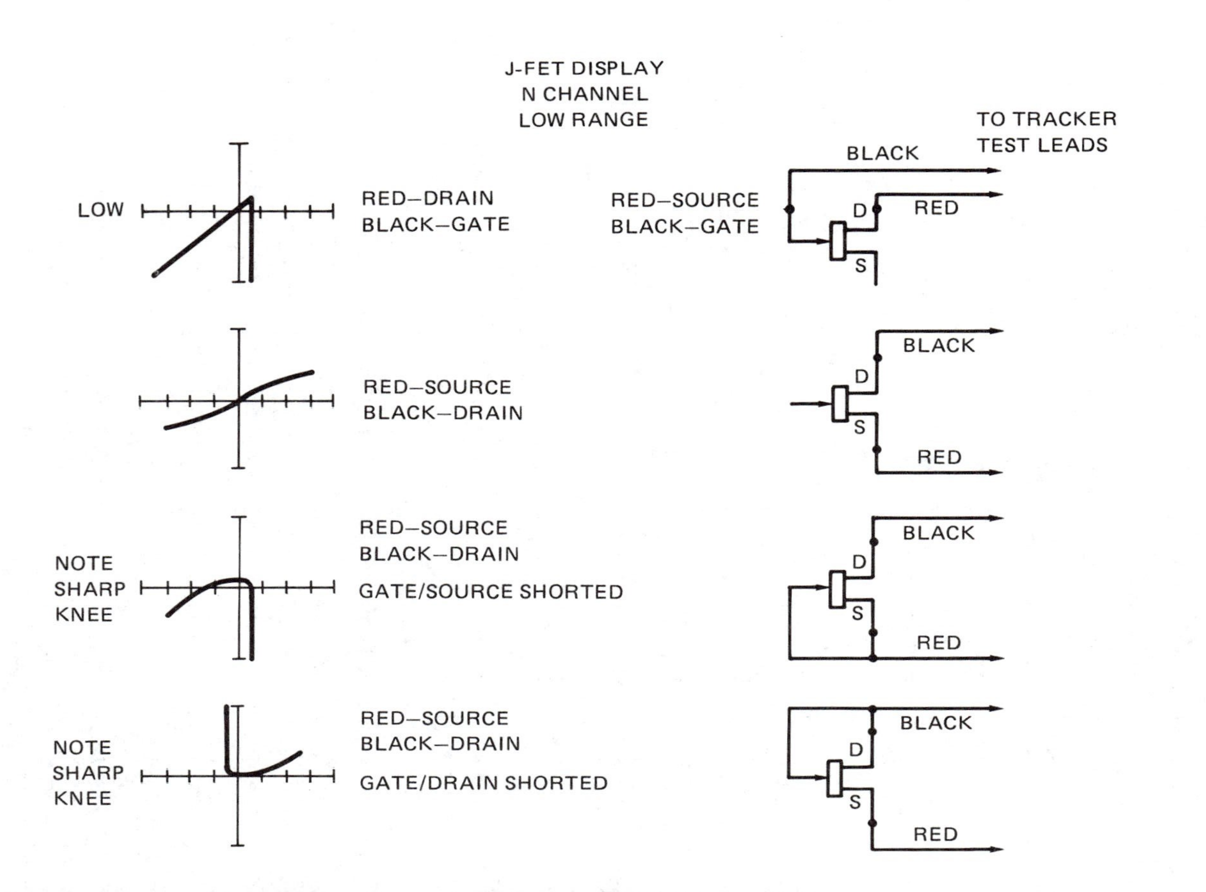
J-FET DISPLAY
N CHANNEL
LOW RANGE
TO TRACKER
TEST LEADS
LOW
RED—DRAIN
BLACK—GATE
RED—SOURCE
BLACK—GATE
BLACK
RED
D
S
RED—SOURCE
BLACK—DRAIN
BLACK
RED
D
S
NOTE
SHARP
KNEE
RED—SOURCE
BLACK—DRAIN
GATE/SOURCE SHORTED
BLACK
RED
D
S
NOTE
SHARP
KNEE
RED—SOURCE
BLACK—DRAIN
GATE/DRAIN SHORTED
BLACK
RED
D
S

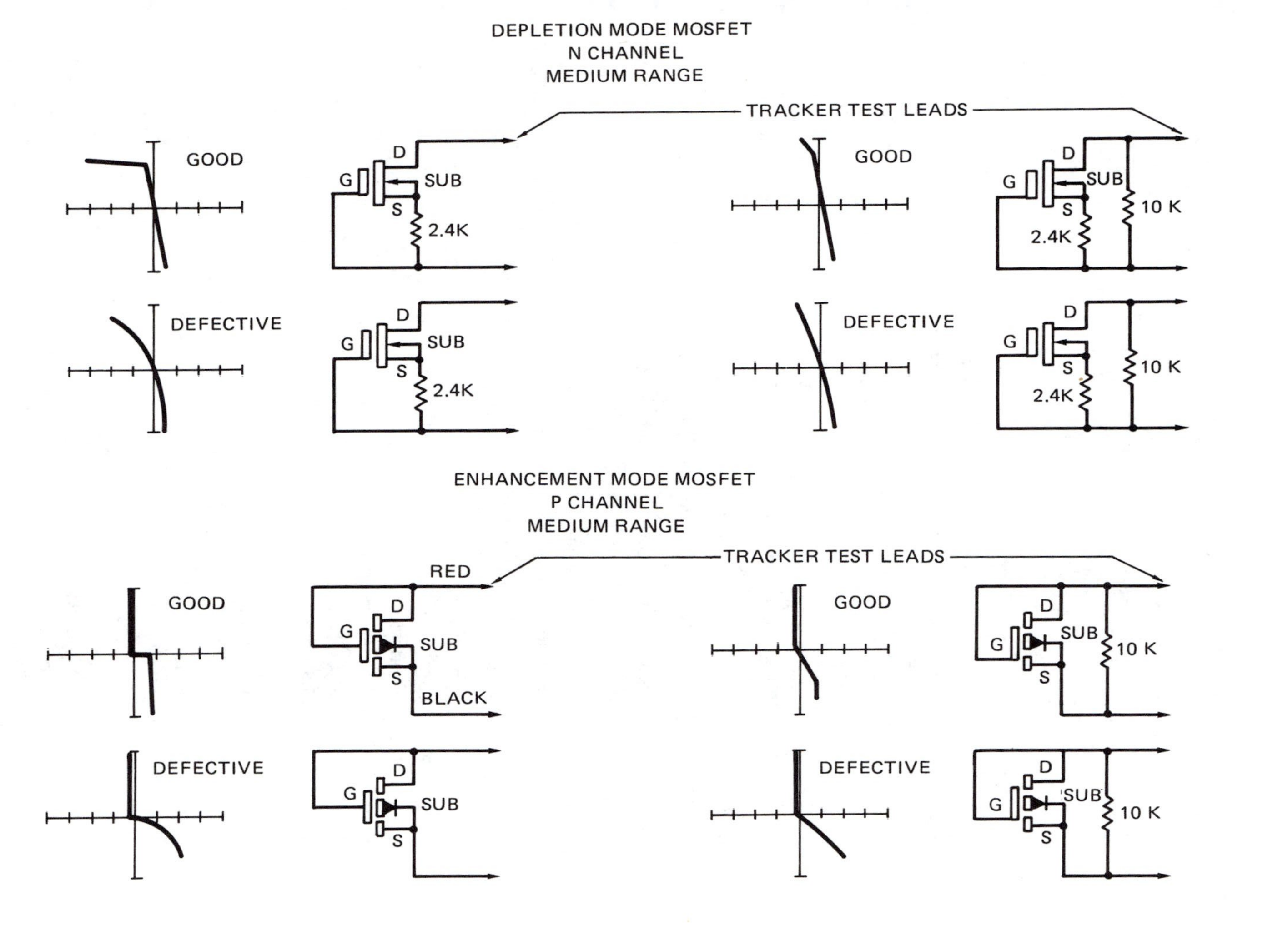

FIGURE 17-17 Testing JFETs (p. 256) and MOSFETs with the solid-state tester. (Courtesy of Huntron Instruments Inc.)

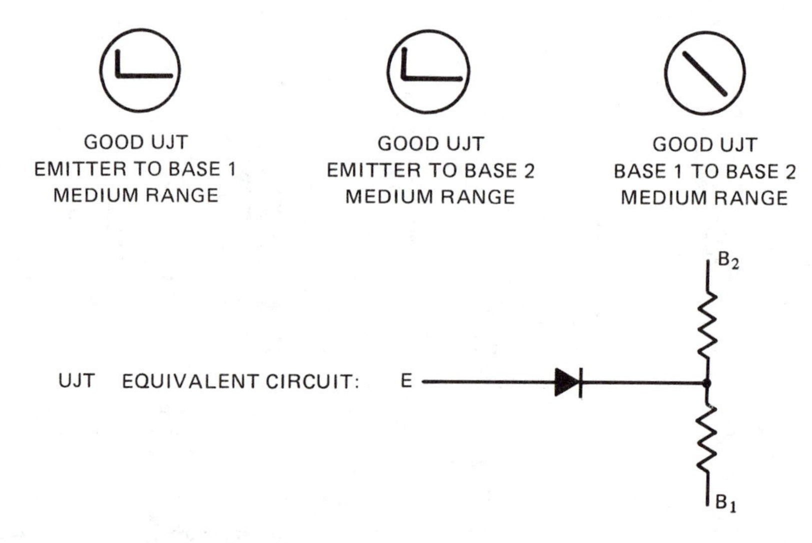

FIGURE 17-18 Unijunction transistor testing with the solid-state tester.

Testing Unijunction Transistors (UJTs)

Solid-State Tester. Check according to instructions provided with the instrument. Typical presentations are shown in Figure 17-18.

VOM. As far as the VOM resistance tests are concerned, the UJT is a resistor with a diode attached to the middle (Figure 17-19). The UJT is tested by taking six readings, just as the bipolar transistor was tested. The UJT should show two resistive readings (through the base or "resistor" section) of roughly 5000 ohms in either direction. The readings from the third lead should show a low resistance in the forward direction to either of the two base leads, and when the leads are reversed should show two very high-resistance readings to the same two leads. The DMM is not recommended for testing UJTs because of too little test current, unless a special function is provided on the DMM for this purpose. Typical VOM resistance readings on several UJTs are shown in Table 17-2.

A diode junction should show a low reading and a high when the leads are reversed, when measuring from the emitter to either base connection.

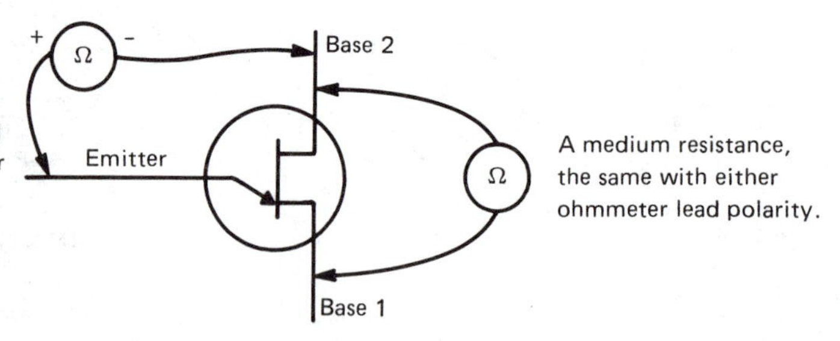

FIGURE 17-19 Testing a UJT with an ohmmeter.

TABLE 17-2 Sample Ohmmeter Readings on Several Unijunction Transistors

OHMMETER POLARITY		UNIJUNCTION TRANSISTORS		
+	−	2N1671B	2N491	2N4870
Emitter	Base 1	4 K	3 K	3 K
Base 1	Emitter	∞	∞	∞
Emitter	Base 2	10 K	7 K	4 K
Base 2	Emitter	∞	∞	∞
Base 1	Base 2	9 K	6 K	5 K
Base 2	Base 1	9 K	6 K	5 K

X1000 Range of VOM used.
∞ = open

Testing Silicon-Controlled Rectifiers (SCRs)

Solid-State Tester. Check according to instructions provided with the instrument; depending on the HUNTRON model used, SCR checks vary. Model 2000, the latest, has provisions on the front panel to fire SCRs during cathode-to-anode testing. The firing pulse width and amplitude are adjustable to better compare the firing characteristics of individual components. See Figure 17-20 for waveform presentations with Model 1005 and Model 2000.

VOM. The SCR is four alternating layers of semiconductor material. See Figure 17-21 for testing SCRs with an ohmmeter. The SCR should indicate open between the anode and cathode with the gate lead open or shorted to the cathode. The gate should test as a diode with respect to the cathode, with one reading on the

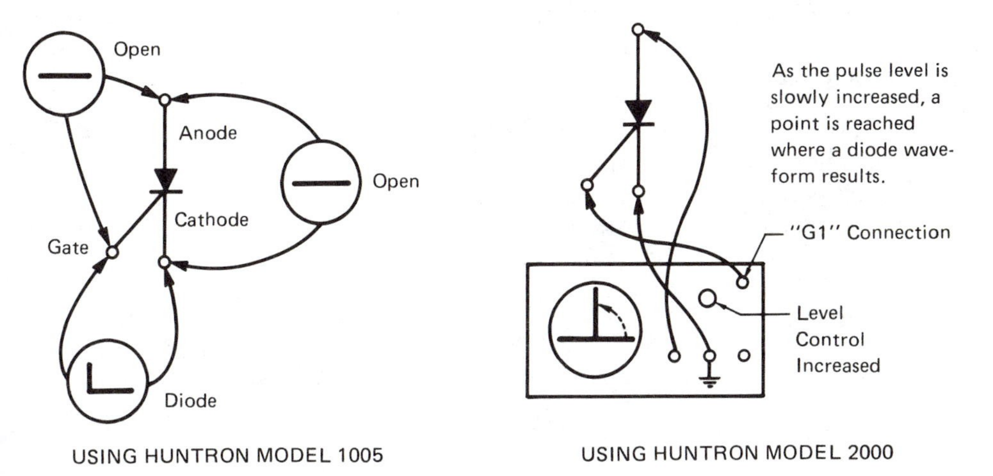

FIGURE 17-20 Solid-state tester waveforms for silicon-controlled rectifiers (SCRs).

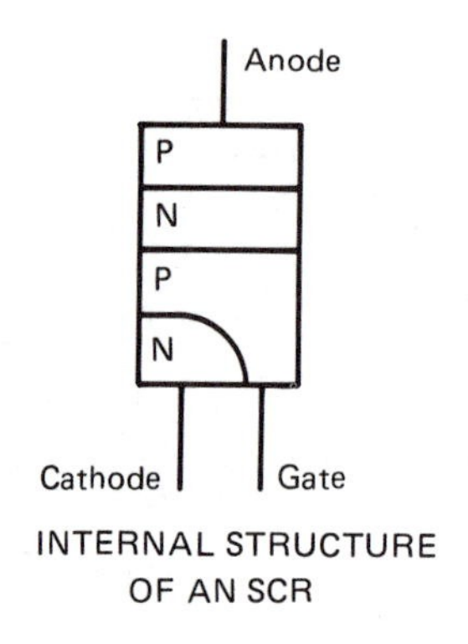

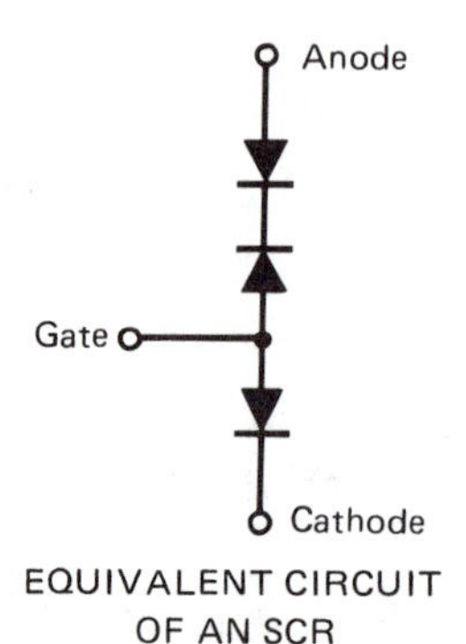

SCR TESTING WITH AN OHMMETER

Ohmmeter Polarity +	–	Reading
G	C	Low
C	G	∞
G	A	∞
A	G	∞
C	A	∞
A	C	∞

Note that a diode junction is evident between the gate and the cathode. All other readings should be open.

FIGURE 17-21 Testing a silicon-controlled rectifier with an ohmmeter.

RX100 scale showing low resistance, then open with the leads reversed. The DMM is not recommended for junction testing unless it is provided with the special junction-testing function. In this case, the DMM should show one diode junction between the cathode and the gate elements and five other open readings for the other test-lead combinations.

Testing Light-Emitting Diodes (LEDs) and Infrared LEDs (IREDs)

Solid-State Tester. Check according to instructions provided with the instrument. (Sample presentations are shown in Figure 17-22). Use the "low" range. The Huntron will actually light the LED, a more conclusive test than simply testing its conductivity throughout the AC voltage curve applied. A defective yet still-operating LED may show leakage in the reverse direction. In circuits that depend upon the rectifying action of the LED this reverse leakage can create problems. This leakage characteristic is not normally noticed with other test instruments.

VOM. The resistance function of the VOM will show a good LED as a fairly low resistance in the forward (lighted) direction and as a very high resistance in the reverse direction. Due to the high voltage drop across the LED in the forward direction when conducting, the higher ranges of the ohmmeter such as RX100 or

FIGURE 17-22 Testing light-emitting diodes with the solid-state tester.

RX1000 should be used. As a note of interest, a *red* LED will show about 1.6 V DC across it when lighted; green and yellow diodes have about 2.0 V DC across them. A DMM is not generally recommended for testing LEDs due to the higher voltage required over the 0.3/0.7 volt of a "normal" diode, and the DMM's insufficient current to properly exercise the junction. An exception is a DMM equipped with a special range for this purpose.

The infrared LED (IRED) is tested just as an LED, but there will be no visible light coming from the device. The normal forward voltage drop is about 1.0 V DC.

Testing Other Display Components

Gas Discharge (Neon) Displays. These tubes are best tested by substitution in an operating circuit. They require voltages much higher (75 to 150 V DC) than those provided by the VOM or solid-state tester for individual element testing. They should test as an open circuit on an ohmmeter.

Testing Photodiodes

The photodiode is operated in the reversed-polarity mode much as a zener (Figure 17-23).

Solid-State Tester. The Huntron will show presentations similar to Figure 17-24(A) when the diode has no light shining upon it, and similar to Figure 17-24(B) when light is provided to the diode's junction. Use the "low" range.

DMM. The DMM may be used to test these diodes for comparison purposes, using the resistance scale. Connect the diode with reversed (normally open) polarity.

VOM. Photodiodes may be tested using the RX100 scale of a VOM if sufficient voltage is available at the probes to reverse-bias the diode and provide a high enough sensitivity to indicate a low resistance when the diode is shown a light. If the RX100 scale is not sufficient, try the next higher range.

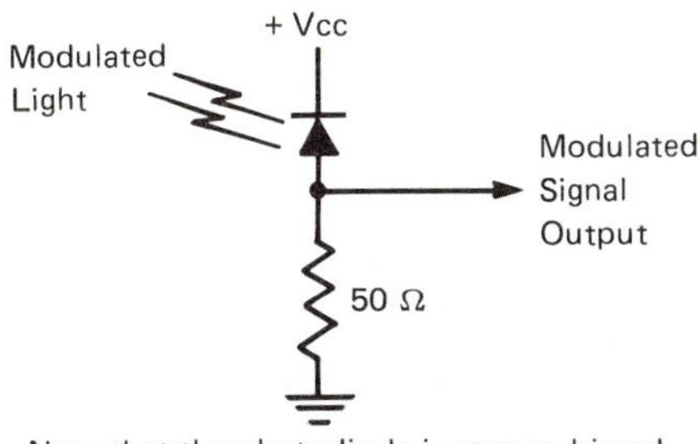

FIGURE 17-23 Sample photodiode operating circuit.

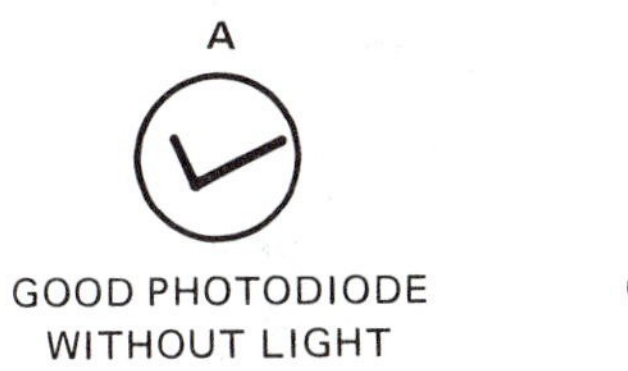

FIGURE 17-24 Testing photodiodes with the solid-state tester.

GOOD PHOTOTRANSISTOR
WITHOUT LIGHT
COLLECTOR TO EMITTER
MEDIUM RANGE

GOOD PHOTOTRANSISTOR
WITH LIGHT
COLLECTOR TO EMITTER
MEDIUM RANGE

FIGURE 17-25 Testing the photo transistor with a solid-state tester.

Testing Phototransistors

The presence of light will cause the resistance from emitter to collector of a phototransistor to decrease when the transistor is shown the wavelength of light to which it is sensitive. Visible-light phototransistors respond to visible light, and infrared phototransistors may require the use of an infrared source to test them. Depending upon the sensitivity curve of the phototransistor, fluorescent light may produce sufficient stimulation of a good infrared phototransistor to indicate whether the device is working or not. Comparison with a good phototransistor of the same type is recommended.

Solid-State Tester. The Huntron can be used to test a phototransistor (Figure 17-25). Use the "medium" range.

DMM. A DMM is very effective in testing the phototransistor, particularly if the diode function is available and used. Be sure to have the right polarity applied to the phototransistor (red to the collector if an npn type).

VOM. A VOM on a high range (RX100) will indicate a very high resistance, approaching an open, when the positive lead of the VOM is applied to the collector lead and the negative lead to the emitter. When light is then shone on the transistor, the resistance should decrease greatly, depending upon the sensitivity curve of the transistor and the type and intensity of the light source.

GOOD TRACE

GOOD TRACE

BAD TRACE
(NOTE "FRACTURED"
KNEE OF CURVE)

These indications are only a few of the possible traces obtainable in troubleshooting. The instruction book with each Huntron instrument should be consulted for specific waveforms.

FIGURE 17-26 Representative indications when testing ICs with the solid-state tester.

Testing ICs

Solid-State Tester. Check according to instructions provided with the instrument. (See Figure 17-26 for representative displays of tests on ICs.) Use the "medium" range.

VOM/DMM. Testing ICs is rarely practical with these instruments. Lacking a Huntron, substitution is the best method of testing. A possible exception is where a short is suspected, in which case an ohmmeter may be used to verify the suspicion.

Testing Resistors

Solid-State Tester. This instrument will give an indication of resistance depending upon the resistance involved and the instrument scale used. It may be used only to approximate resistor values. The DMM and the VOM are better for checking actual resistance values.

VOM/DMM. Simply measure the resistance of the resistor in question. Since the resistor is not a polarized component, lead polarity makes no difference. Some circuits require precise values of resistors. These resistors are best tested for exact value by using a DMM. The use of metal-film resistors in a circuit indicates a possible requirement for precise replacement values.

Remember that most resistor failures are secondary failures, the result of a primary failure. Look for another component which, if shorted, may have caused too much current through the resistor that failed. Shorted capacitors, especially electrolytic types, and transistors are often the primary problem.

Testing Varactors (Tuning Diodes)

See the tests for a standard junction diode.

Testing Capacitors

Solid-State Tester. Check according to instructions provided with the instrument. (See Figure 17-27 for some representative displays of tests on capacitors.) Use the "low" range for large electrolytic capacitors of more than about a microfarad, the "high" range for capacitors of 0.0025 to about a microfarad, and the

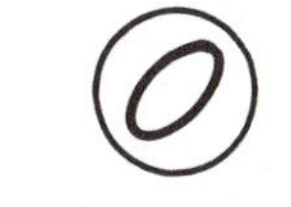

GOOD CAPACITOR

BAD CAPACITOR

BAD CAPACITOR

FIGURE 17-27 Testing capacitors with the solid-state tester.

"medium" range for capacitors between about 0.01 and 4 microfarads. There is considerable overlap in these ranges.

VOM/DMM. Capacitors, other than electrolytic, are not polarity sensitive, and therefore the polarity of the test leads is of no particular consequence. These capacitors should measure "open." With larger capacitors (perhaps a microfarad or so and up) and the higher resistance ranges (RX10,000), it may take several seconds for the needle to come off the "shorted" indication and "bleed" back to an open circuit. This is due to the normal charging of the capacitor. The rate of charge is a function of the size of the capacitor. A very rough idea of capacity may be obtained from comparing a known good capacitor with a questionable one. If of equal capacity, the charging time should be the same. When comparing capacitors this way, be sure to short the capacitors out, then time the charging from the moment the short is removed. It may be convenient to pick an arbitrary value indicated after a few seconds of charging time, rather than wait for the "open" indication.

Very small values, about 0.001 μF and smaller, will charge so fast that about all that can be seen on a VOM is a twitch of the needle. These small capacitors should all test open after the twitch.

Note: When testing capacitors with an *automatic ranging* digital multimeter, it is best to lock the DMM on a specific high-resistance range for comparison testing of capacitor charge times. It is possible for some DMMs to charge a large-value capacitor on the low-resistance ranges very quickly because of the higher current capability of those ranges. This charge may then give a rapid charge indication as the instrument auto-ranges through the higher resistance ranges. This effect may lead the technician to believe that the capacitor has less value than it really does.

Electrolytic capacitors are polarity sensitive. Be sure to apply the ohmmeter test leads with the proper polarity. If in doubt, see Chapter 16 for the ohmmeter-lead-polarity testing procedure. Reverse polarity testing of an electrolytic capacitor may show severe leakage. Electrolytics are similar to other capacitors in their charging action, but they are allowed a certain amount of leakage, with proper polarity applied, without necessarily being "bad." On an ohmmeter, this forward leakage will show up by the charging of the capacitor only to a certain resistance, at which reading the ohmmeter needle settles. This end reading is the forward leakage value of the capacitor. To see if this leakage is acceptable, test a known good capacitor of the same type and preferably from the same manufacturer. If the readings for the good and suspected capacitors are about the same, the suspected capacitor is *probably* all right. Very large electrolytics may have considerable leakage. The larger the capacitor, the more leakage is acceptable.

When replacing electrolytic capacitors with ones that have been on the shelf for some time, it is a good idea to "form" the new capacitors. Application of full voltage to a capacitor that is not "formed" may result in immediate destruction of the capacitor if the supply current is sufficient. The reason is that the capacitor will have an initial leakage resistance that is very low—possibly low enough to cause severe heating and destruction of the capacitor.

To form a capacitor, apply the capacitor's maximum rated voltage through a very-high-value series resistor. The value of the resistor may be estimated by using a simple formula:

$$\text{Series Resistor (in megohms)} = \frac{100}{\text{Capacity (in } \mu\text{F)}}$$

For example, a 20-μF capacitor at 50 V would be supplied the full 50 V through a series 5-megohm resistor. This series resistor will limit the possibly destructive initial current flow into the capacitor to a safe value until the capacitor develops its normal internal high resistance. After perhaps ten or fifteen minutes, the capacitor should reach very nearly the full supply voltage, at which time the capacitor may be discharged and installed.

Testing Microphones

Dynamic microphones are much like a small speaker—a moving coil within a magnetic field. If the test leads of a VOM on the lowest range (RX1) are applied to a good dynamic microphone, there will be an audible "click" when the leads are first applied. A DMM may not produce sufficient current to perform this test. The DMM will, however, give the resistance of the coil. The resistance reading of either the VOM or DMM may be compared to a known good microphone. Note that some dynamic microphones may have a subminiature transformer concealed in the microphone or stand. Either the element (the coil and magnet assembly) or this hidden transformer may be defective. These microphones may also be tested as crystal microphones.

Crystal and ceramic microphones have a piezoelectric crystal supported in such a way that sound waves stress the crystal, which then produces a voltage. These microphones are best tested by using an oscilloscope on a high gain setting and connecting the output of the microphone to the scope input. Whistle past (not into) the microphone or talk (say the word "four," pronounced "Foowwwer"). Blowing or spitting into a microphone can damage it, and in some sound studios could get a technician into a peck of trouble. It is *very* unprofessional to mistreat a microphone in this way.

A carbon microphone is a variable resistor of sorts. A test jig consisting of a battery power supply and other components can be constructed to test it (Figure 17-28). More often, it is preferable to simply try a different microphone than take time to make this test jig.

Testing Speakers

A quick go-no-go test of a speaker is to place two fingers in the center of the speaker, on either side of the central decoration that defines the area of the coil. In small speakers this is usually a piece of felt. Pressing directly in the center of the felt will stretch it; pressing around the edges is the proper place to test. While gently pressing the diaphragm in and releasing, listen for any scratching sound coming from the cone. Scratching indicates that the coil is rubbing on the

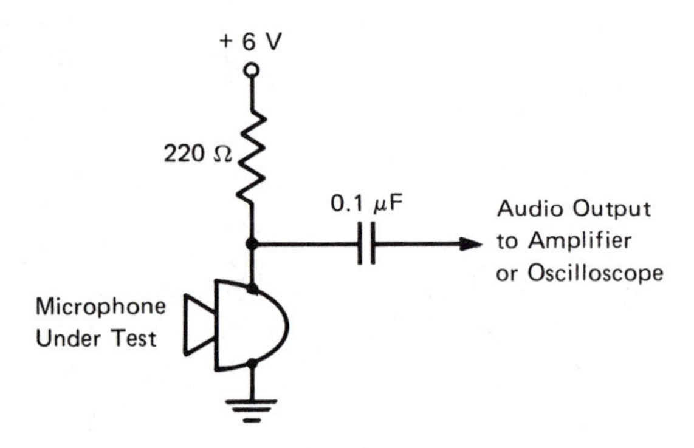

FIGURE 17-28 Carbon microphone test jig.

magnet assembly of the speaker. Replacement is the only reasonable solution for small speakers, but rebuilding is feasible for larger ones, A rubbing speaker will sound distorted and can in some cases cause other electronic problems if the speaker coil carries a DC component with respect to chassis ground.

Solid-State Tester. The Huntron can test speakers, producing a low hum from the speaker and indicating a very low resistance. Use the "low" range.

VOM/DMM. The VOM is a good instrument to test speakers because on the lowest (RX1) ranges there is sufficient current produced by the meter for the speaker to click in response. If there is no response at the terminals on the frame of the speaker, try contacting the coil wires where they connect to the flexible leads between the coil and the frame terminal.

The DMM may not produce enough current to make the speaker thump, but the resistance reading produced should be low, about 2 to 16 ohms, depending on the impedance of the speaker. The resistance of 4-, 8-, and 16-ohm speakers will be about 2.4 to 3.8 ohms, 4.8 to 7.6 ohms, and 9.6 to 15.2 ohms, respectively.

Dry Cell. A small dry cell (1.5 V DC) may be used to test speakers. Momentarily connect the battery across the speaker terminals and listen for the click. Do not leave the battery connected, just make a few quick sweeps across the terminals.

Testing Inductors

The exact value of inductance may be determined with the use of an inductance bridge. This is not a normal instrument to have on hand, so the servicing technician may have to rely on other means of testing inductors.

Solid-State Tester. Inductors show an ellipse and appear similar to capacitors.

VOM/DMM. The VOM and the DMM can give a very rough indication of the condition of an inductor by measuring the resistance of the winding. Comparison with a known good, identical inductor can give an indication of shorted turns, if the number of turns shorted out is many, or if the inductor wire size is small, with a resulting high DC resistance. The best test for low-frequency inductors is to measure the AC across the inductors in a working circuit for good and for suspected inductors. An inductor with shorted turns will have a lower AC voltage than a good inductor. Suspected bad radio-frequency inductors should simply be replaced for comparison of the results.

Very-large-power inductors will take time to build up the magnetic field when tested with an ohmmeter, and therefore will test the opposite way that a capacitor does—with a momentary pause, then the VOM needle goes upscale to a lower resistance reading. *Be careful* when removing the test leads—the inductor may have built up a considerable amount of energy, and there may be enough to shock you if your fingers are across the test leads (your fingers shouldn't be on them anyway!).

Remember that inductor failures are usually secondary, having been caused by the same factors that cause resistors to fail. Some other, primary failure is causing too much current through the inductor.

Testing Batteries

Battery voltage is the universal indicator of condition, but by itself it is almost meaningless. Of course, if a 9-V battery measures 5 V, it is *really* bad. But if a battery measures about the right voltage, it *doesn't* necessarily mean that the battery is good. This can be very misleading to the uninitiated. An important point to remember: *Always test a battery while under load.* Only by putting a load on the battery can its true condition be judged. Since most cells deteriorate by increasing their internal resistance, it is this resistance that must be indirectly checked. Only by passing a reasonable current through the cell can the voltage available at the terminals be meaningful.

Table 17-3 shows some test loads (as recommended by Eveready-Union Carbide) and the minimum acceptable voltages across the loads for common small-battery types.

Testing Transformers

A quick check for open windings or shorts between windings on any kind of transformer may be made using an ohmmeter, either a VOM or DMM. Shorted turns are more difficult to detect.

Power transformers can be tested for shorted turns by actually applying the proper voltage at the proper frequency to the primary winding and measuring the output voltage of each secondary. Caution must be exercised on transformers having secondaries producing high voltages. Shorted turns on the primary

TABLE 17-3 Test Loads and Minimum Voltages for Common Batteries

BATTERY TYPE	LOAD (R)	MINIMUM VOLTAGE
Cylindrical Cells and Batteries		
1.4-V hearing aid	60	1.2
1.5-V carbon-zinc (general use cell)	10	1.1
1.5-V alkaline	1	1.1
1.2-V ni-cad	1	1.1
6-V camera battery	1.2 K	4.4
6-V lantern battery (dry-cell type)	40 Ω	4.4
9-V carbon-zinc	250	6.6
9-V alkaline	15	6.6
Miniature Cells		
1.4-V mercury†		
watch, calculator		
high drain	200	1.0
low drain	3 K	1.0
camera	300	1.1
hearing aid	1 K	1.2
1.5-V general use manganese	200	1.0
1.5-V silver		
watch, calculator		
high drain	200	1.3
low drain	3 K	1.3
camera	300	1.1
hearing aid	1 K	1.2
1.5-V carbon-zinc	200	1.0
1.5-V lithium-iron	3 K	1.0
3.0-V lithium-manganese	600	2.0

Information courtesy Eveready-Union Carbide.
†Caution: watch polarity as it seems reversed from other cells.

could also cause unusually high voltages to appear on the secondaries for a short time until the transformer overheated. A power transformer designed for continuous use should *not* get hot if operated without a load. If it does, it is an indication of shorted turns.

Transformers usually fail due to excessive current flow through them. This implies that associated loads should be carefully checked for shorts or overloads.

Installations using portable AC generators may cause transformer failure if the proper frequency is not maintained. Some 60-Hz transformers will overheat and burn out with 55 Hz or less, depending on the amount of load. A combination of full load and a few cycles low in frequency can, over a relatively short period of time, burn up a transformer.

Audio transformers should be tested with an ohmmeter for open windings and for windings shorted to each other. Substitution is the next best method to use if no defect is shown by the resistance readings, and yet shorted turns are suspected.

RF transformers are first tested with an ohmmeter. If no problem is evident with an ohmmeter, check by substitution. If the transformer is tunable, it may be checked by verifying that it tunes properly. A disconnected or defective capacitor across an RF inductor or transformer can cause the transformer to seem bad because it won't tune properly.

Radio-frequency inductors and transformers are sometimes fitted with adjusting cores or "slugs." These slugs are made of a material that is very easily cracked. If a slug is split in such a way that it is jamming and will not come out with normal means, it may be that the whole transformer will have to be removed from the board and replaced with a new one. If the application is not too critical, however, it is sometimes possible to remove these broken slugs by one of two methods. The first is to break out the old slug by chipping off successive bits, thereby either taking out the entire slug or tapering it in such a way that a pair of hemostats might be able to grasp the remaining piece. A more drastic method is to provide support for the coil form and force the slug out without turning it at all. This method may ruin the component, but occasionally the old slug can be pushed out without too much damage. If need be, the technician might have to carefully drill a hole in the PC board to admit a pushing tool from the bottom of the board. If the old slug will come out without severe damage to the coil form, a single small piece of rubber band should be placed entirely through the form and a new slug inserted. The piece of rubber band should be sufficient to take up any slack caused by forcing out the old slug. When adjusting the new slug, avoid pressing down on it as much as possible.

Testing Quartz Crystals

Quartz crystals are difficult to test outside of the circuit for which they are designed. Differences in circuit capacity of only a few picofarads may make the difference as to whether the crystal will oscillate or not. Substitution and then frequency testing is the only sure method of determining whether a crystal is good or not. The frequency counter is the best instrument to use for checking frequency.

It is noteworthy that some circuits provide a tunable coil or capacitor which *must* be at or above a certain critical value for the crystal to oscillate. Before replacing a crystal (which itself very rarely fails, since it doesn't "wear out"), be sure that the circuit is properly adjusted (swing any associated adjustment back and forth to see if the circuit begins operating) and that the circuit is not overloaded by the following stage to the point where oscillations cannot be sustained.

Testing Switches

VOM/DMM. The ohmmeter should show no resistance on the lowest scale when the switch contacts are closed, and should show an open otherwise. Any heating of the contacts either points to an overload of the switch or indicates that the switch contacts are dirty and have developed too much resistance for the normal current being handled.

Solid-State Tester. The Huntron will show the same symptoms as the VOM: open or shorted, depending on the position of the switch. Any range may be used.

Testing Relays

Relays often have very critical mechanical adjustments. An intermittent or otherwise troublesome relay might possibly be fixed if the technician can see iron filings between the pole piece and the armature. Relays seem to collect these little troublemakers in that particular spot. The iron filings will stand on end in the magnetic field and prevent the armature from coming down close enough to the pole piece for the relay contacts to close. Thus, an intermittent relay. (Of course, the filings sometimes stand on end just right, and sometimes not.) Cleaning or blowing out the relay can clear this problem.

Relays will actuate, or "pull in," at a voltage considerably below their rated voltage. That voltage can then be reduced by a large percentage before the relay will deenergize, or "drop out." The difference between the two voltages ensures that once the relay pulls in, it will stay in until the coil voltage drops to a much lower value. As an example, a 12-V relay might pull in at about 8 V, and stay pulled until the input voltage drops to perhaps 4 V. (These are representative but realistic values to illustrate the point and are not firm statements. Pull-in and dropout values for relays vary greatly.) If a relay will not actuate properly, but will hold in if its armature is pressed by hand, this indicates that too low a voltage is being applied to the relay coil.

Other relay problems are best cured by replacing the relay. The old relay can be tested for an open coil, but relays seldom fail for this reason. Contact alignment and poor electrical contact are the real problems. Attempts to adjust any but the simplest of contacts is a waste of time because it is so difficult to get a bent contact back into precise alignment, sometimes within a thousandth of an inch or two as might be required for small or multiple-contact relays.

Testing Fuses

Fuses were thoroughly discussed in Chapter 6. Routine testing of fuses is done with a VOM or DMM. Look for a good low-resistance element in the fuse of less than an ohm. A few very-low-current fuses may have slightly higher resistance.

Testing Meter Movements

Meter movements are very sensitive and fragile. The basic d'Arsonval meter movement may be tested on the *highest range* (RX10,000) resistance scale of a

VOM. The actual resistance of the movement will be very low, perhaps 50 ohms, but do not attempt to read the resistance with a VOM. Use the high range only to indicate continuity of the meter coil and to verify that the meter needle swings upscale freely. One or perhaps two ohmmeter ranges down (RX1000, RX100) *might* be used, but take care not to make the meter needle strike the upper peg. Testing a meter movement on the lowest resistance ranges of a VOM can easily bend the meter needle and ruin the movement. A DMM can usually be used to check continuity of a meter movement without danger of damaging the meter.

Testing Tubes

When vacuum tubes were in their prime, the tube tester was often viewed by nontechnicians as a final authority. It wasn't. It could be used to give some indication of a tube's condition, but it often failed to detect a tube that was not doing its job. The only reliable test to see if a tube is bad is to replace it with a new one, or at least with a used tube known to be good.

FINAL PRECAUTIONS

The new part to be installed should also be checked to be sure it is good, if the test doesn't take an unreasonable length of time. Resistors, capacitors, and transistors fall into this quick-check category. ICs and other multipin components take a while to check, and the checks would probably not be that conclusive anyway. In these cases, "toothpicking" as shown in Figure 17-9 might be advisable before soldering the IC into the circuit. (Be sure to put the IC into the circuit the proper way!)

SUMMARY

This chapter has covered the removal of the suspected part, the verification and analysis of the failure, and the replacement with a new component. If a replacement for the original part is not available, refer to Chapter 22. Chapter 18 will discuss how to be sure the repair is indeed complete.

REVIEW QUESTIONS

1. Why is it better to cut component leads deep in a pair of wire-cutter jaws rather than near the tips?
2. What is the main advantage of a temperature-controlled soldering iron over one that runs "wide open" all the time?
3. Name two methods of removing solder from printed circuit boards.
4. What is the purpose of "toothpicking" an IC into a circuit rather than just soldering it in?

5. How many resistance readings must be taken with an ohmmeter to determine if a bipolar transistor is good?
6. How many readings must be taken with a solid-state tester to determine if a bipolar transistor is good?
7. Why not use the RX1 scale of an ohmmeter to test small transistor and diode junctions?
8. What two components are inside an optical isolator?
9. What is the internal electrical structure of a UJT?
10. As far as passive testing with an appropriate ohmmeter is concerned, the LED should look like what other common electronic component?
11. Can neon lamps be tested with the common ohmmeter?
12. What instrument is best for testing ICs and for comparing them to good ICs?
13. Is the solid-state tester recommended for testing resistor values?
14. What is the basic pattern expected on a solid-state tester when checking large-value capacitors?
15. What will the needle of a VOM do when testing large-value capacitors?
16. When is it necessary to "form" electrolytics before using them?
17. Why is "spitting" into a microphone to test it discouraged?
18. What might a cone-pushing test reveal about a speaker?
19. What possible fault of an inductor may be undetectable with an ohmmeter?
20. What is the most important thing to remember when testing batteries with an voltmeter?
21. Which is more harmful to a power transformer, a high or low line *frequency*, with the voltage held constant?
22. Would a resistance reading of 25 ohms be acceptable for a closed switch?
23. Generally speaking, is it worthwhile to attempt adjusting bent contacts on a multicontact relay?

chapter eighteen

Final Inspection and Return to Service

CHAPTER OVERVIEW

This chapter covers the alignment, reassembly of the equipment, and final testing. Paperwork that may be required is also discussed.

ALIGNING EQUIPMENT

The alignment of equipment is the process of adjusting DC circuits and tuning AC and RF circuits so that the equipment meets the manufacturer's original specifications. As examples, a DC amplifier might be aligned with internal adjustments so that it provides the original voltage gain of exactly 5.00 times the input signal, or a transmitter might be tuned to provide the specified 20 watts of RF output that it originally produced.

The final checkout phase of a repair will sometimes reveal a problem that can be corrected by aligning appropriate circuits. This is the case when the operator has tampered with the internal adjustments of the equipment. A touch-up alignment is indicated when the equipment almost meets specifications and particularly if the equipment has also seen many hours of service. Alignment of at least one or two stages is required when the technician has had to replace certain critical components such as tunable IF and RF transformers, crystals, or most components in calibrated analog circuits, such as voltmeter circuits or oscilloscope amplifiers.

It is important to have the proper alignment instructions for the equipment being adjusted. There is no way to look at an adjustment and be certain if it is one that is peaked, dipped, or a band-pass amplifier with its own special tuning requirements. Improper handling of some of the more critical adjustments may result in complete inoperation of the equipment, operation at improper frequencies, or even damage to the equipment because of excessive current flow. In summary, know what the adjustment is for before you change it.

A small alignment tool is available that allows temporary changing of a tuning adjustment without turning the inductor core. Sometimes it is desirable to know if changing the adjustment will produce the desired end effect, without having to change the adjustment physically. There is an alignment tool that has a small piece of powdered iron in one end and a piece of brass in the other. Placing this tool into a coil form will increase the inductance when the powdered iron end is inserted, and decrease it when the brass end is inserted. Of course, the coil form has to be large enough to allow inserting the tool. Many coil forms are too small to use this method.

Generally speaking, alignment is similar to signal tracing: The proper signal is introduced into the input, and the load instrument is connected to the output. Whereas in signal tracing the output is observed for a go or no-go indication, alignment consists of making adjustments to the circuit while the signal is passing through it. The adjustments are made for the desired changes in the output indication.

Source Instruments

Source instruments used in alignment are usually existing signals from the equipment itself or external signal generators. In the case of band-pass circuits, a special signal generator called a *sweep generator* is used. This will be covered in detail later in this chapter.

When using a signal generator, it is sometimes difficult to get an exact frequency reading using the dial on the instrument. Alignment often requires precise, resettable frequencies. In these cases, it may be a good idea to connect a frequency counter to the signal generator output to obtain an exact frequency indication. Turn up the output of the signal generator until the frequency counter has a stable count and adjust the generator to the desired frequency. Once the frequency is determined, the output of the signal generator may then be reduced to the proper value for the alignment test. Signal generators of marginal quality will change in frequency when the output attenuator is turned near maximum output settings, so use of a frequency counter may not be advisable in this case.

Load Instruments

Load instruments used in alignment are generally an analog meter (VOM, FETVM, VTVM) and an oscilloscope. The meter is usually connected to a point in the equipment that has a varying DC potential, dependent upon the adjustment in question. In receivers, this monitoring point is usually the AGC (automatic gain control) voltage bus in an AM (amplitude-modulated) circuit or a limiter stage in an FM (frequency-modulated) circuit. The oscilloscope may be used to directly observe the RF changes produced when a stage is adjusted, therefore the instrument should be connected to the output of the stage *following* the stage being tuned. The indicating instrument should always be at least one stage removed from the one on which adjustments are to be made because of the detuning effect of the instrument and the cable going to it.

If connecting the indicating instrument to the tuned stage is necessary for some reason, connect it to a low-impedance point where the loading and detuning effect is minimal, such as the emitter-resistor of an RF amplifier stage. This is a low-impedance point that is not affected by the attachment of an oscilloscope probe. In this case, tuning is a matter of peaking the DC level at the emitter.

When making any tuning adjustments, it is preferable to use a detecting instrument that will give a visual, analog readout. A VOM needle is a good example of such a readout. Sometimes it is tempting to tune a receiver by using the speaker, listening for changes in the strength of the receiver signal from the speaker as a tuning indication. Although this will work to some extent, the human ear is not sensitive to very small variations in amplitude and thus does not make a good "peaking" indicator.

The strength of nearby (strong) RF signals that are already radiated into space (like those of a transmitter) may be monitored very nicely with a field-strength meter. This is a meter to which is connected a detector diode (RF rectifier) and sometimes a simple LC tuning circuit (Figure 18-1). If the radiated signal is not quite strong enough for the field-strength meter to be used, a communications receiver may be used to detect the presence of weaker signals. If the communications receiver also has a signal-strength meter, that meter may be put to good use as an indicator of relative strength. This might work well in an application such as tuning up intermediate stages of a transmitter, if the receiver is capable of receiving the intermediate frequencies being produced.

There is one variable to consider in measuring the strength of signals produced within a shop area. Such signals are often influenced by the physical position of the technician around the equipment and especially of the hands. So, if you are monitoring the strength of a signal and tuning for maximum or minimum, be sure to keep the position of your body and hands as constant as you can during the adjustment procedure.

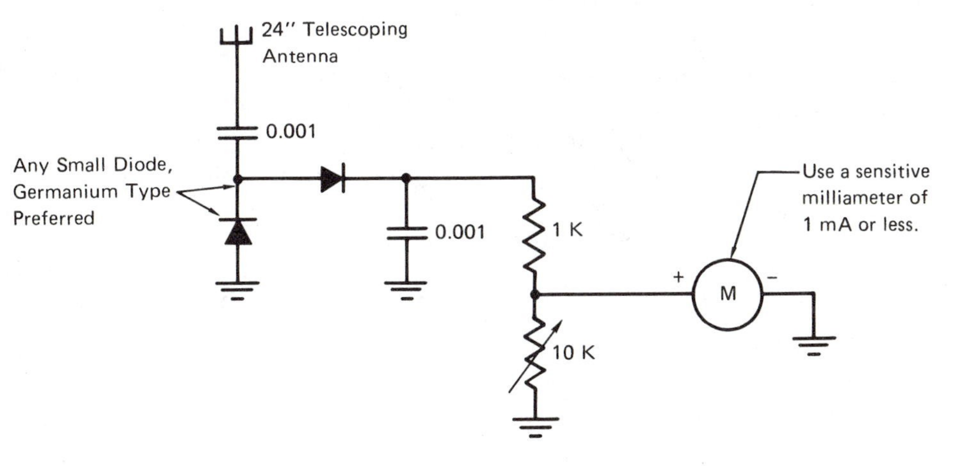

FIGURE 18-1 A voltage-doubling field-strength meter.

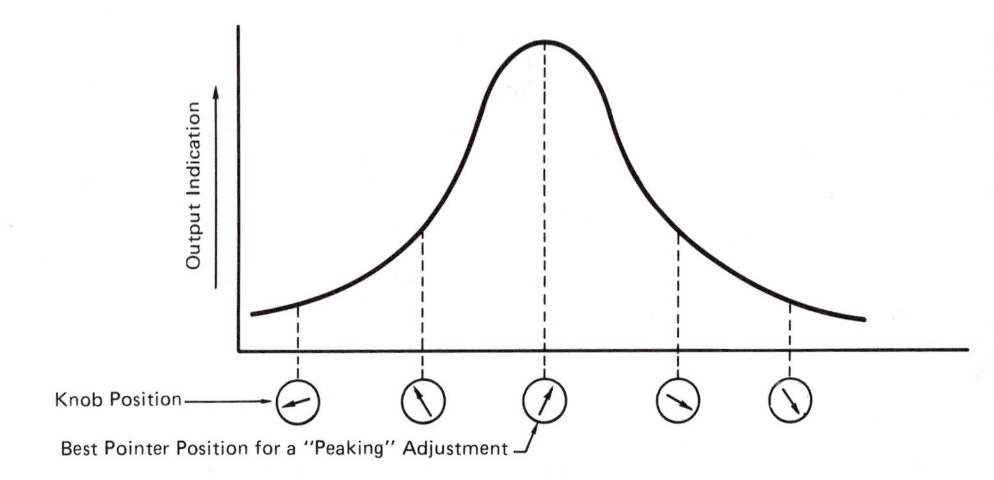

FIGURE 18-2 Adjustment of a peaking control.

Peaking Alignment

Peaking alignment is a procedure whereby an input signal is applied to a circuit much in the manner used in signal tracing. The output is also monitored in a similar way. Adjustments are made within the equipment between the input and output points. The adjustments are then made to get the maximum reading on the output circuit. This procedure progresses until all of the adjustments within the equipment are made. In technical jargon, the equipment has been "peaked up."

Existing sources are often used for peaking signals, such as the oscillator of a transmitter. Many adjustments within a transmitter are simply peaked for maximum power output to the antenna circuit. An important point to be made here is that when peaking adjustments for maximum output, *keep the input signal low.* Overdriving an input circuit produces an effect of broadening the adjustment so that the point of best adjustment is not definite. By keeping the input signal low enough that the control is responsive, the best point is easily sensed while swinging the control back and forth (Figure 18-2). When making adjustments such as these, the technician should swing the control back and forth in a wide range to get the feel of the adjustment. Small, jerking adjustments cannot give the proper feel for the best adjustment at which to stop.

An analog meter or oscilloscope is preferable to using a digital multimeter as an output indicator. The interpretation of rapidly changing numbers is more difficult than watching the smooth swing of a needle or an expanding pattern on an oscilloscope.

Dipping Alignment

Dipping adjustments are made in the same way as the peaking adjustments just covered, but the output is tuned for minimum indication (Figure 18-3).

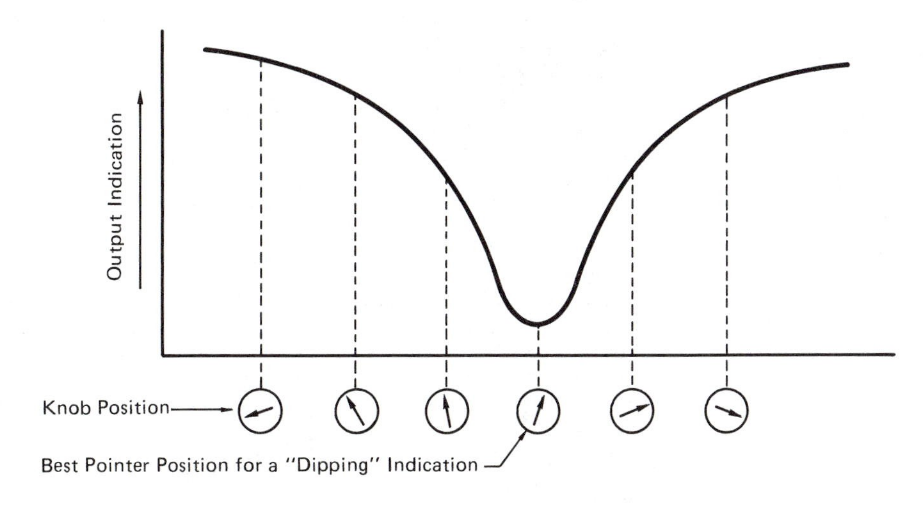

FIGURE 18-3 Adjustment of a dipping control.

The Nulling Adjustment

Nulling a voltage means to take it to zero. Such adjustments can be overdone to produce negative voltages, so nulling is rather like "tuning out" all voltages entirely. This is a common adjustment in DC circuits.

Band-pass Amplifier Alignment

The band-pass amplifier makes use of the overcoupling characteristics of a transformer to increase the band of frequencies that the amplifier will amplify. It is a common circuit in television and radar receivers. Alignment of these circuits must be done using a signal generator that changes frequency constantly over the band of frequencies used in the circuit. The response at the various frequencies is shown on an oscilloscope and adjustments made to respond properly. Proper response is a matter of consulting the manufacturer's alignment instructions. Figure 18-4 shows such an alignment setup. In order to observe the output waveform, the fluctuating amplitude of the RF waveform must be rectified and filtered. This is a form of AM detection and is accomplished in a special probe designed for this use. It is called a *demodulator probe* (Figure 18-5).

Again, it is important to keep the input signal low enough to get a good response curve. The alignment instructions will often specify the amount of input signal to apply.

REASSEMBLING THE EQUIPMENT

There isn't much to putting things back together if you're the person who took them apart. If someone else took the equipment apart some time ago, don't be surprised if the bolts are not there or other small hardware is missing. It often

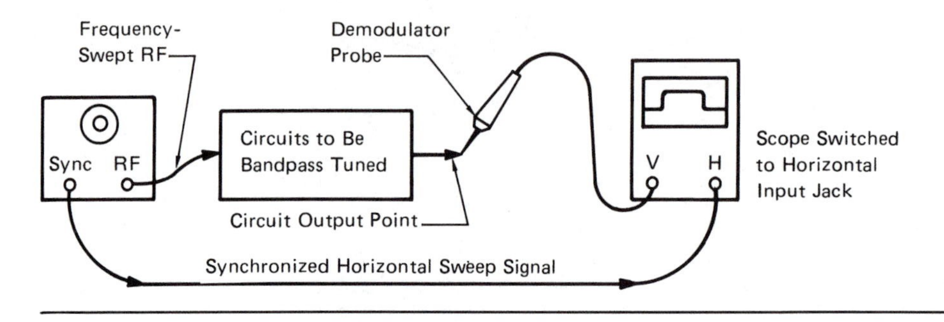

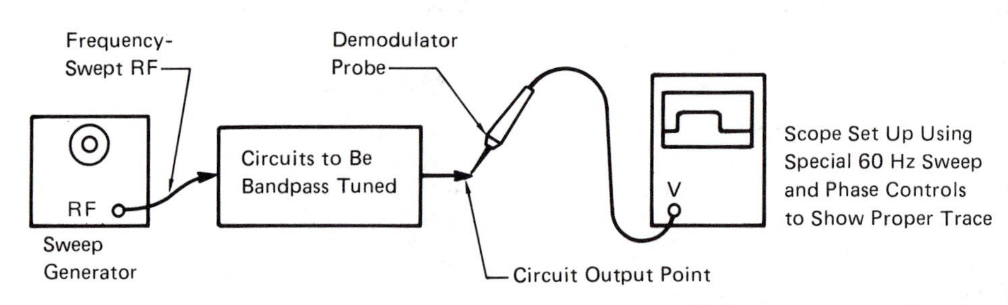

FIGURE 18-4 Two methods to set up band-pass alignment instruments.

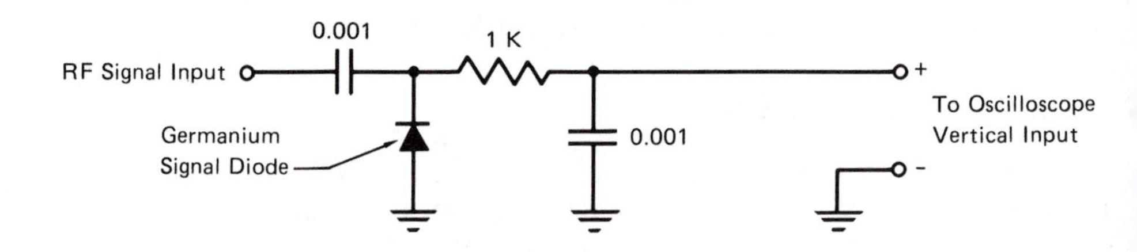

FIGURE 18-5 Schematic of an oscilloscope demodulator probe.

happens. Beware of putting long bolts into holes not intended for them. This can sometimes cause problems if the bolt is long enough to contact parts of the circuit board. Also beware of pinching wiring when tightening cover screws.

A professional touch that can impress the customer very favorably is to give the front panel and the knobs a bath. When equipment is returned looking strikingly cleaner than when it came into the shop it helps relations immensely. Common household cleaners will work fine, a favorite being Formula 409. The grooves on most knobs will collect finger oils and dirt. This is cleaned off by removing the knob, saturating it with the cleaner, and sweeping out the grooves with a toothbrush.

Another, often overlooked, touch is to apply a *very little* lubrication to appropriate places—just a half-drop on the control shafts where they go through the front panel, for instance. The new "feel" of the control is impressive to the operator used to squeaky operation of the knob.

THE PRACTICAL RF DECIBEL FOR TECHNICIANS

Decibels involve logarithms. A technician in the field working on a transmitter may need to convert RF power ratios into decibels, but there certainly won't be a table of logarithms handy. The technician probably won't even have a calculator. With my luck, I wouldn't even have a pencil!

Decibels are really quite easy to juggle in your head! If two powers have a 2:1 relationship, they are 3 dB apart (actually, 3.01 dB). The larger power is 3 dB stronger than the weaker. The weaker is *down* 3 dB, or −3 dB, from the stronger. One more rule: If two powers are related by 10:1, they are 10 dB apart. The stronger is *up* 10 dB from the weaker, and the weaker is *down* 10 dB (−10 dB) from the stronger. With these simple facts in mind, let's take an example and see how these rules enable us to do dBs in our heads.

Suppose we want to find what power is 16 dB above a watt. The first 10 dB is an increase of ten times, so our power is now 10 watts rather than the original 1 watt. That leaves 6 dB to account for. Take 3 dB of that 6, and double the 10 watts to 20. For the remaining 3 dB double again, and the final power is 40 watts. To recap, ten plus three plus three accounts for the full 16 dB, and the power was one watt times ten, times two, and times two.

Let's do the same problem using 17 dB instead of 16 dB. This is even simpler: Take the one watt times ten to account for the first 10 dB. That gives 10 watts. Again take the power up ten times, as though the dB figure were 20 dB. That would give 100 watts, but we need to *back up* 3 dB to get the 17. So just *halve* (remember, *down* 3 dB is half) to get the final power of 50 watts.

This may be a bit confusing at first, but after juggling a few mental problems, it becomes very easy to work out decibel problems to within a decibel, at worst, of the final figure.

FINAL TESTS

It is necessary to conduct final tests on equipment that has been repaired in order to ensure that all faults have been corrected. The sections that follow give examples of common equipment and the specific items that should be checked before returning to service. For any equipment, the manufacturer's specifications must be met before the equipment will give proper service.

Transmitter Checkouts

Frequency. Before it leaves the shop, every transmitter should be checked for proper frequency of operation with a frequency counter. Remember to terminate the transmitter in a proper (52-ohm) load. Use whatever attenuation is required to prevent overloading the front end of the counter with the high power available from some transmitters.

The frequency counter, incidentally, can be very easily checked for calibration by comparing its internal crystal frequency standard with that of the Na-

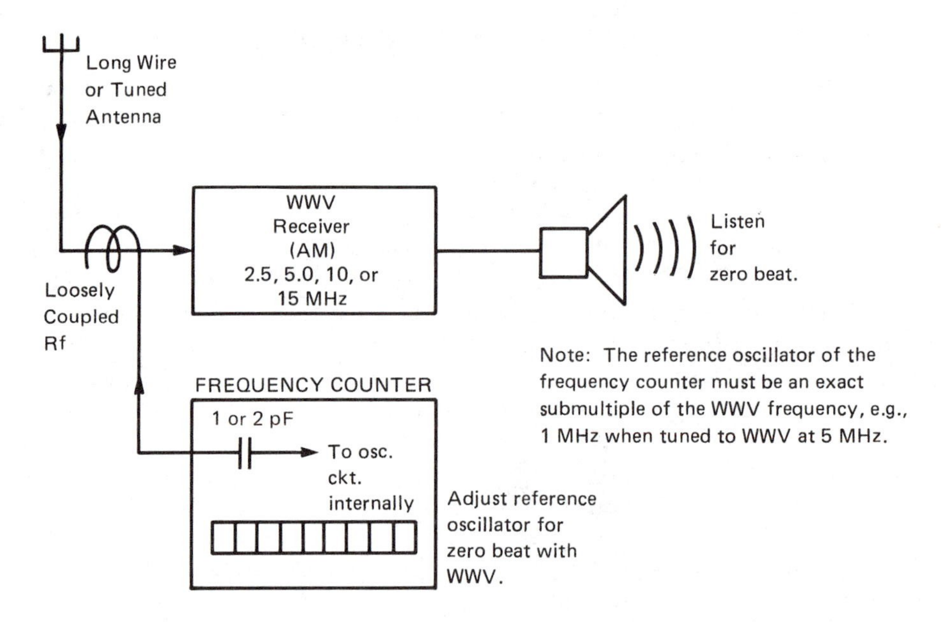

FIGURE 18-6 Calibrating the frequency counter with WWV.

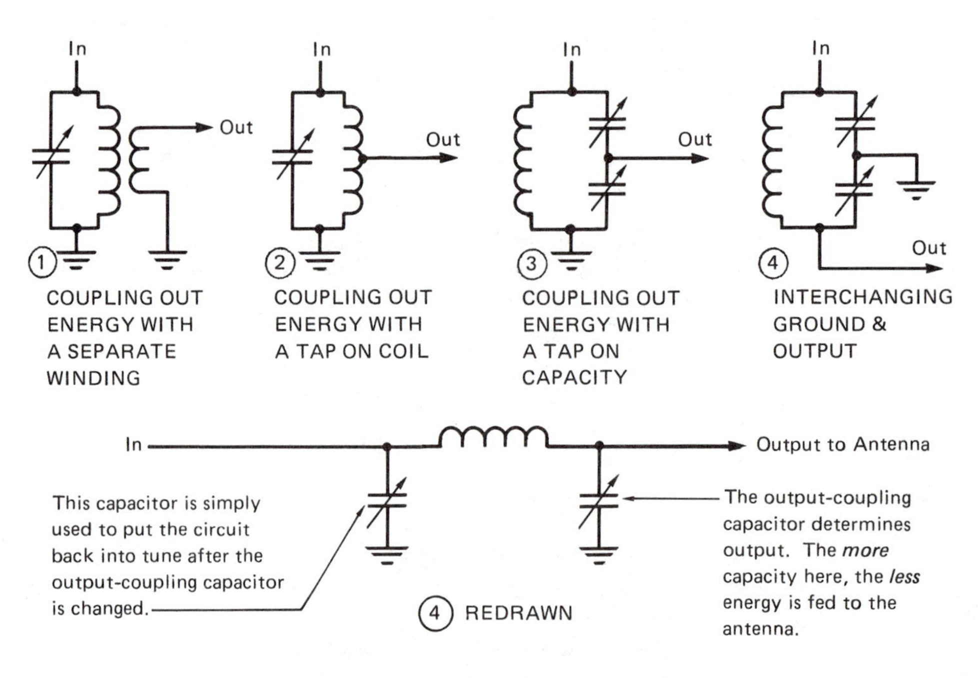

FIGURE 18-7 How the pi-network circuit is derived.

tional Bureau of Standards' transmitting station WWV in Boulder, Colorado. If the crystal used as a standard in the frequency counter uses a 1-MHz or 10-MHz standard oscillator, a beat should be heard if WWV is tuned in and a sample of the frequency counter standard signal is also introduced into the antenna circuit. A very loose coupling, often a few inches of bare wire in the room, is sufficient to obtain a beat. Since WWV broadcasts simultaneously on 5 MHz, 10 MHz and 15 MHz, the frequency providing the loudest beat can be used. The frequency standard can then be zeroed in with WWV, becoming a "secondary standard" for the shop (Figure 18-6).

Power Output. The power output of the transmitter should also be checked. The usual way is to provide either a terminating wattmeter (one with a termination of 50 ohms built in) or a through-line wattmeter and separate dummy load.

Some older transmitters required tuning of several stages from the front panel of the equipment. The general rule was that the input (also called grid, excitation, or drive control) was to be peaked. This was usually monitored by watching a front-panel meter. The output control (called plate, output, or final control) was tuned for a dip or decrease in the same meter that was previously peaked. One of the more difficult adjustments to be made on these transmitters was to tune the output stage. This entailed manipulating two controls, sometimes three. The most difficult pair was the output tuning and the loading controls. Many equipment operators never really understood what these controls did, and why. See Figure 18-7 for an explanation of the two controls, and why they act as they do.

Single-sideband transmitters are tuned by temporarily introducing just enough RF feed around the balanced modulator to provide good indications of peak-and-dip, as previously described. Power-output measurement of an SSB transmitter requires a different method than the simple reading of a steady-state RF output. In order to provide a check of the linearity of the modulating and RF systems of an SSB transmitter, a two-tone test is used to evaluate the power output. A special oscillator produces two audio tones of 900 Hz and 1300 Hz

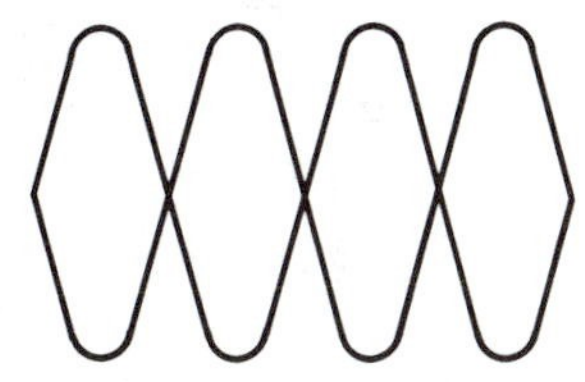

NORMAL RF ENVELOPE
OF TWO-TONE TEST

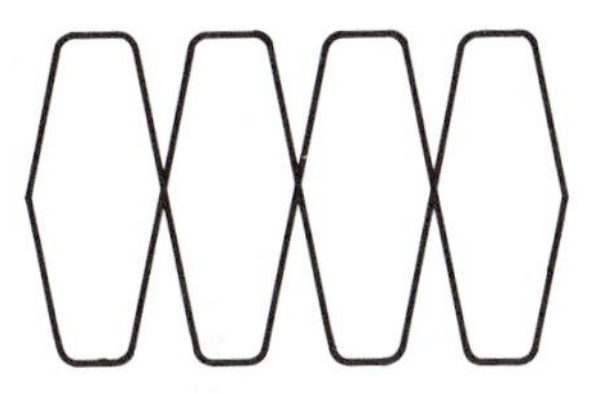

PEAK FLATTENING DUE TO
INSUFFICIENT ANTENNA
COUPLING, TOO LITTLE RF
DRIVE INTO THE AMPLIFIER
OR POOR REGULATION OF
THE DRIVER STAGE

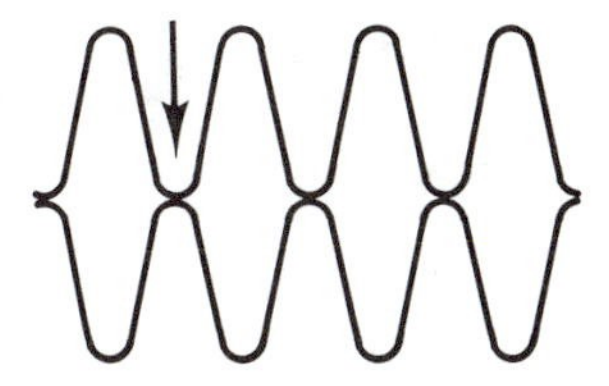

POOR CROSSOVER POINTS
INDICATE PROBABLE
MISADJUSTMENT OF THE
QUIESCENT (IDLING) BIAS
ON THE OUTPUT STAGE

FIGURE 18-8 The meaning of two-tone linearity tests on single-sideband transmitters.

simultaneously, which are applied to the modulator input. The transmitter output will look like one of the outlines in Figure 18-8.

The peak-envelope power of a signal produced by a transmitter can be calculated from the two-tone transmitted output as shown in Figure 18-8. This is done by determining the circuit output impedance (usually 52 ohms), and noting the peak-envelope voltage on a calibrated oscilloscope (from the midpoint, or zero crossover point, to either positive or negative peak value), which value is multiplied by 0.707 to obtain the rms voltage. Then it is simply a matter of squaring this voltage and dividing by the value of the resistance. Here's the formula:

$$PEP = \frac{(Erms)^2}{R},$$

where PEP is peak-envelope power, Erms is rms voltage, and R is resistance of the circuit.

The power input to an SSB transmitter will be between 1.4 and 2 times this output power value, depending greatly upon the efficiency of the power-amplifier stage. A typical value that may be used often is 1.57 times the output power equals the input power.

Another check to be made on a transmitter should be a check of the spectrum that is transmitted. The instrument used is called a spectrum analyzer. This is particularly important with today's crowded frequency spectrum. The instrument to be used will have instructions as to how much signal may be applied to the front-panel connector without damage. Normally, an attenuator must be used to prevent overloading the spectrum analyzer. Analyzer checks are extremely important during and after aligning VHF and UHF transmitters. Some of these circuits are highly critical: A fraction of a turn on some of the interstage adjustments can cause the production of drastic interference signals from the transmitter.

Modulation Tests. A check should also be made of the amount of transmitter modulation. The AM modulation on lower frequencies (under about 30 MHz) can be made with the use of a high-frequency oscilloscope. A 30-MHz oscilloscope would do this job very nicely. Figure 18-9 shows proper and improper indications of 100-percent AM modulation.

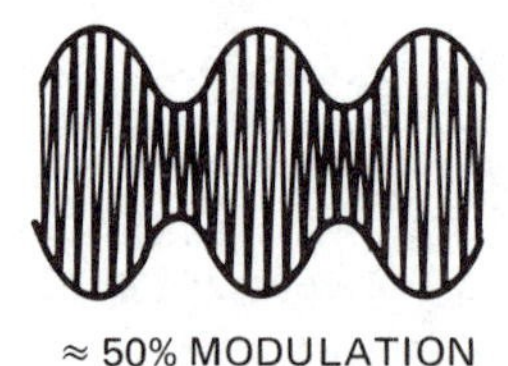

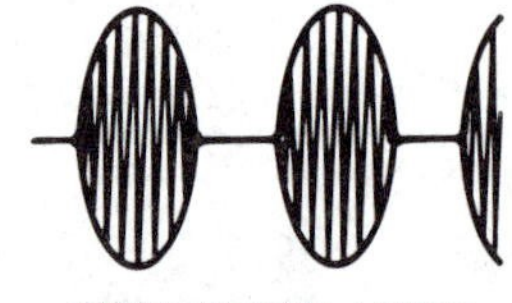

FIGURE 18-9 Amplitude-modulating percentages.

FM transmitters will require a special instrument, a deviation meter, to measure the amount of frequency modulation. The calibration methods vary with the type of meter used, so the deviation meter instruction manual must be consulted for setup procedures.

Receiver Checkouts

Sensitivity. Every receiver leaving the repair shop should be checked for sensitivity. FM communications receivers often use the 20-dB signal+noise-to-noise test. Checking an FM communciations receiver is quite easy. Monitor the speaker output with an AC voltmeter. Turn off the squelch (let the speaker roar), and at a conveniently loud setting of the volume control, note the audio voltage at the speaker. Now increase the RF signal from the signal generator, at exactly the right frequency. The noise from the speaker should reduce greatly. When the audio voltage is 1/10 of original, read the amount of RF entering the receiver. This is the sensitivity of the receiver.

Frequency. The FM communications receiver must be set exactly on the proper channel for optimum sensitivity and range. This is often determined by monitoring the DC output of the discriminator while supplying a precise, on-channel signal, preferably from a crystal-controlled oscillator known to be exactly correct or from a synthesized signal generator in proper calibration. The frequency vernier adjustment of the receiver will probably be a small trimmer capacitor or inductor that changes the frequency of the crystal-controlled local oscillator, or a single adjustment to be made in a frequency synthesizer. The equipment instruction book should be quite specific as to how this adjustment should be made.

There are few high-frequency AM communications these days, having been replaced by SSB receivers. SSB receivers must be set exactly on frequency for the received signals to be natural sounding. Off-frequency operation of an SSB receiver (or a poorly adjusted transmitter) will result in abnormally high- or low-pitched signals. If far enough off frequency, the signals can be totally unintelligible.

Tunable receiver circuits may require alignment at both ends of the tuning dial. When this is necessary, the general rule to put the dials back into calibration is to adjust the inductance at the low-frequency end of the dial and adjust the capacitance of the circuit at the high-frequency end.

Audio Amplifier Checkouts

Audio amplifiers may be checked for distortion. In noncritical applications, a simple test by ear may suffice. In critical applications, measurement of the total distortion may be required. In this case, a sampling of pure tones across the audio spectrum is applied to the amplifier input, one at a time, and the output analyzed with a distortion analyzer for compliance with manufacturer's requirements. When these measurements are required, specific sets of criteria must be provided, usually by the manufacturer.

Audio sensitivity is another measurement that may be required in some applications. Again, specific test criteria must be provided for the technician to verify proper sensitivity.

Other electronic equipment will have similar requirements as those given above. With any equipment, the manufacturer should state the minimum acceptable performance, and with the proper test procedures provided by the manufacturer, the technician will be able to learn the test procedures using the test equipment at hand.

Printed Circuit Tests

Printed circuit boards are seldom complete in themselves. They are usually part of a larger system or equipment. Because of this, they usually have many connections for inputs and outputs. This makes their checkout considerably different from that of a complete unit.

There are several ways to check a PCB for proper operation. The first is to insert the card into the equipment of which it is a part and verify that it performs its proper functions. In some cases a technician may have to repair a card and send it back to the customer to insert and test in a working system. In this case, the technician should attach a note to the card saying something like *"Test this card before using as a spare."*

The second way to check circuit cards is to make a mockup, the external circuitry necessary to test the circuits on the card. This will require connectors to mate with the card, miscellaneous wiring, and signal-input sources. Monitoring of the output of the card will probably be done with oscilloscopes, lights, or other appropriate monitoring indicators. This approach for testing PCBs is recommended only if there are enough boards being repaired to make the time spent in making the mockup worthwhile.

THE PAPERWORK

Equipment History Entries

Some companies require that detailed equipment histories be kept on all of their electronic inventory. In this case the technician is the only one qualified to make these entries because of his or her detailed knowledge of the failure.

Equipment histories can sometimes be used to indicate trends of failures. As an example, the failure of a particular capacitor during the summer months might suggest the need for changing to a capacitor of different rating or manufacture to prevent future failures. This sort of planning will generally be done by the shop supervisor, but it is the technician who will probably note the trend and who should call it to the attention of the proper person to evaluate the situation.

The history entries should be done neatly so that others may read them. In addition to the name or initials of the servicing technician, it may be required that the technician place in the record his or her FCC license number. Although not required by the majority of employers, it is a good idea for the technician to keep his personal log of equipment worked on. These logs can come in very handy when questions arise weeks later in reference to a specific job. Minimum information is all that is needed, such as the date, the equipment type, the problem, and particularly serial numbers. Since it is a personal log, it might also be a good place to make other notes, such as symptoms and cures.

Billing the Customer

Local policies will indicate how to do this billing. The time spent on the repair and the numbers of the part used should be noted at the completion of each job. Perhaps a piece of paper with this information is handed to the company clerk who handles it from this point. In some cases the technician makes out a job order sheet with this information, and this document is used to bill the customer. Ask your shop supervisor how the paperwork for billing is handled.

Ordering Parts

When a repair uses parts from stock, that stock must be replaced. The procedures consist of noting the part number and quantity and contacting the supplier. The supplier can be a local "generic" parts house or the manufacturer for proprietary parts. Proprietary parts will not be available anywhere but through the manufacturer. The manufacturer either makes these parts for its equipment or procures "generic" parts and removes the identification, affixing its own. In this way the manufacturer will get your return business for parts, priced at what the market will bear. Often an identical part might be obtained at a parts distributor for a fraction of the manufacturer's price, if you know the identity of the part.

In any case, the shop stock must be kept up so that repairs in the future can be made without undue delay. This inventory of parts may be kept manually or, as in some of the more progressive shops, on computer data-base programs. This latter method is better. It is sure and accurate *as long as the stock changes are given to the computer!* A single omission of information makes the inventory system that much weaker.

Shop policy should be established for the occasion when the technician wants to borrow a part for a few minutes for a quick check, then return the part to stock.

PACKING FOR SHIPMENT

Pack electronic equipment for shipment with care. Almost any packing that will cushion the equipment may be used, but one exception is the styrofoam packing materials that are now in common use. These are very static-producing and should not be used with equipment having CMOS chips. This is very important to note when packing printed circuit cards. These cards have many input and output leads coming to the edge of the card without additional protection.

Wrap the equipment and tape the wrapping in place. Place the bundle into a suitable cardboard box along with any paperwork required to be inside the box, and seal it with nylon tape. Attach the mailing label and any shipping documents to the outside of the box as indicated by shop policy. Don't forget to keep a log of the shipping of the item in case there are inquiries as to the date and place shipped.

SUMMARY

This chapter concludes the matter of repairing equipment and returning it to service. The next chapter will cover some aspects of preventing failures.

REVIEW QUESTIONS

1. What is a peaking adjustment?
2. What special instrument is required for a band-pass alignment job?
3. What is the dB difference between 10 watts and 60 watts?
4. What is the standard impedance of an RF wattmeter?
5. What is the RF frequency standard for the United States?

chapter nineteen

Routine and Preventive Maintenance

CHAPTER OVERVIEW

Twenty-five years ago preventive maintenance was very necessary. The failure rate of vacuum tubes and associated components made frequent checking of electronic equipment good sense. Component failures were in large part due to the stresses of high temperatures and high voltages used with vacuum tubes.

Today we have the semiconductor with its high reliability. Transistors and integrated circuits run on very little power and produce very little heat. The voltages used are low and components are subjected to less voltage stress. All in all, reliability has taken a great leap forward. It is not unusual for equipment to operate 24 hours a day for years with no maintenance at all.

WHY MESS WITH IT?

Working inside equipment is really quite hard on it. Wiring is pulled, circuits are probed with test leads, and the equipment is bumped around. In the very process of trying to prevent failures, failures can be caused. Serious thought should be given to doing nothing with transistorized equipment that is working. If it is performing as it should, seriously consider leaving it alone.

On the other hand, some equipment *must* work as reliably as possible. Critical equipment such as that used in aviation must undergo periodic tests to forestall any possible malfunction. The technician will be guided by shop policy in preventive maintenance decisions. When working on critical equipment, it is imperative that the technician not introduce problems into good equipment. This is avoided by handling, transporting, and working on the equipment with unusual care. Particular attention should be paid to careful opening of the equipment.

In a few cases government regulations require routine maintenance. As equipment reliability increases, regulations may change to lengthen the intervals of maintenance or perhaps eventually cease to require periodic tests at all. At this time, the Federal Aviation Administration (FAA) requires periodic testing of specific equipment used in aircraft applications.

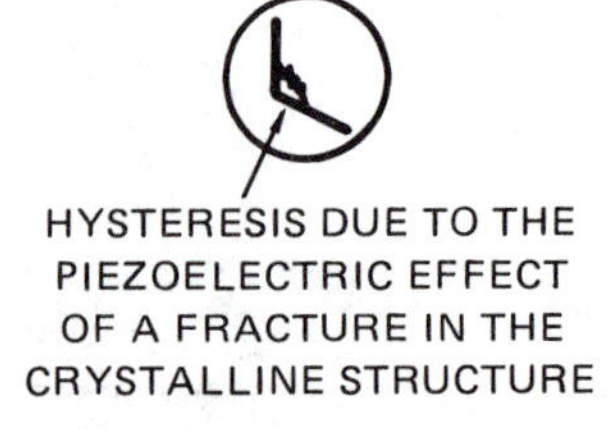

FIGURE 19-1 Using the solid-state tester to predict failures by early detection of abnormal responses.

USING THE SOLID-STATE TESTER TO PREDICT FAILURES

In those applications that require the greatest reliability, the solid-state tester has definite value as a predicting instrument. Semiconductor failures may not occur all at once. A fractured or contaminated crystal structure may function quite normally in a circuit for some time with such a defect. The solid-state tester can detect many of these failures early, sometimes before the circuit ceases to function.

Figure 19-1 shows several defective waveforms that might not be evident during the normal operation of the circuit. As Mr. Hunt of Huntron Instruments so aptly put it, "It can't get better. . . ." So the logical thing to do is to replace the component before it causes circuit failure. The use of the solid-state tester as a predictive instrument is the same as its use for troubleshooting any dead circuit board. Check the edge connectors for proper waveforms and then sector the board and check component by component as covered in Chapter 16.

TDR FOR CABLE MAINTENANCE AND TROUBLESHOOTING

TDR is an acronym for Time-Domain Reflectometry. Despite its impressive title, the principle involved is extremely simple: Pulse a cable with a voltage and watch for reflections, coming back like radar, due to discontinuities in the cable. There are two methods used to pulse a cable. Coax cables are pulsed with a square wave. Only the initial negative-to-positive transition is used to detect reflections. After a relatively long time, the voltage is dropped to zero and another positive-to-negative transition is sent down the cable to repeat the cycle (Figure 19-2). The second method is to use a very short positive pulse down cables of the twisted-pair variety.

FIGURE 19-2 Sample presentations when using time-domain reflectometry (TDR) as a troubleshooting aid on coaxial cable runs.

The advantage of TDR is that small discontinuities, even the presence of cable connectors, can be easily "seen" on the scope face. If large reflections are shown, a coax run probably requires maintenance. The time from the initial pulse to the discontinuity is shown along the horizontal scale of the CRT. This has a direct relationship to the physical distance from the instrument to the problem. Sophisticated equipment is available that will pinpoint problems out of thousands of feet of cable to within a few inches.

Besides the obvious advantage of the TDR instrument for preventive maintenance by spotting slowly deteriorating cables, the instrument can also be very useful in identifying and locating troublesome problems when they occur. This ability is extremely valuable when dealing with long cables strung in inaccessible locations, such as onboard ships or in aircraft.

If the technician has access to an oscilloscope with a frequency capability of 30 MHz or better, he or she can use with the oscilloscope the simple circuit of Figure 19-3 as an inexpensive TDR generator. It is capable of checking cables of about 6 feet or more for large discontinuities such as impedance changes from 50 ohms to 72 ohms, shorts, and opens. It is not capable of the resolution (ability to discern details) of the commercial TDR generator, but it is quite serviceable for routine use.

To use the TDR generator circuit, install it as close to the oscilloscope front panel as practical. Use external triggering and be sure that the scope doesn't have a trace with the trigger disconnected. Attach an unknown cable to the point indicated on the schematic, which should also be as close to the oscilloscope vertical input connector as practical. The scope sweep speed should be set to the fastest sweep available. The vertical amplifier should be set so that the waveform is about half the height of the screen. Typical presentations of the TDR generator are shown in Figure 19-4.

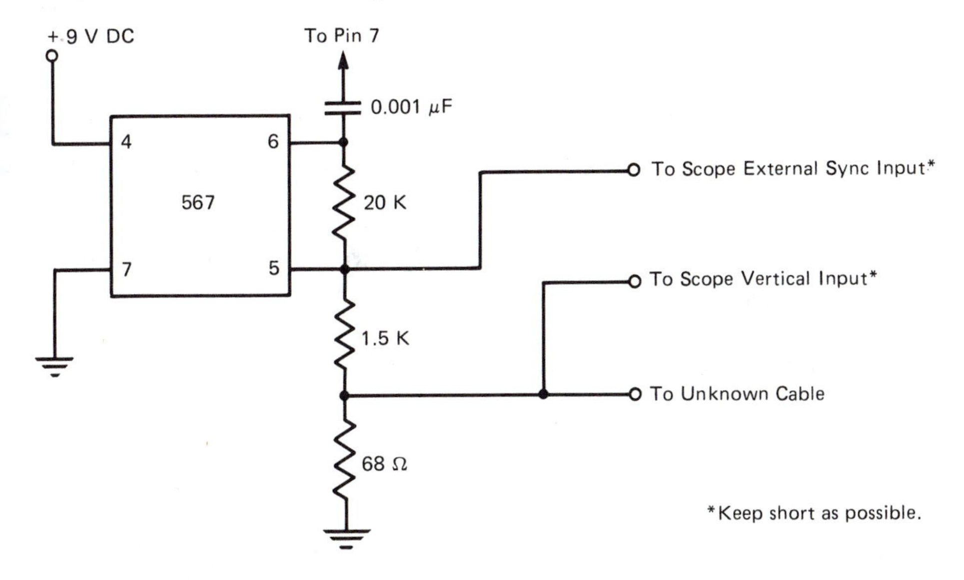

FIGURE 19-3 Circuit for inexpensive TDR generator.

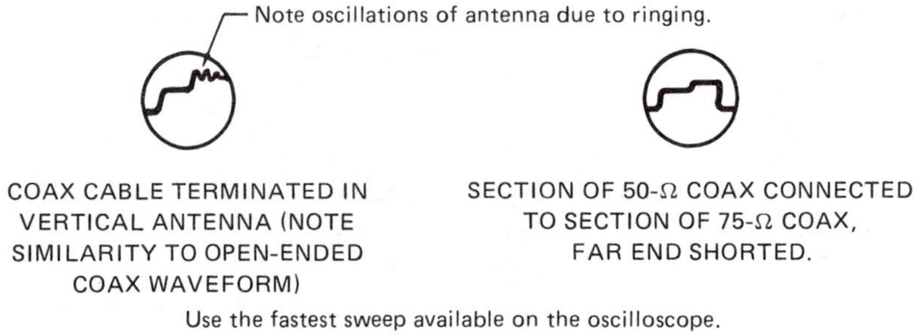

COAX CABLE TERMINATED IN VERTICAL ANTENNA (NOTE SIMILARITY TO OPEN-ENDED COAX WAVEFORM)

SECTION OF 50-Ω COAX CONNECTED TO SECTION OF 75-Ω COAX, FAR END SHORTED.

Use the fastest sweep available on the oscilloscope.

FIGURE 19-4 More TDR waveforms using circuit of Figure 19-3 and a 35-MHz oscilloscope.

MECHANICAL MAINTENANCE

There are a limited number of things that might require lubrication in electronic circuits. Blower motors might need a drop of oil now and then, but they often have sealed bearings that cannot be lubricated. When the bearing squeals, it is time to remove the motor and replace the bearings. With the proper tools and replacement bearings the job can be done with a little thought by any technician.

Front-panel controls can occasionally use a half-drop of oil where the shafts go through the panel sleeve. Controls should operate smoothly without grinding or squealing. Some equipment is also provided with drawer-type rails so that the equipment may be pulled from its cabinet without falling to the floor. These rails may be lubricated *sparingly* with heavy grease.

Cooling louvers should be kept clear of airflow obstructions. Loose paper has a tendency to flip up in front of suction louvers, cutting off airflow. Keep liquids off the tops of electronic equipment. Coffee cups are a common problem around electronics—A spill of coffee, particularly if well sugared, can make quite a mess on a circuit board.

SOLVENTS FOR CLEANING

When using any cleaner other than mild soap and a damp rag, there is a possibility of damage from the solvent. Plastics are susceptible to immediate and long-term damage from solvents. Denatured alcohol seems to be about the safest solvent to use, but it should be used on plastics only as a last resort, and even then it should be used on a test spot first.

WEATHERED ELECTRONICS

Antennas and equipment subjected to the weather have special problems. Moisture and heat are the big ones. Moisture seems able to get into anything! Sealing against rain and moisture is difficult, but it can be done. The main problem in sealing against moisture is that temperature changes cause the expansion and contraction of enclosed air spaces, which produces a suction that draws in rain and moist air. If equipment is to be operated outdoors, the only reasonable solution is to put it into a pressure-tight enclosure, preferably under a low positive pressure.

Protection from heat is in direct conflict with the above procedure. Heat is usually dissipated by providing louvers or blowers to carry away warm air. Blowers operated outdoors are particularly failure prone. An alternative to the louver approach can be used if large heatsinks are provided outside equipment for the power-handling components; in this case heat may be passed right through the side of the cabinet to the heatsinks.

The Megger

Antennas accumulate corrosion and dirt. Antennas that are supposed to be an open DC circuit, such as a long-wire antenna, can be monitored for dirt accumulations by regular readings using a special ohmmeter called a *megger*. This instrument supplies a high voltage, typically 500 volts, to the antenna and measures any leakage at that high voltage. This voltage is sufficiently high that many megohms of resistance can be detected (Figure 19-5). The megger is capable of producing quite enough current to shock a careless operator and can be a hazard to anyone touching the antenna while measurements are being taken.

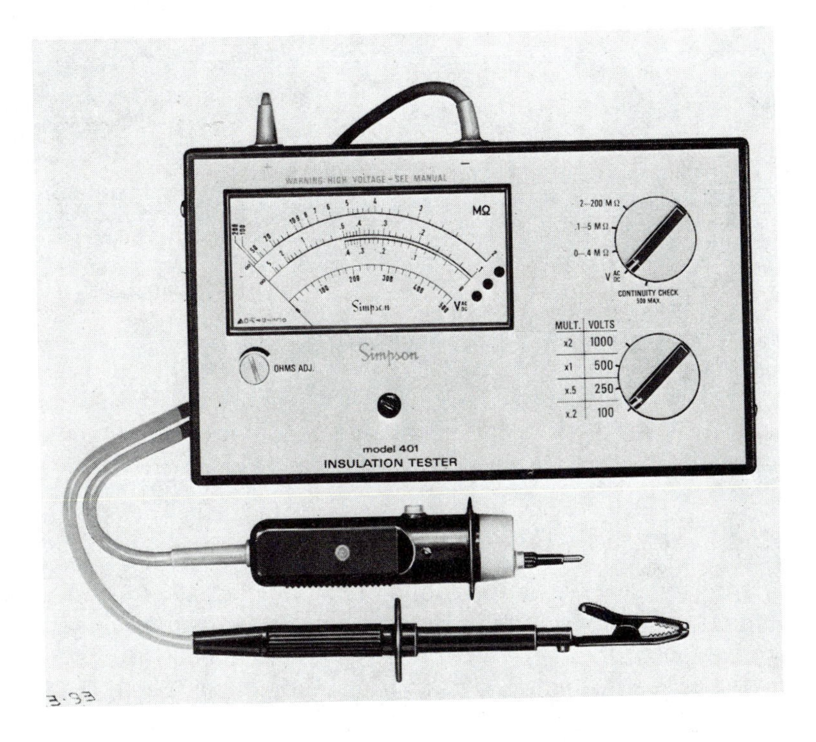

FIGURE 19-5 A megger or high-voltage insulation tester in use. (Courtesy of Simpson Electric Company, Elgin, Illinois)

FIGURE 19-6 Fluke DMM with conductance scales to measure extremely-high-resistance paths. (Reproduced with permission of the John Fluke Mfg. Co., Inc.)

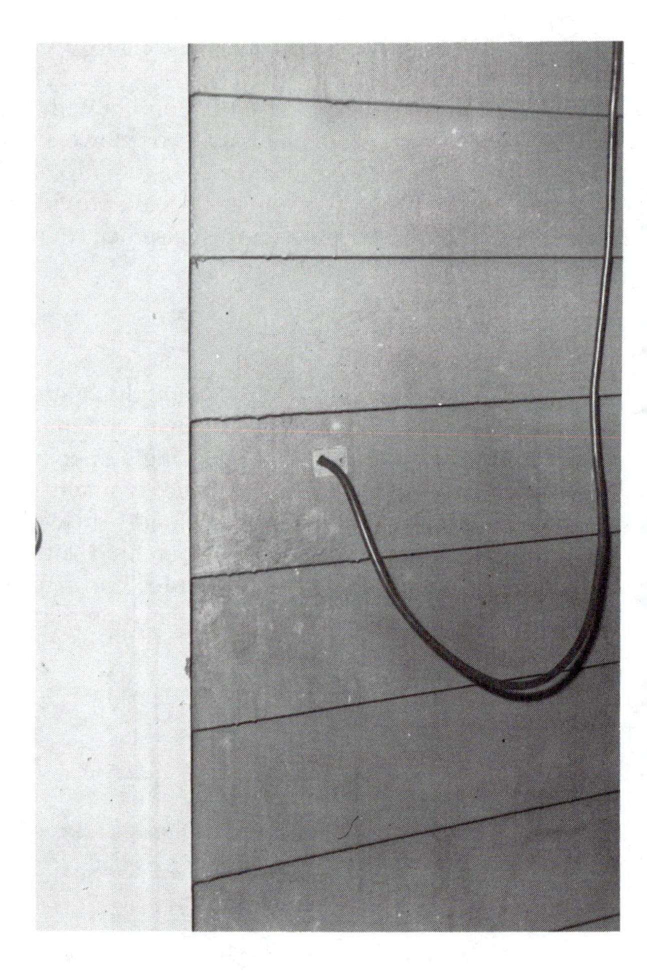

FIGURE 19-7 A drip loop can help prevent water from following the coax inside.

In order to keep track of the gradual degradation of an antenna system, records can be kept to monitor leakage trends over a long period of time.

Measuring Leakage by Other Means

An ohmmeter can give an indication of antenna leakage too. The readings will be limited by the maximum resistance that can be read on the ohmmeter. An alternative method to using the megger or an ohmmeter is to use the ultra-sensitive conductance function available on some of the Fluke DMMs. Conductance is the reciprocal of resistance and is measured in *siemens* (pronounced seamens). The obsolete term for conductance is "mho."

The Fluke Model 8020B, as an example, will measure extremely small conductance (high resistance) using the 200-ns (nanosiemens) range (Figure 19-6). This range will measure from 200 ns (5 megohms) to 0.1 ns (10,000 megohms)! These scales are direct in their approach, not requiring any mental reversal to visualize a leakage problem. The larger the number, the worse the leakage. When using resistance scales, one has to remember that the higher numbers mean less of a problem. When an antenna system reaches a predetermined level of leakage, the antenna is scheduled to be taken down and the insulators cleaned.

Coaxial cables are also subject to failure due to the weather. A coax cable extended upward without sealing off the end of the cable will encourage rainwater to "wick" down the braided covering. What started out as a coax can end up a water pipe! Two things should be done to prevent this: Seal the end of the coax with suitable insulating material, such as tar and/or tape, and provide a "drip loop" (Figure 19-7). Note that water flowing down the outside of the cable drips off at the drip loop at the bottom of the cable, and thus does not enter the building.

SUMMARY

This chapter concludes the maintenance portion of this book. Shop policy and a good measure of common sense will dictate just how much preventive maintenance should be done in a given situation.

REVIEW QUESTIONS

1. What is the operating principle of TDR?
2. What two general indications can TDR give the operator about a cable failure?
3. What is the main difference between an ohmmeter and a megger?

chapter twenty

Techniques for Live Vacuum-Tube Circuits

CHAPTER OVERVIEW

The vacuum tube has fallen into disuse. There are only a few applications left where the tube has no competition from the transistor. Special applications such as cathode-ray tubes (CRTs) and circuits of very high power and/or very high frequencies are still the domain of the vacuum tube.

An excellent reference to have on hand when working with vacuum tubes is RCA's *Receiving Tube Manual*. A copy of their *Transmitting Tube Manual* may also be helpful. These books give pin connections and other information needed to service vacuum-tube equipment.

SPECIAL TOOLS

Some special tools are needed to work on this kind of equipment. Some form of tube puller is handy, to remove tubes placed between other tall components. Miniature tubes also have a tendency to develop bent pins, so a tube-pin straightener is handy for straightening the pins quickly and easily (Figure 20-1).

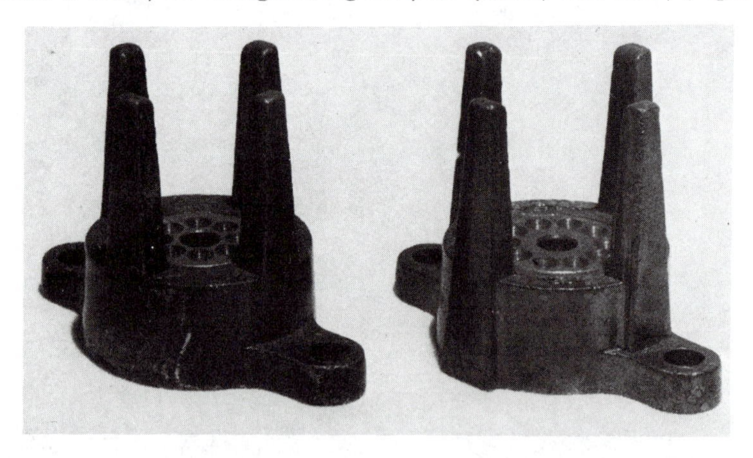

FIGURE 20-1 Miniature tube-pin straighteners for use with seven- and nine-pin miniature vacuum tubes.

SAFETY NOTES

Vacuum tubes require the use of voltages much higher than those used with semiconductors. Familiarity with the following safety rules will help avoid a shock hazard.

1. When using a voltmeter on high voltages, be sure to attach the ground lead to the chassis or common point before checking voltages. Failure to do this first may produce a "hot" negative lead, which can be dangerous if touched.
2. When measuring voltages up to perhaps 200 V, remove any metal jewelry and keep your left hand (assuming you are right-handed) in your back pocket. This helps avoid the possibility of accidental current flow through the chest. It is current flow through the chest that is most dangerous.
3. In addition to the above rules, when measuring voltages higher than about 200 V, other precautions become prudent. Turn off the equipment, discharge the filter capacitors, attach the meter, then turn on the power and take the reading. Do not touch the meter or the leads until the power is again turned off and the capacitors are discharged.
4. The use of an isolation transformer is recommended if working on transformerless equipment operated from the power line. If one is not available, the technician should determine the proper plug polarity to ensure that the chassis of the equipment is connected to the neutral side of the line rather than to the elevated side of the 120-V line.
5. Electrical insulating rubber matting on the floor to stand on in front of the bench is also advisable.

CHASSIS WIRING COLOR CODE

Tube equipment was sometimes wired with a standard color code. Here is that code:

Supply Voltage (B+)	Red
Ground	Black
Plate Leads	Blue
Grid Leads	Green
Cathode Leads	Yellow
High Side Heater	Brown
Low Side Heater	Black
Screen Leads	Orange
AVC Bus	White

This color code was not always followed, but it is sometimes handy as a good place to begin wire checking.

Power transformers also had a standard color code, more often followed than not. When servicing old transformers, *do not* depend on this color coding, for safety reasons. Never trust color coding or the size of a wire to indicate whether or not it may be supplying dangerously high voltage. Assume every winding of an energized transformer is dangerous unless proven otherwise. Low-voltage windings also can be at high potentials with respect to ground, because of deliberate or accidental connection to other windings. Here is the color code for power transformers:

Primary Power	Black
Tap, if any	Black/Yellow or Black/Red
High-Voltage Plate	Red
Tap, if any	Red/Yellow
Rectifier Heater (5 V)	Yellow
Tap, if any	Yellow/Blue
Filament #1	Green
Tap, if any	Green/Yellow
Filament #2	Brown
Tap, if any	Brown/Yellow
Filament #3	Slate
Tap, if any	Slate/Yellow

Audio transformers had a code, too:

Tube Plate Lead	Blue
High Voltage (B+)	Red
Plate Center Tap	Brown or Blue
Grid High	Green
Grid Low	Black

Intermediate-frequency transformers used:

Plate Lead	Blue
High Voltage (B+)	Red
Grid High	Green
Grid Low	Black

TYPICAL VACUUM-TUBE VOLTAGES

Figure 20-2 shows the elements of vacuum tubes and their names. Each element will be discussed in the following paragraphs.

Heater Voltages

In order for tubes to work, they must have a source of heat to "boil off" electrons and begin them on their way through the tube elements. Heater voltages are commonly 6 or 12 volts AC or DC, but may be anywhere from a volt and a half up to as high as 120 volts in a few cases. There are two ways to operate

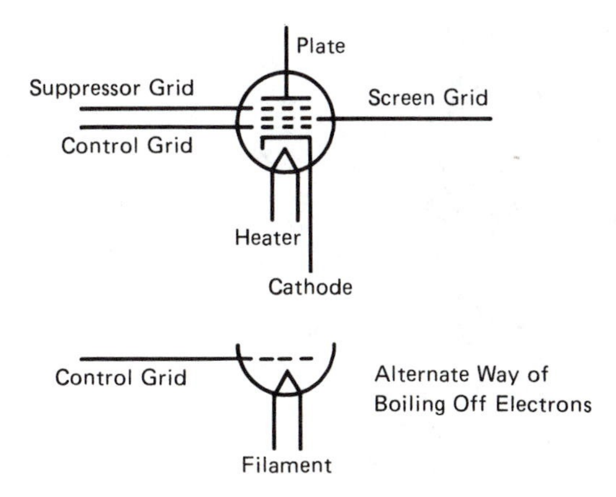

FIGURE 20-2 Vacuum-tube elements.

tube heaters: in series like Christmas-tree lights, operating directly from the 120-V AC power line, or in parallel at much lower voltages, commonly 6 or 12 V AC. The latter method requires a transformer to reduce the line voltage from 120 V to the required heater voltage.

The first check of vacuum-tube heater equipment should always be to see if all the tubes are lit. Glass tubes make it possible to see if the tubes are lighting. If a tube is not obviously lit, a few minutes of operation should make the tube warm to the touch because of the heater and the plate current flowing through the tube as a result of proper operation. If the tube becomes hot, it can be assumed that the heater is working properly.

Heaters often fail by opening. In a parallel heater arrangement a suspected open heater is best checked by removing the tube and checking with an ohmmeter at the tube pins for continuity through the heater element. In a series arrangement the open heater is easiest to find by looking for the full line voltage (120 V AC) across the heater pins of the defective tube. All other tubes will indicate 0 voltage (Figure 20-3). Note that this is the same basic method as described earlier in this book for finding open fuses.

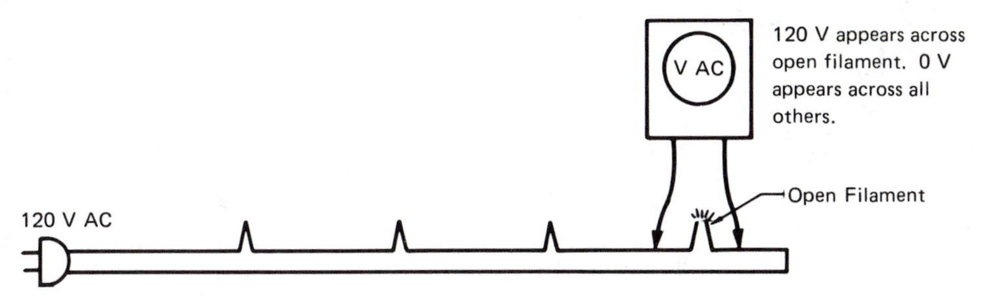

FIGURE 20-3 Finding an open heater in a series string.

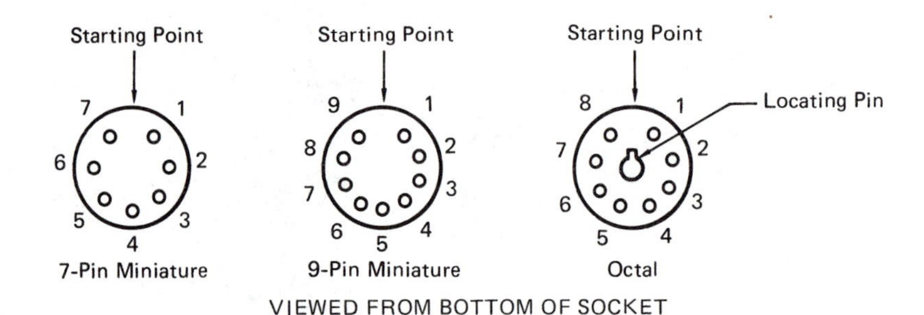

FIGURE 20-4 Tube-pin numbering schemes.

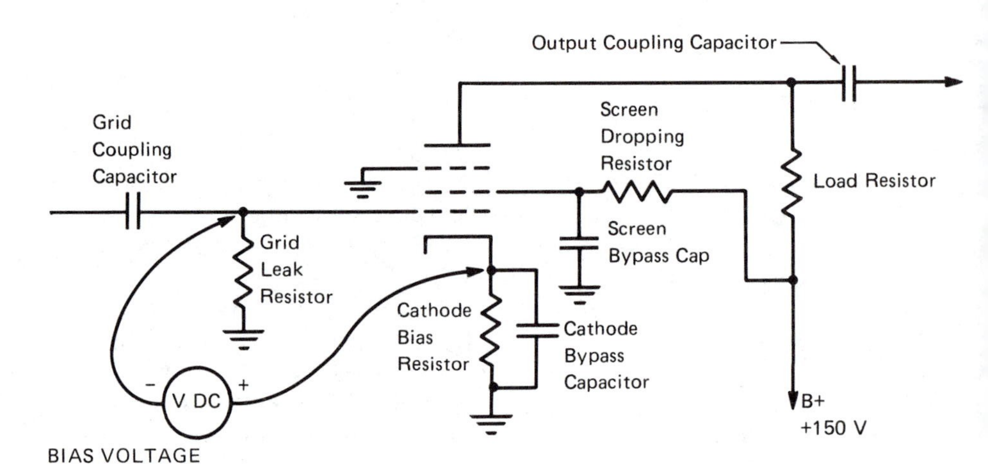

FIGURE 20-5 A typical circuit to obtain grid bias.

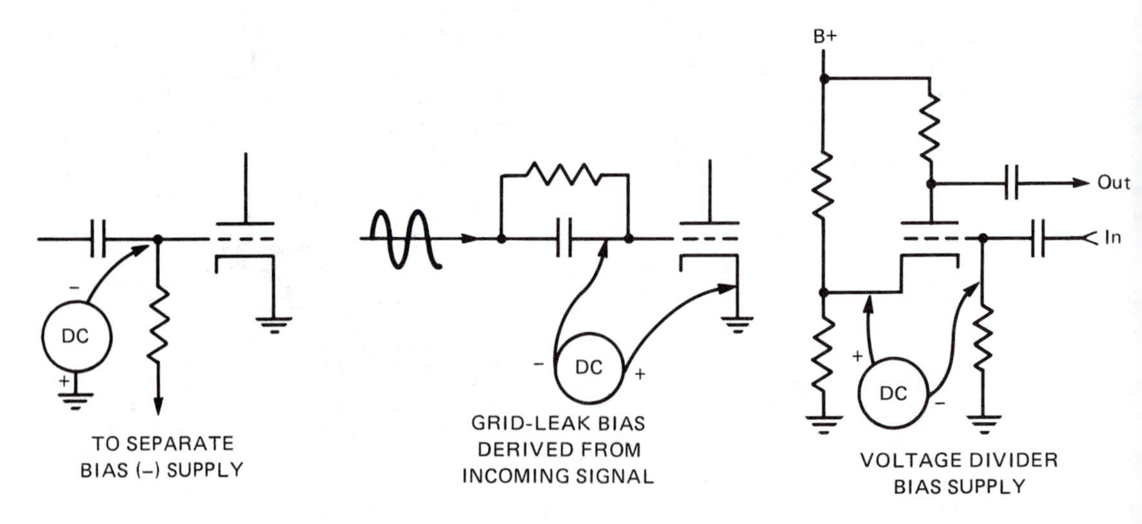

FIGURE 20-6 Alternative ways of obtaining grid bias voltages.

Tube-Base Diagrams

Figure 20-4 shows how tube pins are numbered, as viewed from the *bottom* of the tube socket.

Vacuum-Tube Bias (Grid Voltage)

The vacuum tube will conduct too much current if no bias is applied to the tube. The bias of a tube is a negative voltage on the grid with respect to the cathode (Figure 20-5). This negative voltage may be acquired in several different ways (Figure 20-6).

Class A Operation

The cathode biasing circuit of Figure 20-5 shows the cathode connected to a resistor. Normal plate current flow will produce a voltage across this cathode resistor, putting the cathode a few volts above ground potential. If the grid is operated in reference to ground, the net result is that the grid has a negative voltage with respect to the cathode. This is the most common way of biasing tubes for Class A operation. This makes the tube operate "half-on, half-off." An incoming signal may then vary the output voltage up or down, depending on the grid voltage polarity, negative or positive. This is the most common circuit for the voltage amplifiers in audio amplifiers, oscilloscope amplifiers, and other applications where the low distortion of Class A operation is necessary. The voltage across the cathode resistor is a very good check of the tube's operation. Proper voltage at the cathode indicates that the tube plate current is about normal. This cathode resistor should have a high-capacity electrolytic across it for audio work and a ceramic or paper capacitor for higher frequencies. This capacitor averages out the pulsations of current flowing through the resistor, giving a constant voltage for the cathode.

Class B Operation

A more negative bias voltage than that required for Class A will reduce tube current to Class B operation. The tube can only be turned more "on," since it is already almost "off." This class of operation is used mostly in audio circuits where two of these circuits are operated together to amplify the positive and negative halves of an incoming waveform independently, then put back together again with a transformer. This is called "push-pull" operation (Figure 20-7). The bias for this stage is sometimes derived from a separate circuit rather than using cathode current as in a Class A stage.

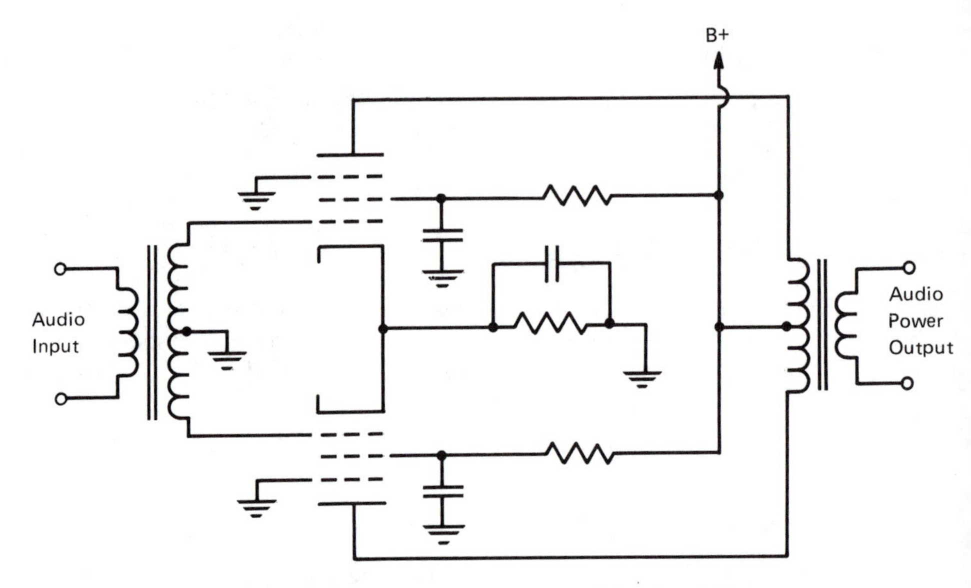

FIGURE 20-7 Push-pull vacuum-tube amplifier.

Class C Operation

A very negative bias voltage of several times that necessary for Class B is used in some circuits where it is desirable to have short, high-powered pulses in the plate circuit. These pulses are reconstructed into a sine wave by inductance and capacitance provided in the plate circuit. The result is an amplifier with high efficiency. Figure 20-8 shows such an RF circuit. Note in this circuit that the bias comes from a separate source.

Figure 20-9 shows some *very* general voltages used in tubes such as those used in receivers. These voltages may be used as an *indication* of proper volt-

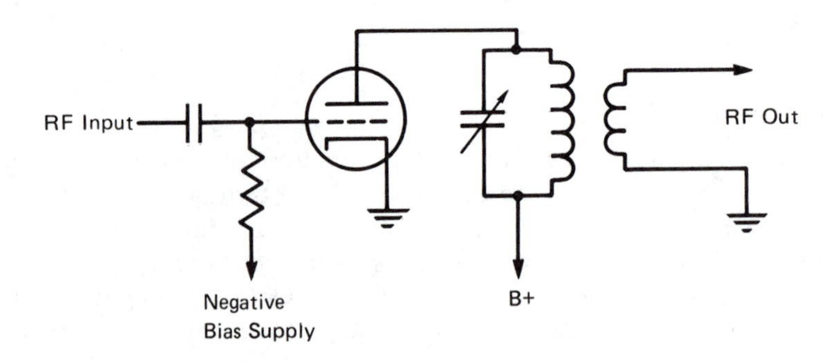

FIGURE 20-8 The radio-frequency Class C amplifier.

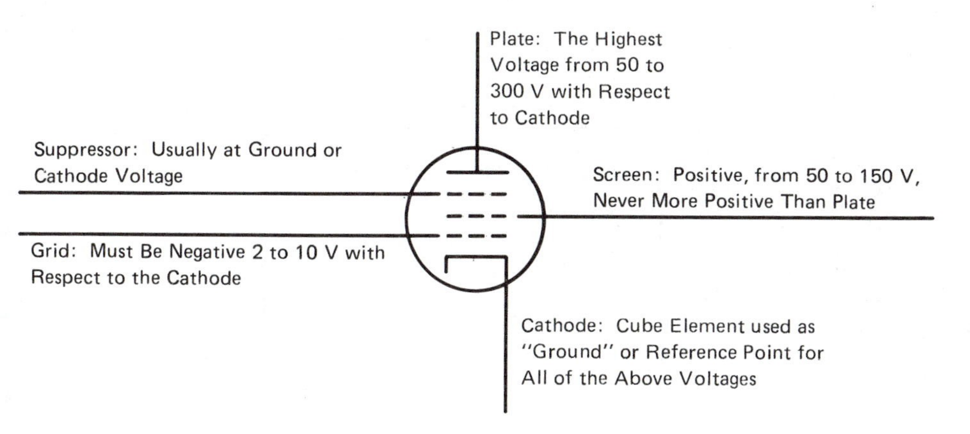

FIGURE 20-9 Typical pentode-tube voltages.

ages in a circuit. Wide voltage variations are to be expected in actual circuits. The complete lack of a voltage is certainly cause for further investigation.

Cathode Voltages

The cathode is the element used as a reference for tube voltages. With the cathode as the reference, the bias on the grid must be negative and all other voltages positive or at 0 volts. The heater circuit is the only one that has no particular reference to the cathode.

Plate and Screen Voltages

Plate and screen voltages are generally quite high. Receiving circuits will have plate and screen voltages from about 100 volts DC up to several hundred. Caution must be used when measuring these potentially hazardous voltages. Note that the screen voltage is high, and the plate voltage is the highest.

Suppressor Grid Voltages

The suppressor is usually connected to either the cathode connection or to ground.

Actual voltages that are normal for equipment must be taken from the equipment instruction book. Voltages for many vacuum-tube circuits can vary 10 percent or 20 percent or more in noncritical applications like audio amplifiers without noticeable changes in performance. Of course this cannot hold true for precision circuits like those used in oscilloscope amplifiers.

Some tubes are made with more than one tube within the envelope. In this case treat each tube as a separate one as far as troubleshooting is concerned.

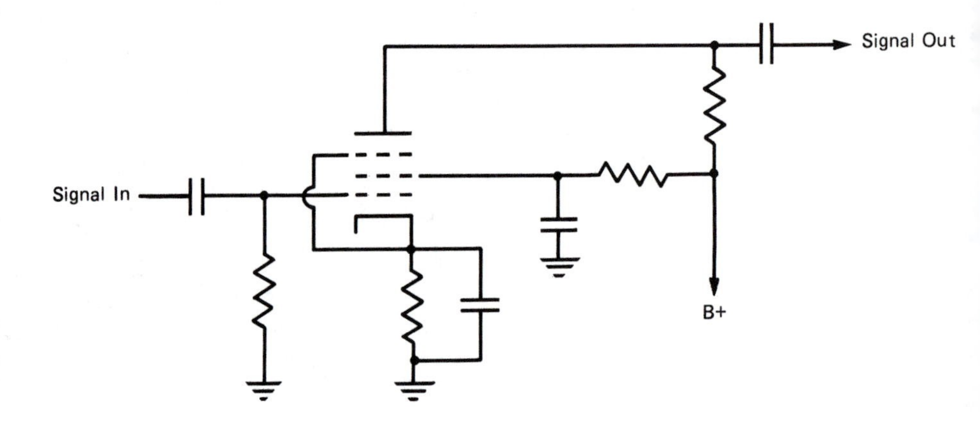

FIGURE 20-10 A typical RC-coupled tube amplifier circuit.

TROUBLESHOOTING VACUUM-TUBE CIRCUITS

The RC-Coupled Amplifier

The vacuum-tube voltage amplifier of Figure 20-10 is easy to troubleshoot with the voltmeter. Assuming that the tube is showing a glowing heater, the first step in checking for failures in this stage is to check the cathode bias. This will be on the order of a few volts, perhaps 3 to 8 volts for small receiving-type tubes. Absence of any voltage would indicate an open, bad tube or lack of screen or plate voltage because the cathode resistor has no current flow through it. Check the screen (if there is one in the tube) and the plate voltages. Whichever is first in the electron stream as it flows from cathode to plate through the tube is responsible for the current to the plate circuit. The lack of either voltage will indicate a problem, which may be found using the voltage troubleshooting techniques of Chapter 5.

If the cathode voltage is too high, there is a good chance that the capacitor that should be blocking plate voltage from the preceding stage from appearing on the grid circuit has become shorted or leaky. Even a high resistance or leakage is sufficient to put a positive voltage on the grid, causing excessive current flow through the tube, hence the high cathode voltage indication. Replacement of the grid capacitor will usually cure this problem. This failure will often damage the tube. If the cathode voltage is proper, a check of the screen, suppressor and plate pins is in order. It is possible to have low plate voltage and a near normal cathode voltage in pentode (five-element) tubes.

All of the DC voltages may be correct, yet the stage will not amplify enough: This is one of the cases where DC troubleshooting cannot detect all of the possible problems. An open cathode bypass capacitor will produce these symptoms. The incoming signal is merely changing the cathode bias voltage, producing very little variation of the plate current. Replacing the cathode capacitor to hold the bias constant will cure this problem.

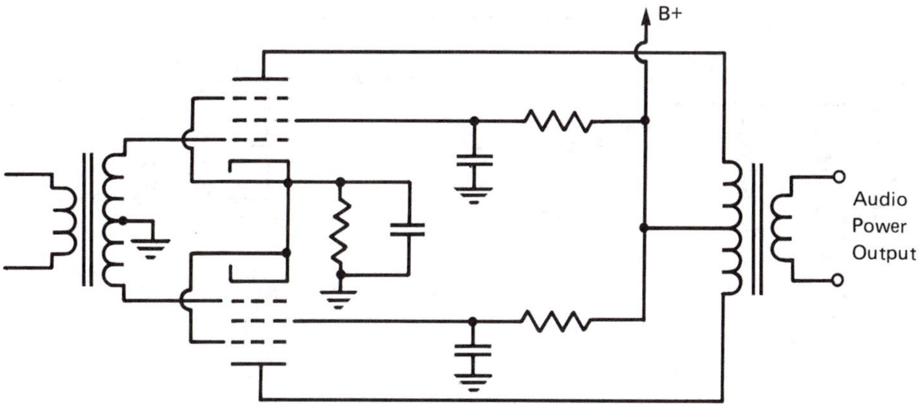

FIGURE 20-11 A typical push-pull amplifier circuit.

The Push-Pull Audio Amplifier Stage

The circuit shown in Figure 20-11 is commonly used in vacuum-tube audio amplifiers, supplying the output audio power to the speaker. Failures of this circuit often take the form of intolerable distortion when one of the two tubes stops working. This results in a "half-wave" output instead of a "full-wave" output. Replacement of the bad tube is the obvious answer.

Distortion can also be caused by improper biasing of both the tubes. The normal bias is produced by average plate currents of the tubes flowing through the common cathode bias resistor and is smoothed out by the cathode bias capacitor. If the capacitor shorts, there will be no bias and the tube currents will be far too high, producing much distortion and making the tubes run very hot. Again, checking the cathode bias is a good place to start.

Each of the tubes must have the proper screen and plate voltages to operate. The lack of a plate or screen voltage is traced to its cause by using a DC voltmeter as described in Chapter 5.

The RF or IF Amplifier

Figure 20-12 shows two common circuits used for the amplification of RF and IF frequencies. Just as with other vacuum-tube circuits, begin troubleshooting these stages by checking the cathode bias first. Using the previous guidelines, finding the problem with a DC voltmeter is easily accomplished. Move on to checking the plate, screen, and suppressor voltages.

The Vacuum-Tube Series Regulator

The regulator circuit of Figure 20-13 is representative of many possible variations. It consists of a series voltage-control element (variable resistor in essence)

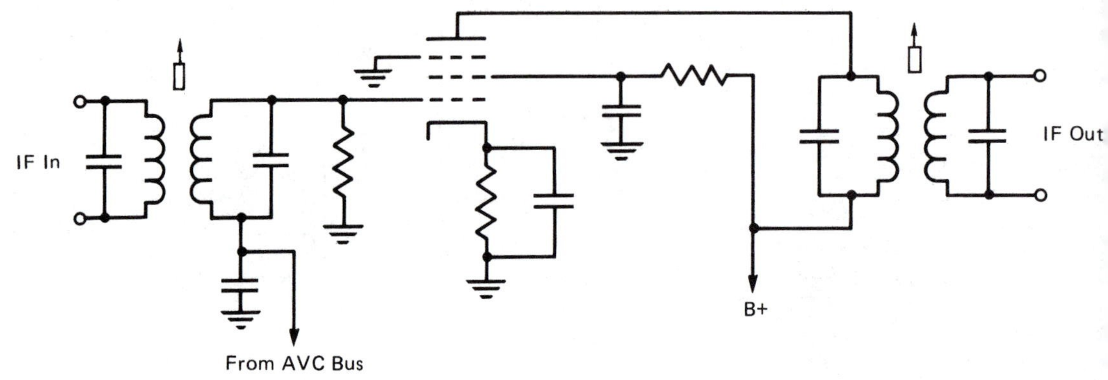

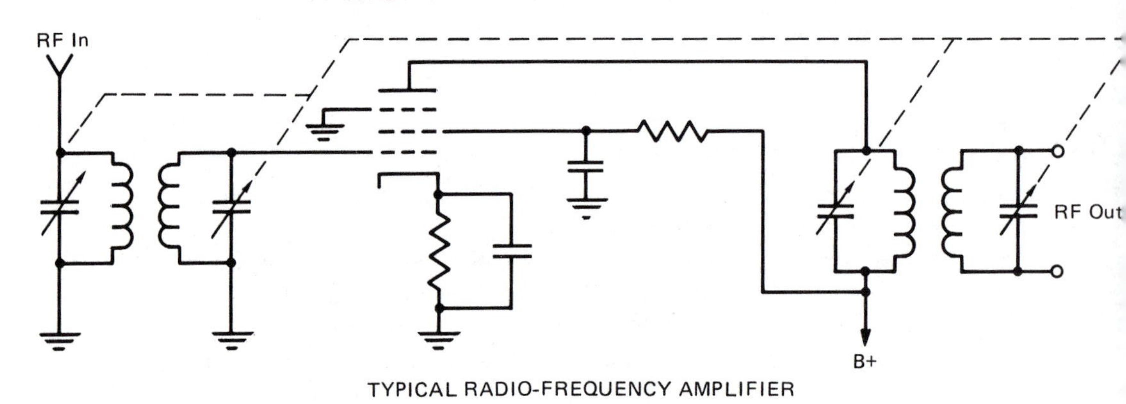

FIGURE 20-12 Typical IF and RF amplifier circuits.

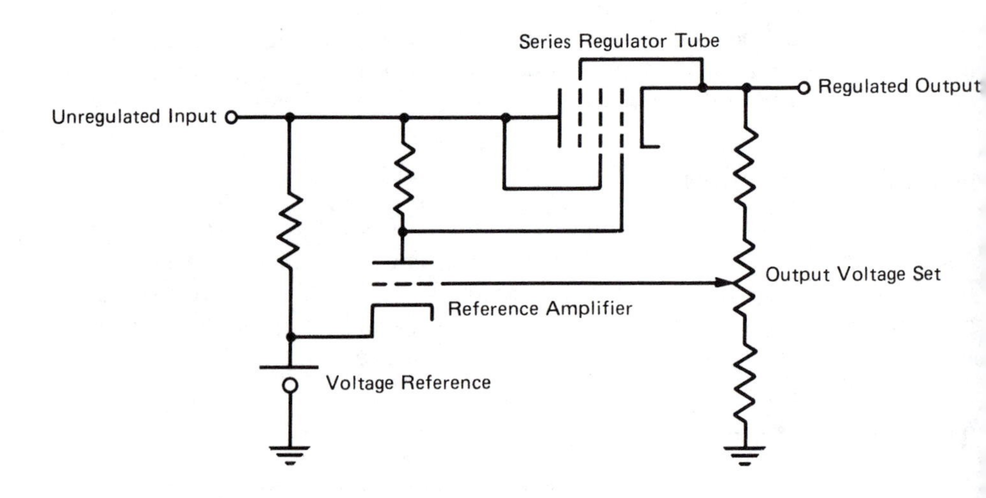

FIGURE 20-13 A series regulator circuit with reference amplifier.

and a reference against which the actual output voltage is compared. A difference between the reference and the output voltage is amplified and applied to the series element to vary its resistance in the proper direction to bring the output voltage back to normal.

As with any series regulator, the heater voltage, the output voltage, input voltage, and reference voltage should be checked first. Since the series element has to have some voltage drop, the input voltage must be higher than the output voltage. For instance, if the output voltage is supposed to be 150 V DC, then one would expect perhaps 180 V DC at the input to the regulator. Anything less than this is not enough for the regulator to work properly. The reference voltage must be correct and stable, too.

The Vacuum-Tube Shunt Regulator

Figure 20-14 shows a common tube shunt regulator. This regulator circuit normally has a current flow through the shunt tube of between 5 and 30 milliamperes. If the input voltage across the tube is sufficient, the tube will fire, causing a visible glow between the elements. Due to the characteristics of the glow tube, the input voltage must momentarily go *at least 15 percent higher* than the regulating voltage of the tube, in order to fire the tube initially.

THE CATHODE-RAY TUBE

The cathode-ray tube (CRT) still plays an important part in electronics. It is the interface component between electrical signals and human beings. *Be careful when measuring the voltages associated with these tubes.* Read the following information carefully before proceeding with any voltage measurements, as high voltages may be present where they are not expected!

CRT Cathodes

The CRT cathode produces the electrons that are eventually splashed into the front screen where they produce light. The cathode of the CRT serves the

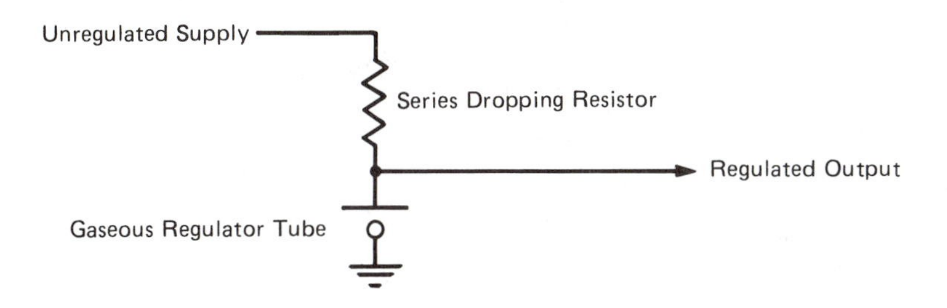

FIGURE 20-14 A typical shunt tube regulator.

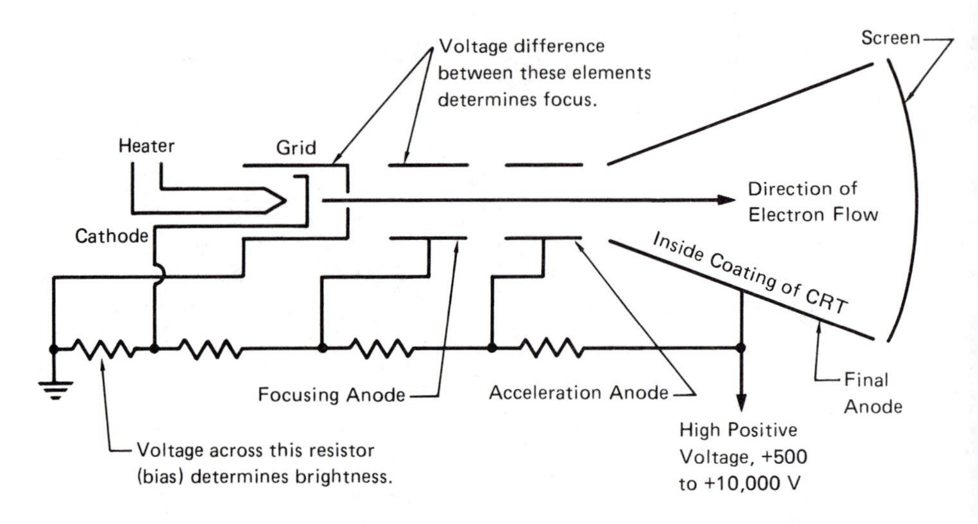

FIGURE 20-15 Basic CRT voltages.

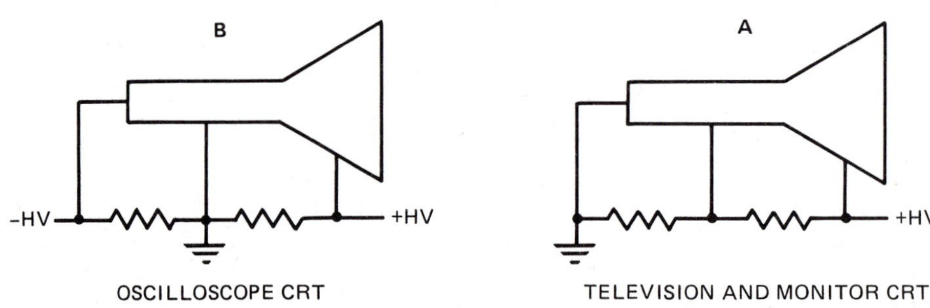

OSCILLOSCOPE CRT

Cathode at high negative voltage with respect to chassis, final anode at high positive.

TELEVISION AND MONITOR CRT

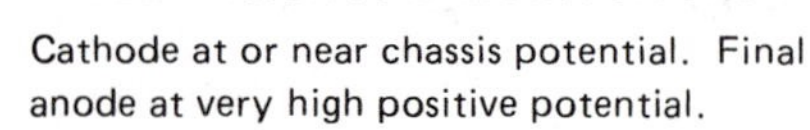

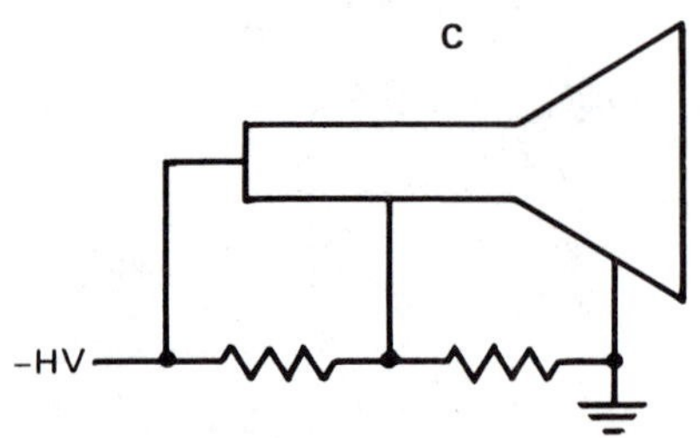

A few oscilloscopes operate the cathode with single, very negative supply.

FIGURE 20-16 CRT high-voltage supplies and cautions to observe.

same purpose as that of the common vacuum tube. The electrons exit the cathode area through a hole in the cylindrical CRT grid (Figure 20-15). Note carefully the voltage difference between the cathode and the last anode. There are very high voltage differences involved. These can be dangerous. Figure 20-16 shows how this can happen. For those technicians who may be accustomed to having the cathode of a tube at ground, the circuits of B and C of Figure 20-16 could hold a nasty surprise.

Oscilloscopes often operate the CRT with the cathode highly negative, the deflection plates near zero volts, and the final accelerating anode highly positive. This enables the deflection plates to be near ground, avoiding design problems interfacing with the amplifier circuits connected to them. Video monitors and TVs often operate the cathodes near ground, since the input to these CRTS involves putting a signal into the grid or cathode circuit to turn the CRT beam on and off. In this case the accelerating anodes are run at very high positive voltages with respect to ground.

Normal CRT Voltages

Normal CRT voltages require that the tube-element voltages increase in positive steps from the cathode to the screen, with the exception of the grid. The grid is held at a negative voltage with respect to the cathode, just as in receiving tubes. This voltage is made adjustable to control the intensity of the spot on the screen. The voltage difference between the focusing and the accelerating elements determines the focus of the spot. Troubleshooting focusing problems should begin with these two voltages. Some CRT circuits may have regulators on several voltage sources to hold critical voltages constant. Failures of these regulator circuits can cause brightness fluctuations and focusing instability of the spot.

The higher the final accelerating voltage, the brighter the spot. If the spot on a CRT is dim and unfocused, the high-voltage supplies should immediately be suspected as being too low or zero. This may be confirmed if the CRT shows images too *large*. The lack of final accelerating voltage makes the image larger because the electron beam is moving slower and is more easily deflected.

SUMMARY

This chapter has very briefly covered the troubleshooting techniques and general voltage expectations for small vacuum-tube circuits. Although tubes are fading from importance, there are some technicians who can use this information to maintain old equipment out of necessity or nostalgia.

REVIEW QUESTIONS

1. What is the first check to be made of inoperative vacuum-tube equipment?
2. Name all of the elements of a vacuum tube that has three grids.

3. What is the polarity of a vacuum-tube grid with respect to its cathode?
4. What would happen to current through a tube if the bias voltage were reduced to zero?
5. What element of a vacuum tube is the reference point for other tube-element voltages?
6. Name two tube elements normally at a high voltage. Name two at a low voltage. Name one at a negative voltage.
7. What circuit component is a common cause of excessive tube plate current?
8. If a shunt gaseous regulator tube is rated at 105 volts but has 115 volts across it and is not glowing, what is the most likely cause?
9. Are CRT cathodes usually at ground or near ground potentials?

chapter twenty-one

Basic Troubleshooting of Power-Line Circuits and Motors

CHAPTER OVERVIEW

This chapter covers the fundamentals of troubleshooting power-line circuits and motors. This is the work of an electrician, but an electronics technician should understand basic wiring, particularly as it may pertain to getting primary power to the equipment that is being repaired. Motors are a part of electronics, often used as blowers or as loads for electronic control circuits.

SINGLE-PHASE POWER DISTRIBUTION AND TROUBLESHOOTING

Most of the equipment and test instruments a technician uses will be operated from batteries or the 60-Hz power line. For this reason, the technician should be familiar with the basics of power distribution and the problems it may present. To work on power circuits as a profession, a thorough knowledge of the National Electrical Code is essential. These standards ensure that electrical wiring is installed and maintained in a safe, professional matter.

Electrical outlets for 120 V AC should be provided with three connections. These three connections are for the "hot," the neutral, and earth ground. Older installations of electrical wiring provided only two connections, the "hot" and neutral. These older outlets should be replaced with the newer ones whenever practical. Of the three connections, the narrower of the two blade connections should be the "hot" connection, but by no means depend on this (Figure 21-1).

The 120 V AC provided at an outlet is one of a pair of "mirror-image" voltages of 120 V AC. To assist in understanding this arrangement, consider how the transformer on the electrical power pole outside the house is connected to the power panel (Figure 21-2). By connecting a load from point A to point B, 120 V AC is utilized. Connecting from point B to point C also results in 120 V AC. Connecting from A to C, however, allows the *sum* of this to be used, 240 V AC.

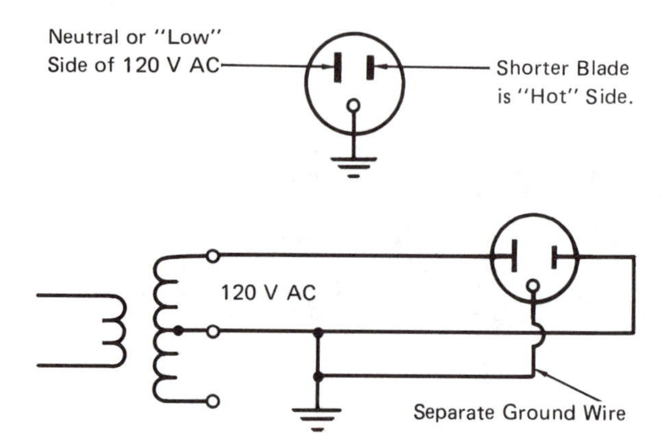

FIGURE 21-1 The 120-V AC outlet.

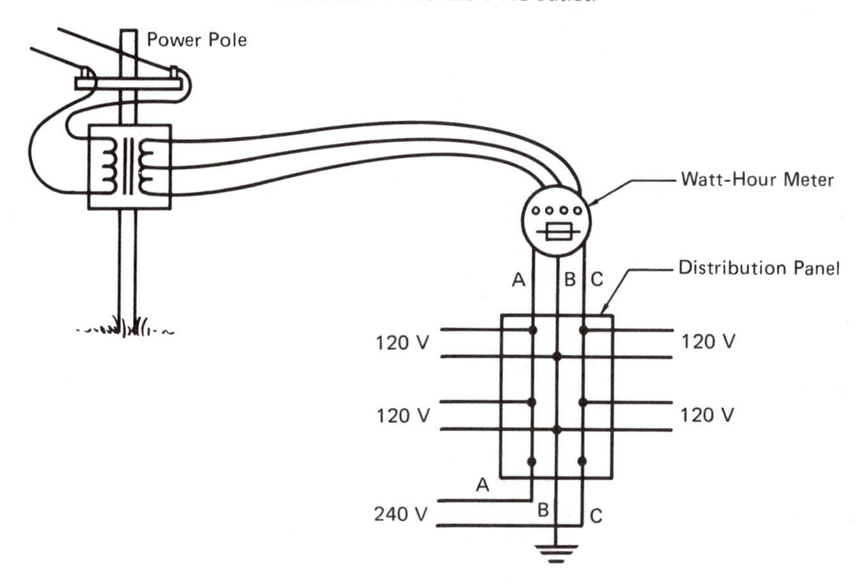

FIGURE 21-2 Residential power circuits.

Using this scheme of three wires from the power transformer feeding the breaker panel, one of two voltages may be selected. For very heavy loads such as hot-water heaters and cooking stoves, connection is made to the 240-V AC connections, resulting in half the current draw that would be required if the appliance were operated from only one side of the supply transformer, 120 V AC. Using less current means that smaller wiring can be used.

THREE-PHASE POWER SYSTEMS

A three-phase supply line is used when a large amount of power is being used. Large commercial and industrial users have huge heating requirements and large

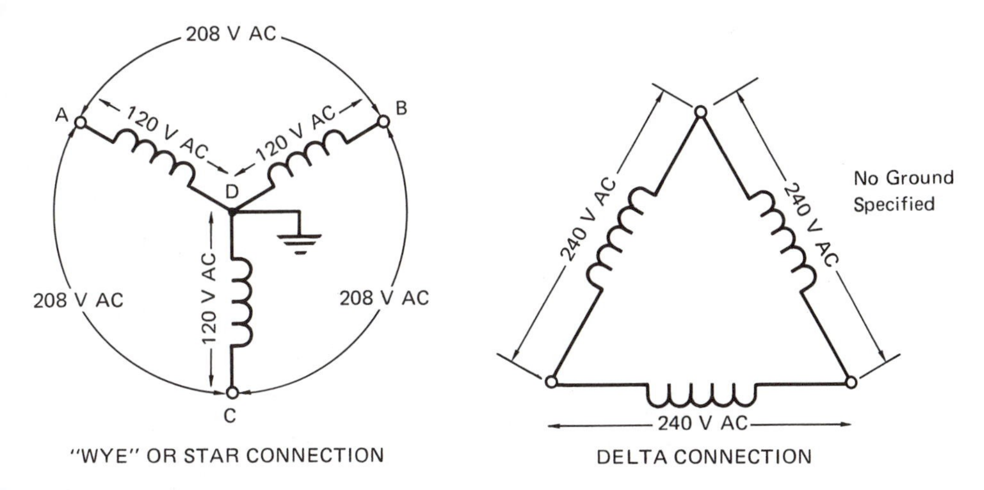

FIGURE 21-3 Delta- and wye-connected three-phase power distribution transformers.

motors to turn, which require such circuitry. Figure 21-3 shows two ways of connecting these three-phase systems, the delta- and wye-connected configurations.

A SPECIAL CIRCUIT

Figure 21-4 shows a circuit in common use in AC lighting circuits, whereby one load may be controlled by two switches. Although a very simple circuit, it is one that is hard to find in reference books.

DC MOTOR BASICS

The Permanent-Magnet Motor

The small permanent-magnet (PM) motor has many uses: driving tape transport mechanisms in DC-operated recording equipment, as a blower in DC applications, and in a reversed application, as a generator of DC voltages for monitoring shaft rotational speed (Figure 21-5).

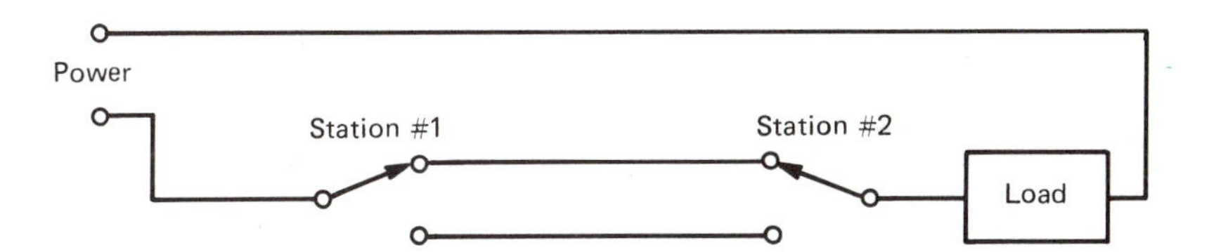

FIGURE 21-4 Two-station control of a single load.

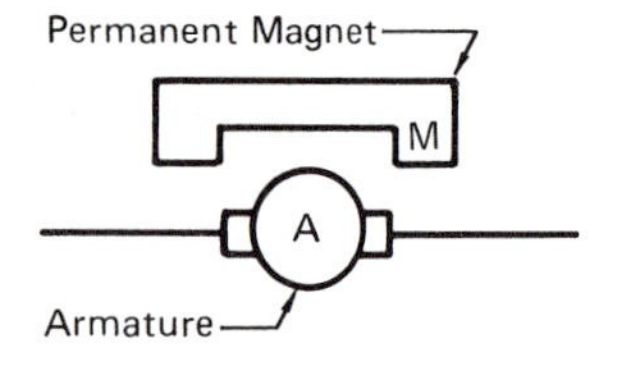

FIGURE 21-5 Schematic diagram of a permanent-magnet motor/tachometer generator.

FIGURE 21-6 A "pancake" permanent-magnet DC motor (or generator).

PM motors come in many sizes and shapes. A modern PM motor may take the shape of a pancake (Figure 21-6). This is a particularly efficient motor.

A permanent-magnet motor is easily reversed in direction by reversing the current flow through the armature. Speed control can be accomplished by varying the voltage applied to the armature or by applying a voltage with a varying duty cycle (Figure 21-7).

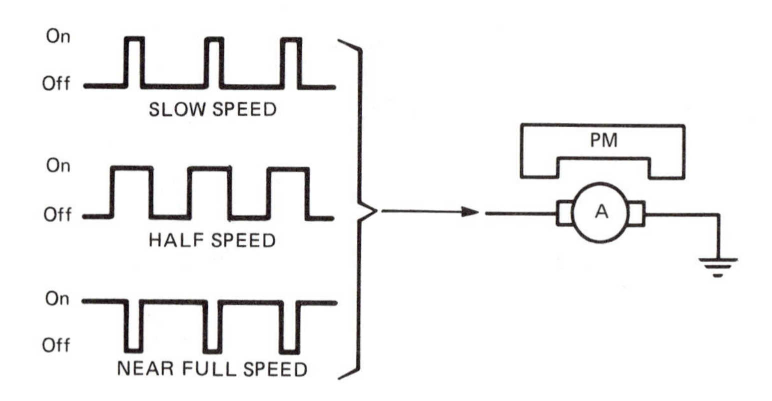

FIGURE 21-7 Speed control of a DC motor by varying input duty cycle.

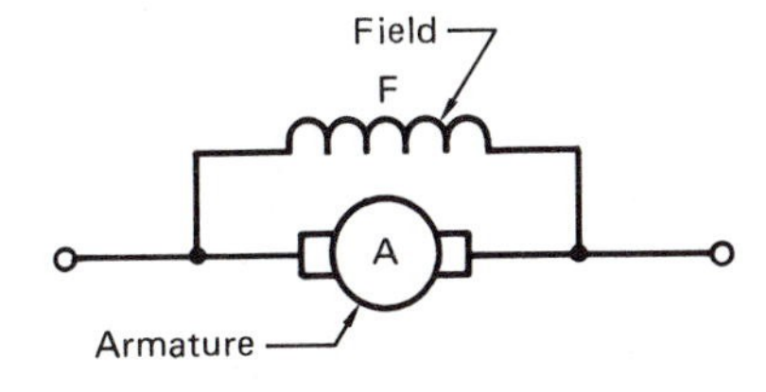

FIGURE 21-8 Schematic of a shunt motor.

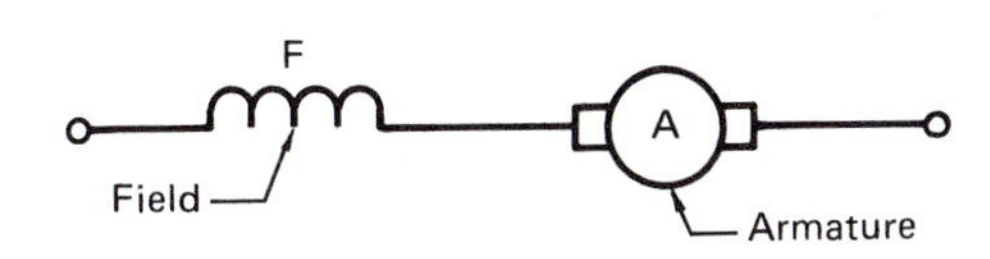

FIGURE 21-9 Schematic of a series motor.

Shunt Motors

Shunt motors have a field winding, used to produce the stationary magnetic field in which the armature turns. The term *shunt* refers to the method of connecting the field in parallel with the armature (Figure 21-8). The shunt motor will run at fairly constant speed with a given voltage. This made it attractive at one time as a motor to drive generators that produced other voltages. This motor still finds some use in commercial applications because its speed is easily controlled with the application of varying voltage. In such an application, the DC applied is derived from AC lines, rectified by high-powered SCRs.

The direction of rotation of a shunt motor can be changed by reversing either the armature polarity *or* the shunt field. Because of the lower current required by the field, it is usually the one reversed.

Large DC shunt motors will *increase* in speed with a *decrease* in the field magnetic strength. This made a special circuit necessary to turn off armature power in the event of field-current failure. Without a field, the motor can "run away" and throw itself and the load apart if sufficient power is available to drive the motor. Armature current increases tremendously with field failure.

The shunt motor can also be used as a generator if field current is supplied. The small residual magnetism of the field pole is often sufficient to produce enough armature output to reinforce the field, resulting in a self-buildup of proper field current. Varying the strength of the field current controls the armature output current in direct proportion. Automotive generators were once commonly of this type. They have been largely replaced by the more efficient alternator.

Series Motors

Series motors have the field in series with the armature (Figure 21-9). The series motor may be operated on DC or AC if designed for that purpose.

The series motor is in very common use in household appliances. It is a good motor for such high-speed, low-torque applications as vacuum cleaners and food blenders. The series motor has a high torque at very low speeds. This is because the armature current is high before the armature comes up to speed. This causes heavy field current because the two are in series. Since there is high current, the field winding is heavier than that of a shunt motor, with fewer turns. With fewer turns, the motor is able to operate on AC current since there is less inductance. In AC applications, this motor is often called a "universal motor."

The series motor can be reversed in direction only by reversing the current flow in either the armature or the field. Reversing both does not reverse the motor, thus its ability to operate on AC. Since it operates at high speed, the series motor often fails due to bearing failure. Inexpensive motors have bushings rather than ball bearings; the whole motor is often replaced rather than repaired. The speed of a series motor varies widely with load. Speed control is possible but not entirely satisfactory by varying the applied voltage.

Compound Motors

Compound-wound motors are made to combine the merits of both the shunt and series motors: good speed regulation with varying load and high starting torque (Figure 21-10). The failure of either of the fields of a compound motor may still enable the motor to operate, but without one of the two advantages just mentioned. To reverse such a motor, both of the fields or just the armature may be reversed. Speed control is possible by varying the input voltage.

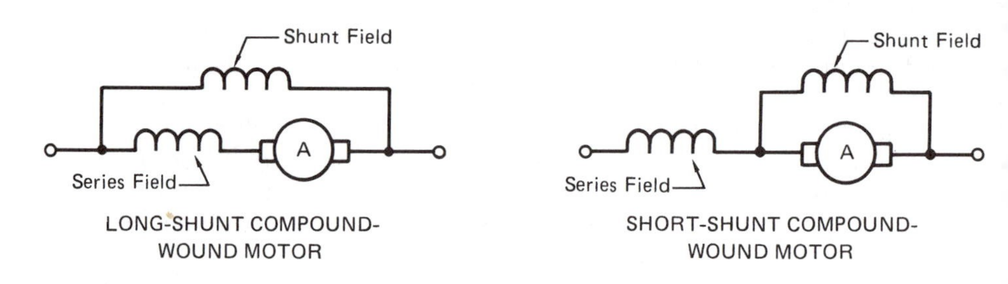

FIGURE 21-10 Schematics for the compound-wound motor.

THE ALTERNATOR

The alternator is not a motor to drive a mechanical load. It is only a generating device whereby mechanical rotation is converted into electrical current. The alternator has a rotating field within a three-phase, heavy-current winding (Figure 21-11). It will produce a large current output at relatively low shaft speeds, compared to the older DC generator. Output-current regulation of the alternator is accomplished by varying the strength of the rotating field. Polarity output of the alternator can be changed only by reversing the leads coming from the rectifier stack.

Common failures of an alternator are the bearings and the brushes that carry current into the rotating field. The regulator is also often a problem; the output of the regulator controls the alternator, so troubleshooting any failure

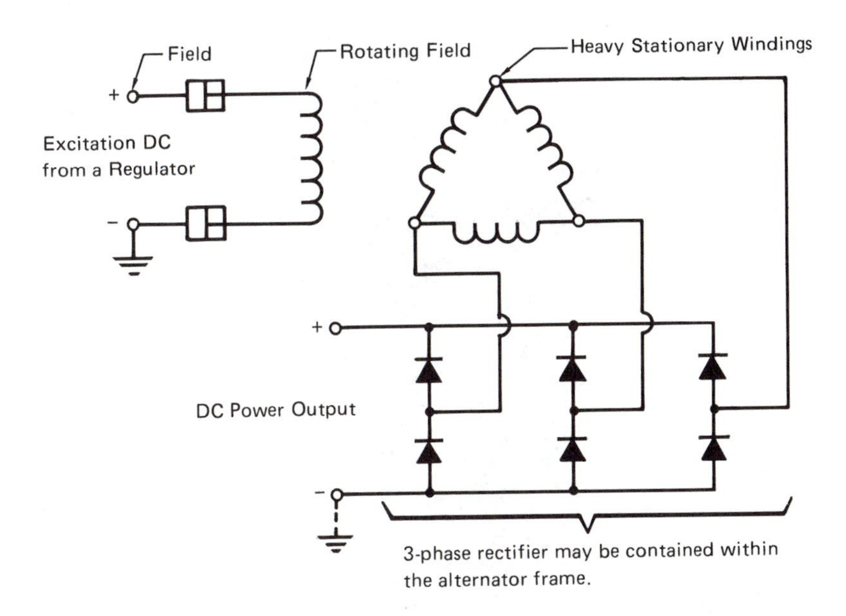

FIGURE 21-11 Schematic of an alternator.

should include the regulator. If voltage is being supplied to the field of an alternator, it should provide an output current. If it does not, the alternator is defective.

AC MOTOR BASICS

AC motors are generally dependent upon the frequency of the AC line to determine their rotational speed. Varying the line voltage serves only to change the torque of the motor, not the speed. Elaborate circuitry is necessary to change the speed of these motors. An AC-motor speed control usually converts the AC line to a DC voltage, then "chops up" the DC into AC again, at a variable frequency (Figure 21-12).

The Split-Phase Motor

This AC motor is in common use in washing machines and refrigerators. One winding is used only to provide the direction and torque for starting the motor under load. Once the motor is up to near operating speed, the starting winding must be turned off (Figure 21-13).

When this motor is first turned on, it has high torque. As it nears operating speed, the "click" of the centrifugal switch controlling current flow through the starting winding should be heard. If the switch does not make contact when the motor is first turned on, the motor probably will not be able to turn the normal load and will just hum. Eventually, it may turn itself off if it is provided with a thermal overload switch, as most of these motors are. If the centrifugal switch

FIGURE 21-12 A high-quality commercial three-phase motor speed control is complex. This cabinet contains two such drives. (Photo courtesy of Eaton Corporation)

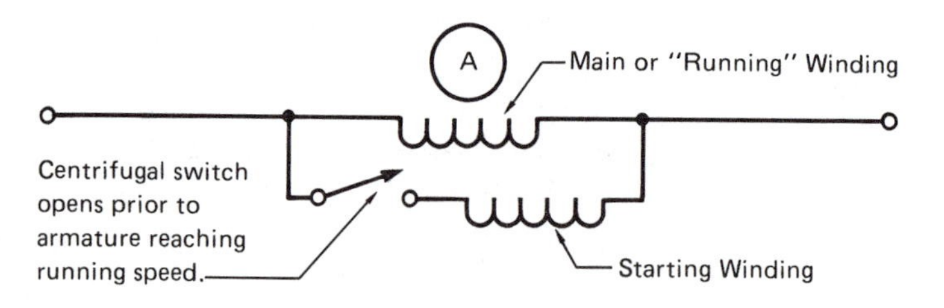

FIGURE 21-13 Schematic of the split-phase motor.

does not open before operating speed is attained, the motor draws excessive current. Again, the thermal overload of the motor should act to protect the motor from burnout. There will be a louder hum in this case than in the case above. These same symptoms will be evident if the motor is prevented from turning by an excessive torque load.

The direction of rotation of the split-phase motor may be reversed by reversing the connections to the starting winding.

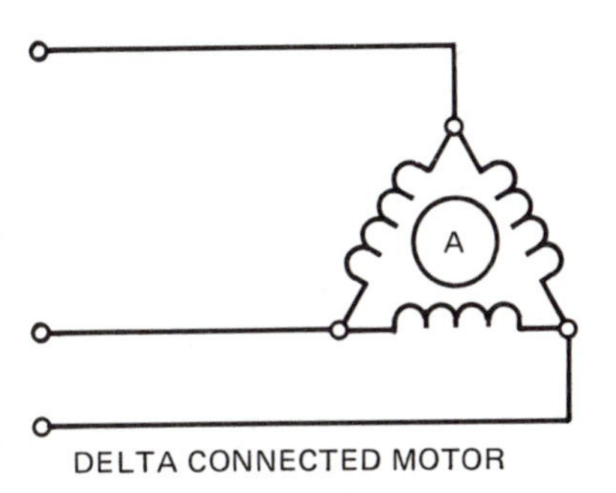

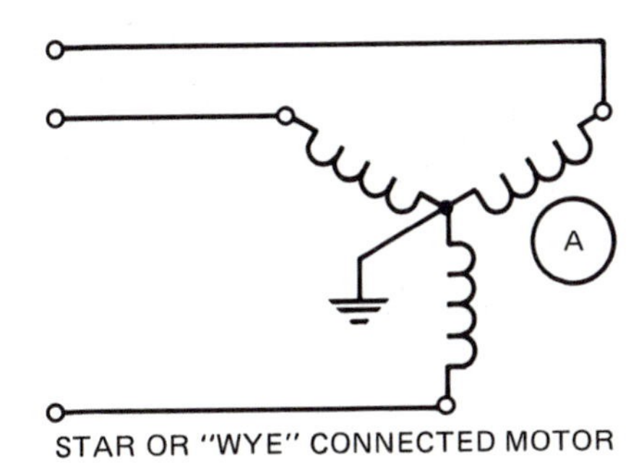

FIGURE 21-14 Schematics of three-phase motors.

Three-Phase Motors

Three-phase motors are used when torque and speed loads are heavy, i.e., in commercial and industrial applications (Figure 21-14). The direction of rotation of a three-phase motor is determined by the direction of rotation of the phases. Reversing any two of the three leads to the motor will reverse its direction.

TROUBLESHOOTING MOTOR CIRCUITS

The most common instruments used to test motor circuits are the voltmeter and the clamp-on ammeter. The AC clamp-on ammeter is an important instrument for monitoring and troubleshooting AC motors of all sizes. It is particularly handy to check if a three-phase motor is drawing current from all three lines.

Another instrument of interest is the DC ammeter, which does not need to be inserted into the circuit. This is an ingenious device that attaches a pointer to a small magnet (Figure 21-15).

When this instrument is placed next to a wire carrying DC current, the magnet will be deflected by the interacting magnetic field of the wire. Approximate amperage is indicated on the scale. This is a particularly good instrument

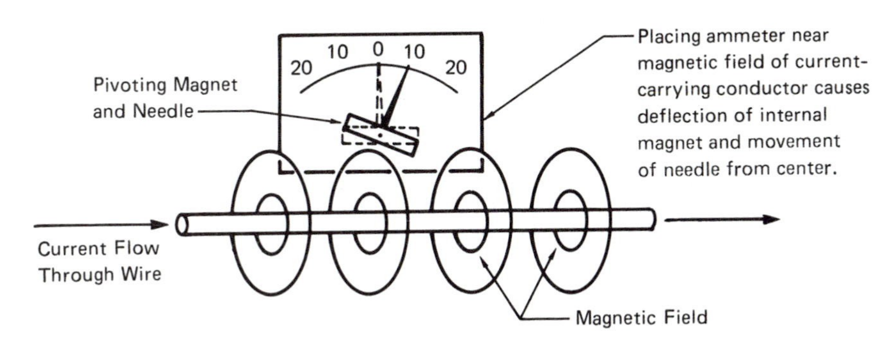

FIGURE 21-15 A special DC ammeter that does not require electrical connection into the circuit.

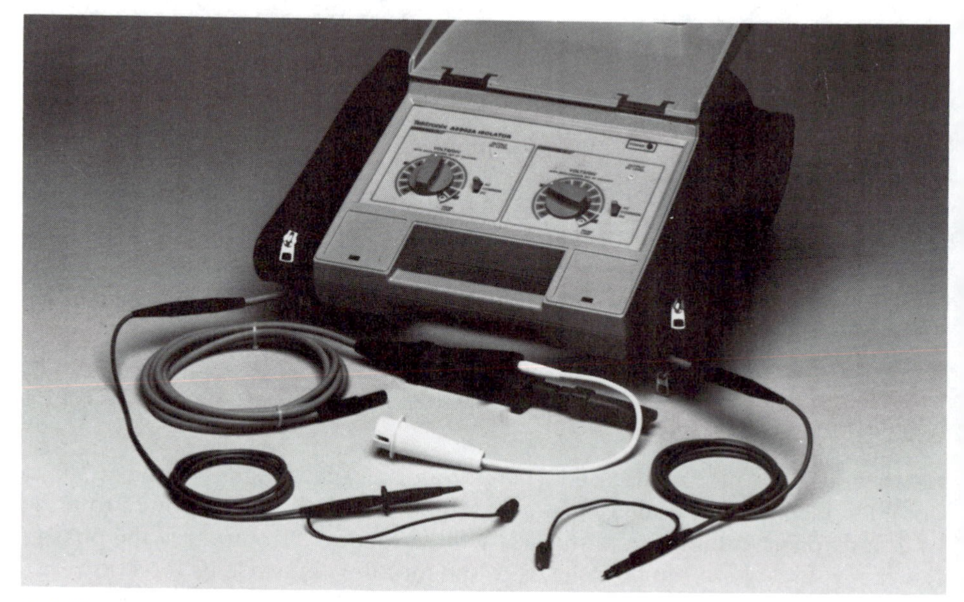

FIGURE 21-16 The isolator allows safe measurement of small voltages superimposed on much higher voltages by keeping the oscilloscope frame at ground potential. Its most common use is to observe voltage not referenced to ground.

for automotive applications. It is simple, reasonably accurate (perhaps within 20 percent, good enough for most is-it-there-or-not troubleshooting), and rugged. DC clamp-on ammeters can also be used to monitor motor currents in DC circuits.

USING THE OSCILLOSCOPE ON AC LINE PROBLEMS

The chassis of an oscilloscope is normally connected to earth ground through the power plug. This connection is the same as the grounding clip on the end of the probe. Before connecting this grounding lead to *any* circuit, be sure that the resulting grounding will not cause problems. The practice of cutting off the ground lug of the power cord, or otherwise isolating the chassis from earth ground is a dubious practice at best and can result in very dangerous voltages on the oscilloscope chassis. It could also cause failure of the insulation in the oscilloscope power supply or damage to the circuit under test. Recognizing this problem, one company has solved it with a new accessory, the Isolator™ (Figure 21-16).

SUMMARY

This chapter has presented some of the basic facts necessary for an electronics technician to deal with common power-line problems and a few other pertinent facts and warnings. However, it is no substitute for the knowledge and experience of a qualified electrician.

REVIEW QUESTIONS

1. Will a permanent-magnet motor produce electrical output if the shaft is turned?
2. How can a permanent-magnet motor be reversed?
3. How can the direction of a shunt motor be reversed?
4. How may the direction of rotation of a series motor be changed?
5. If the field of a large shunt motor weakens, what happens to the armature speed?
6. Can a series motor be used on AC?
7. Can an alternator be used as a motor?
8. How is the output of an alternator controlled?
9. How is the direction of rotation of a three-phase motor reversed?
10. Does an AC clamp-on ammeter require an internal battery?

chapter twenty-two

If the Right Part Isn't Available

CHAPTER OVERVIEW

The best policy is to replace a part with one identical to the part that was originally installed. Sometimes this is not possible for one reason or another. Perhaps the company who made a special part is out of business, or perhaps the equipment must be operational *now,* but the right part can't be obtained for a few weeks. In such a case, the technician may be called upon to make substitutions and should be prepared to make the suggestion when appropriate.

There are many possible mechanical considerations in making substitutions, all of which could not possibly be covered here. Mounting hole patterns, clearance to other components, etc., are left to the ingenuity and common sense of the technician.

PARTS SUBSTITUTIONS

Get the whole story on both the old and new components before installing them. Substitution of parts different from those originally installed may not meet with success because of subtle differences such as frequency response. (A transistor is an example of a frequency-limited component.) Other differences, such as the identification of the leads of a transistor, might be compensated for with success if the technician is paying attention to detail. Lead identification is a very important consideration when substituting types of transistors.

The following sections describe components that are eligible for emergency replacement with parts other than the original.

Substituting Discrete Semiconductors (Transistors, Diodes, etc.)

There is a great deal of interchangeability of transistors, diodes, and other semiconductors. The major manufacturers publish books of cross-references among thousands of kinds of transistors and their own line of semiconductors. The RCA "SK" series, the Motorola "HEP" series, and the Sylvania "ECG" series are good examples of interchangeable semiconductors. These publications will also suggest replacement transistors for many other types for which exact replacements may no longer be available.

Substituting Integrated Circuits

ICs may be interchangeable from one IC number to another without circuit changes. A replacement IC with the original IC number should be used whenever possible. Some manufacturers use slightly different numbering systems for the same IC. Motorola, for instance, uses a 14013 IC for the industry standard IC number 4013. The IC is the same circuit, but with a different number. Considering the family differences between the TTL Schottky and other subfamilies, it might be that substitution of a different subfamily will accomplish the same end result in *noncritical* applications. Take the specifications of Table 13-1 into account when considering these possible substitutions.

Substituting Resistors

A workable substitute for a resistor may be made in one of three different ways: in the form of several resistors in series adding up to the original resistance, by several resistors of values in parallel to equal the original, and a temporary method of substituting a variable resistance.

Series Resistors. When substituting a series of resistors for an original resistor, the sum of the resistances must equal the original value. For example, if the original resistor was 100 ohms, the substitute resistors must add up to 100 ohms (Figure 22-1). The power ratings of the substitute resistors must not be exceeded, either. If there is any doubt, a good rule of thumb is that none of the substitute

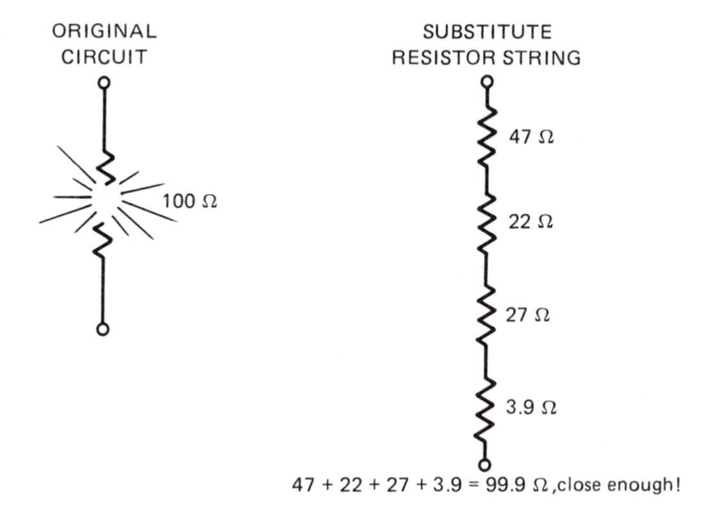

FIGURE 22-1 Substituting series resistors for a specific value.

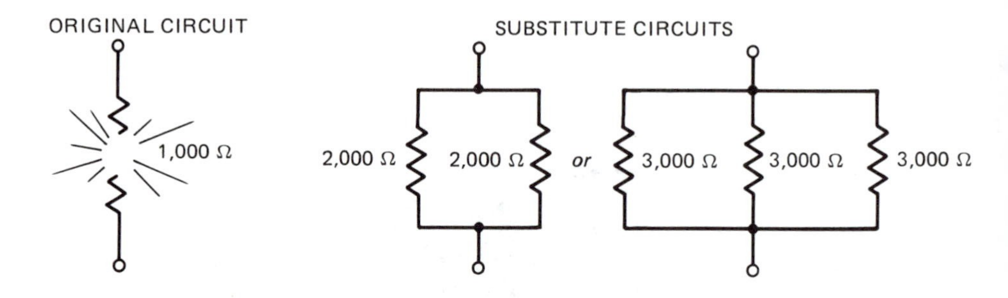

FIGURE 22-2 Substituting parallel resistors for a specific value.

resistors should be smaller physically than the original one. Smaller ones could be used if the actual individual resistor power dissipation is calculated and not exceeded.

Parallel Resistors. Although more complicated to calculate, parallel resistors may also be substituted for the original. The easiest way to figure the value of a pair of resistors to substitute is to make each of the *two* substitutes exactly *twice* the value of the original. That way, two of them in parallel equal the original (Figure 22-2). As an alternative, *three* resistors of exactly *three* times the original could be used. Using four or more resistors to substitute can get a bit ridiculous, but it can be done following the same rule.

A harder way to use a parallel combination is to obtain a resistor of the same power rating as the original with a value as close as possible to but not less than the original. All that remains is to find a paralleling resistor to "trim down" this resistance to the proper value, that of the original resistor. Considering Ro as the original resistor and Rp as the primary replacement resistor, the Rt or trimming resistor needed is found by the formula:

$$\text{Rt} = \frac{1}{\dfrac{1}{\text{Ro}} - \dfrac{1}{\text{Rp}}}$$

For example: If a 456-ohm resistor is needed, and the next closest higher value is 480 ohms:

$$\text{Rt} = \frac{1}{\dfrac{1}{456} - \dfrac{1}{480}} = \frac{1}{(0.00219) - (0.00208)}$$

$$= \frac{1}{(0.00011)} = 9120 \text{ ohms}$$

Thus, a precision resistor of 9120 ohms placed across 480 ohms comes very close to the original value of 456 ohms.

Using a Variable Resistance. A potentiometer having a maximum resistance of *more* than the resistor to be replaced may be used in temporary replacement applications. The closer the total potentiometer is to the value of the resistor to be replaced, the less critical will be the adjustment. This method is satisfactory for low-wattage applications only.

Substituting Capacitors

Two 0.01-μF capacitors in parallel will do the same job as a single 0.02 μF capacitor in situations where temperature and precise value are not a major consideration. Individual voltage ratings must be equal to or greater than that of the original capacitor.

Electrolytics. Electrolytic capacitor circuits have a bit more to consider when substituting capacitors different from the original. The polarity of installed electrolytics must be observed. In addition to the voltage rating, the leakage-resistance characteristic of these capacitors makes some special steps necessary.

When using substitute paralleled electrolytic capacitors, add the individual values of the substitutes to equal the original. The voltage rating of each must be equal to or more than but reasonably close to the original (Figure 22-3). A new voltage rating too far in excess of the original will not allow the capacitor to "form" properly, with a resulting high leakage current through the capacitor.

When substituting electrolytic capacitors in series for an original capacitor, the individual capacities interrelate much like resistors do in parallel. Each capacity must be *more* than the original. Two 20-μF capacitors in series replace a single 10-μF capacitor. Three 100-μF capacitors in series will look like about 33 μF, and so on. The voltage rating of each series capacitor may be added to obtain the original capacitor voltage rating. For instance, two 100-volt capacitors in series can be used in lieu of a 200-volt capacitor, subject to the limitations of the next paragraph.

Resistors must be placed in parallel across each of the series capacitors used. These are necessary to assure equal distribution of the applied voltage across each capacitor. Without these resistors, the DC voltage would divide according to the highly unpredictable internal leakage of the capacitors and could result in at least one being stressed to an overvoltage. The resulting failure of one capacitor by shorting will often cause the remaining capacitors to fail by increasing the voltage across them, a kind of "domino effect."

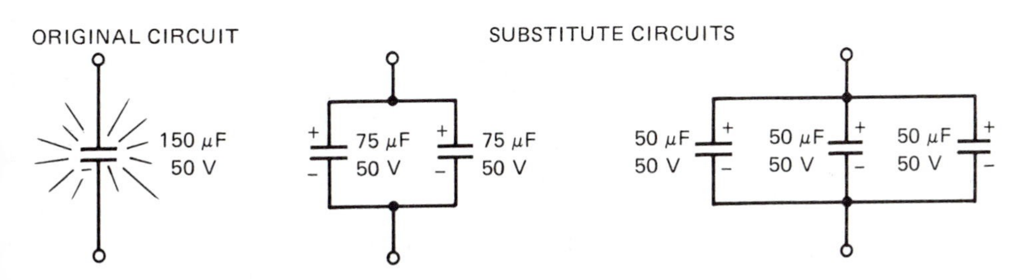

FIGURE 22-3 Substituting parallel capacitors for a specific value.

Substituting Microphones

Microphones may be substituted pretty freely within each type. In other words, a carbon microphone will substitute for any other carbon mike. Likewise a crystal for a crystal and a ceramic for a ceramic. Dynamic microphones may substitute for one another if the associated matching transformer is set for the same impedance with its output winding selection.

Substituting Speakers

There is little noticeable difference to the ear between the popular 4- and 8-ohm speakers. In noncritical applications, such as automobile radios, the two may often be used interchangeably. In more critical high-fidelity use, two 4-ohm speakers may be substituted for a single 8-ohm speaker. Conversely, two 8-ohm speakers may be wired in parallel to load a 4-ohm amplifier. Since the larger the size of the cone, the better the speaker will respond to low frequencies, substituting a larger speaker will often improve the low-frequency sound of the speaker installation.

Whenever speakers are operated together, such as in a stereo or multiple-speaker installation, the bass (large) speakers should be phased. This is simply ensuring that the speaker cones move in the same direction, in or out, upon application of a specified voltage. Speakers may be phased in the following manner:

1. Apply, for a moment, a 1.5-volt flashlight battery to the connections of the speaker. Note whether the cone moves in or out. Put a spot of paint or nailpolish to the terminal that has the positive end of the battery connected.
2. Apply, again for just a moment, the battery to the second speaker. Look for the same motion of the cone that you had in the first speaker. Reverse the battery leads if necessary to obtain this motion.
3. Now that the motions of the speaker cones are the same, put another spot of paint on the second speaker terminal, on the terminal that has the positive side of the flashlight battery connected.
4. Connect the "hot" side of the audio amplifier to the terminals marked with the paint spots, the grounds to the remaining terminals.

Substituting Batteries

A battery of a physically larger size and identical voltage will generally last longer than a smaller one. There is a difference between the carbon-zinc (1.5 V) and the alkaline types (1.4 V) as to the maximum amount of energy available, but so far as the terminal voltage is concerned they are interchangeable. The ni-cad battery has a lower voltage (1.2 V) than either the carbon-zinc or the alkaline. Some applications may require more cells when using ni-cads than other types, but the rechargeability of the ni-cad is a big advantage. Table 22-1 compares the types of cells and the main advantages of each.

TABLE 22-1 Important Characteristics of Common Types of Cells

TYPE OF CELL	ADVANTAGES	DISADVANTAGES
Carbon-zinc	Inexpensive	Produces corrosive leakage some time after complete discharge
Alkaline	Long life at moderate drain	More expensive than carbon-zinc
Mercury	Long shelf life	Expensive
Lithium	Very long shelf life	Expensive
Sealed ni-cads	Rechargeable	Require 14-hour recharge time
Wet ni-cads	Rechargeable; high drain rate	Expensive, electrolyte corrosive, explosive gases while charging
Wet lead-acid	Rechargeable; high drain rate	Electrolyte corrosive, explosive gases while charging
Lead-acid, gelatinous	Rechargeable	

Substituting Transformers

Replacement power-line transformers must have the same primary and secondary voltage ratings as the original. The current rating of the replacement windings must be equal to or greater than the original. Extra windings on a substitute transformer should be left disconnected and taped off.

Transformer windings may be series connected to provide proper primary or secondary voltages if necessary. Be sure to phase them properly. Two secondary windings of 6 V AC may be connected, in series for example, to provide 12 V AC. Connect them and test for 12 V AC across the pair. If there is no voltage, reverse *one* of the windings, which should take care of the problem. It does not strain or damage the transformer to misphase them in series connections, you just won't get what you expect.

Substitute AF and RF transformers should have ratings the same as the originals.

Substituting Switches

There are many factors to consider when substituting switches:

1. the current capability
2. the voltage rating
3. spring-loading, if any
4. normal-open or normal-closed arrangement
5. number of contacts
6. make-before-break or break-before-make contacts

It is permissible in some cases to substitute a different spring-loading arrangement as a matter of necessity. The voltage and current rating of the substitute switch must be equal to or better than the original.

A switch may be rated as a break-before-make type. This will ensure that when changing positions, no two selecting contacts will be connected together by the common wiper. A make-before-break type, on the other hand, will short two adjacent positions together while changing positions. In some applications, this is a major consideration.

Multiple-deck rotary switches can sometimes be repaired by changing a single deck with another switch having an identical contact arrangement. Be careful that the shaft indexing is the same on both switches, however.

Substituting Relays

Relay substitutes must have the original configuration of normally-open or normally-closed contacts, of the same or better current and voltage rating. Extra contacts may be left unconnected or, better yet, paralleled with similar contacts to help share the current load.

Relays intended for use on AC voltages have a small copper pole added to the armature. This provides a more constant magnetic flux to hold in the armature. These relays will also work quite well on DC voltages. However, DC relays will not work satisfactorily on AC. They buzz and will not hold "in" unless a large AC voltage is applied to them. If necessary, a DC relay may be made to work on AC if a rectifying diode is placed in series with the coil, and a filtering electrolytic capacitor is placed across the coil to help smooth out the resulting pulses of DC (Figure 22-4).

The new DC relay will require a certain amount of current to operate satisfactorily in this AC circuit. Calculate the DC current required with Ohm's Law:

$$\text{Current} = \frac{\text{DC Coil Voltage}}{\text{Coil Resistance}}$$

Use this current and the required coil voltage to calculate the series resistance needed to limit the supply AC current:

$$\text{Series Resistor} = \frac{(\text{Peak Supply Voltage}) - (\text{Coil Voltage})}{\text{Current Needed}}$$

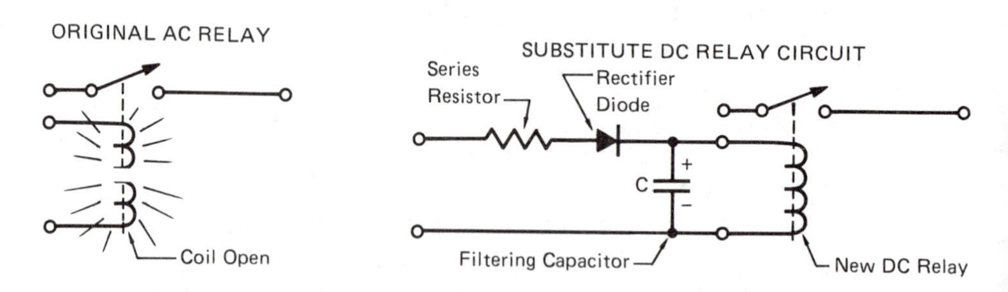

FIGURE 22-4 Substituting a DC relay for a defective AC relay.

The power dissipation of the resistor is found by multiplication:

$$\text{Power} = (\text{Peak Supply Voltage} - \text{Coil Voltage}) \times \text{Current Needed}$$

As an example, let us suppose that we have a 24-V DC relay that we must substitute for a 120-V AC relay that has burned up. The resistance of the new relay coil, measured with an ohmmeter, is 2400 ohms. The current required for normal operation of this particular relay is then 24/2400 = 0.01 ampere. Subtracting the 24 V from the available peak voltage (120 × 1.414) or 170 V, we get 146 V to "get rid of" with the series resistor. This will be the voltage drop across the ideal resistor when it is conducting the 0.01 ampere. We calculated the resistance required by dividing the voltage by the current, or 96/0.01 = 14,600 ohms. The power requirement for the resistor is the voltage times the current, or 146 × 0.01 = 1.46 watt. We could safely use a 2-watt resistor.

A DC relay of the same voltage rating as the original will work on AC voltages if the AC is first rectified and then filtered with an electrolytic (Figure 22-5). The indicated diode and capacitor should work well for DC relay coil currents up to an ampere of current with a 120-V AC input.

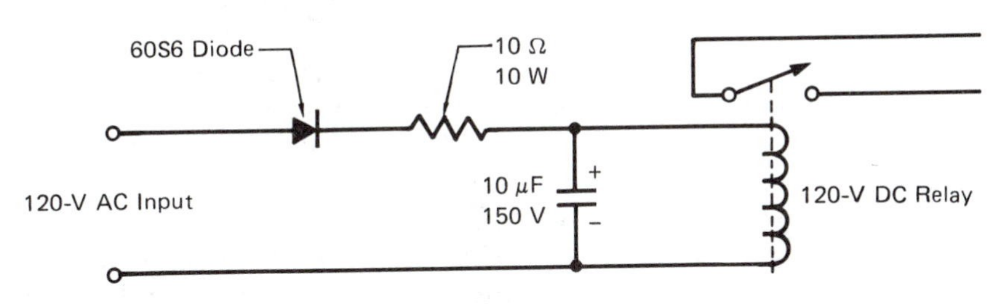

FIGURE 22-5 Operating a DC relay from an AC source.

Substituting Fuses

Replace with only the same type (fast- or slow-blowing) and with current rating equal to or slightly *less* than the original. For instance, it would be permissible to substitute a 3-ampere fuse for an original 3.2-ampere fuse. Do not substitute a larger fuse than the original unless continued operation is more important than the damage that will probably result.

Substituting Meters

D'Arsonval meters having the same basic sensitivity and the same coil resistance may be substituted for originals, even though the scales may not be the same. For instance, a 1-milliampere meter movement with a scale of 0 to 10 would work for a blown 1-milliampere meter with a scale of 0 to 100. A small notation next to the meter would be appropriate to indicate that the meter reading should be multiplied by a factor of ten to obtain the correct reading. Sometimes the meter scales can be changed from the old meter to the new.

Substituting Tubes

Whole booklets have been published on the interchangeability of vacuum tubes. The reader is referred to these publications, available from Howard Sams Publishing Co., for further information on the subject. There are numerous interchangeability possibilities within tubes.

SUMMARY

Substitution of parts as covered in this chapter is to be used only when absolutely necessary or in noncritical applications. Indiscriminate substitution of nonstandard parts can cause problems if not done properly, and can in some cases reflect unfavorably on the technician. When in doubt as to the advisability of a substitution, wait for the original part if at all possible.

REVIEW QUESTIONS

1. Can a physically larger resistor of the same resistance value be substituted if the original size is unavailable?
2. Two series resistors of 10,000 ohms, 1 watt are replacing a 20,000-ohm resistor, ½ watt. Is this an acceptable replacement?
3. Two paralleled 10-K, ½-watt resistors are replacing a 5-K, 2-watt resistor. Is this an acceptable substitution?
4. You must find an emergency replacement for a .01-μF capacitor. All you have is a pair of 0.02-μF capacitors. How would you connect them?
5. You have substituted two 20-μF electrolytic capacitors in series for a 10-μF, 100-V DC capacitor in a filtering circuit. Each of the capacitors you have used is rated at 50 V DC, and the total voltage across the series is 100 V DC. What further consideration must you give to the voltage ratings?
6. What four voltages may be obtained from the output windings of a transformer having two secondaries, 6 volts and 10 volts?
7. What is a break-before-make switch?
8. What series resistor would be required to substitute a 12-V DC relay for a 24-V DC relay, if the new coil requires 60 milliamperes? What power must the resistor throw off in heat?
9. Can a quick-blowing fuse be substituted for a slow-blowing fuse?

appendix I

Resistor Color Codes

Two-Significant-Figure Color Code

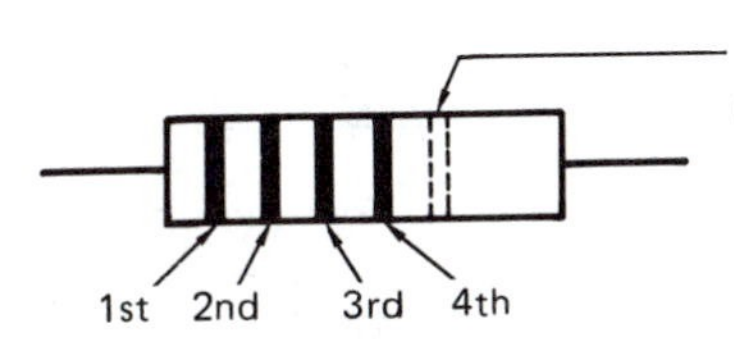

COLOR BANDS				
	1st	2nd	Multiplier 3rd	Tolerance 4th
Black	0	0	1	
Brown	1	1	10	
Red	2	2	100	
Orange	3	3	1,000	
Yellow	4	4	10,000	
Green	5	5	100,000	
Blue	6	6	1,000,000	
Violet	7	7	10,000,000	
Gray	8	8	100,000,000	
White	9	9	1,000,000,000	
Gold			0.1	5%
Silver			0.01	10%
No Color				20%

Three-Significant-Figure Color Code*

	COLOR BANDS			
	1st	2nd	3rd	Multiplier 4th
Black	0	0	0	1
Brown	1	1	1	10
Red	2	2	2	100
Orange	3	3	3	1,000
Yellow	4	4	4	10,000
Green	5	5	5	100,000
Blue	6	6	6	1,000,000
Violet	7	7	7	10,000,000
Gray	8	8	8	100,000,000
White	9	9	9	1,000,000,000
Gold				0.1
Silver				0.01

**All one-percent tolerance resistors.*

appendix II

Electronics Schematic Symbols

COMMON ELECTRONICS SYMBOLS

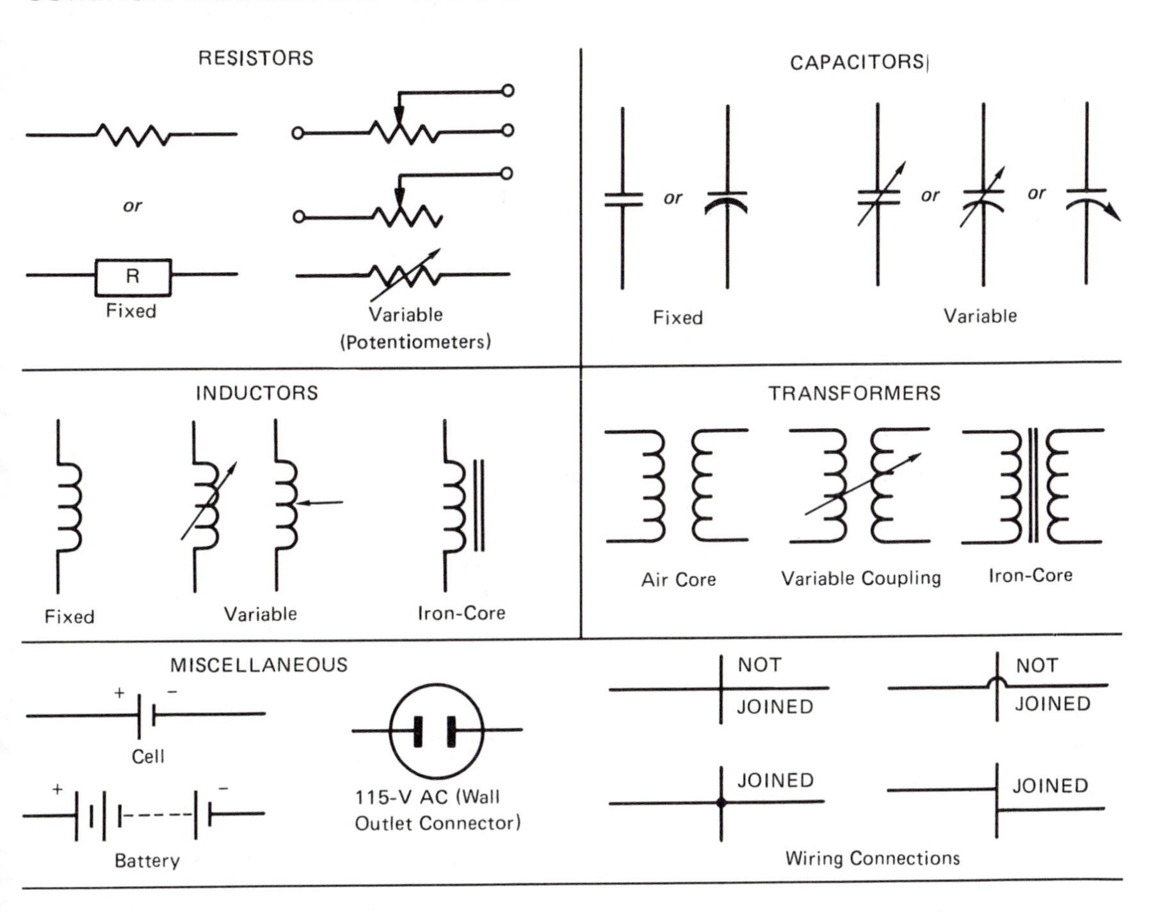

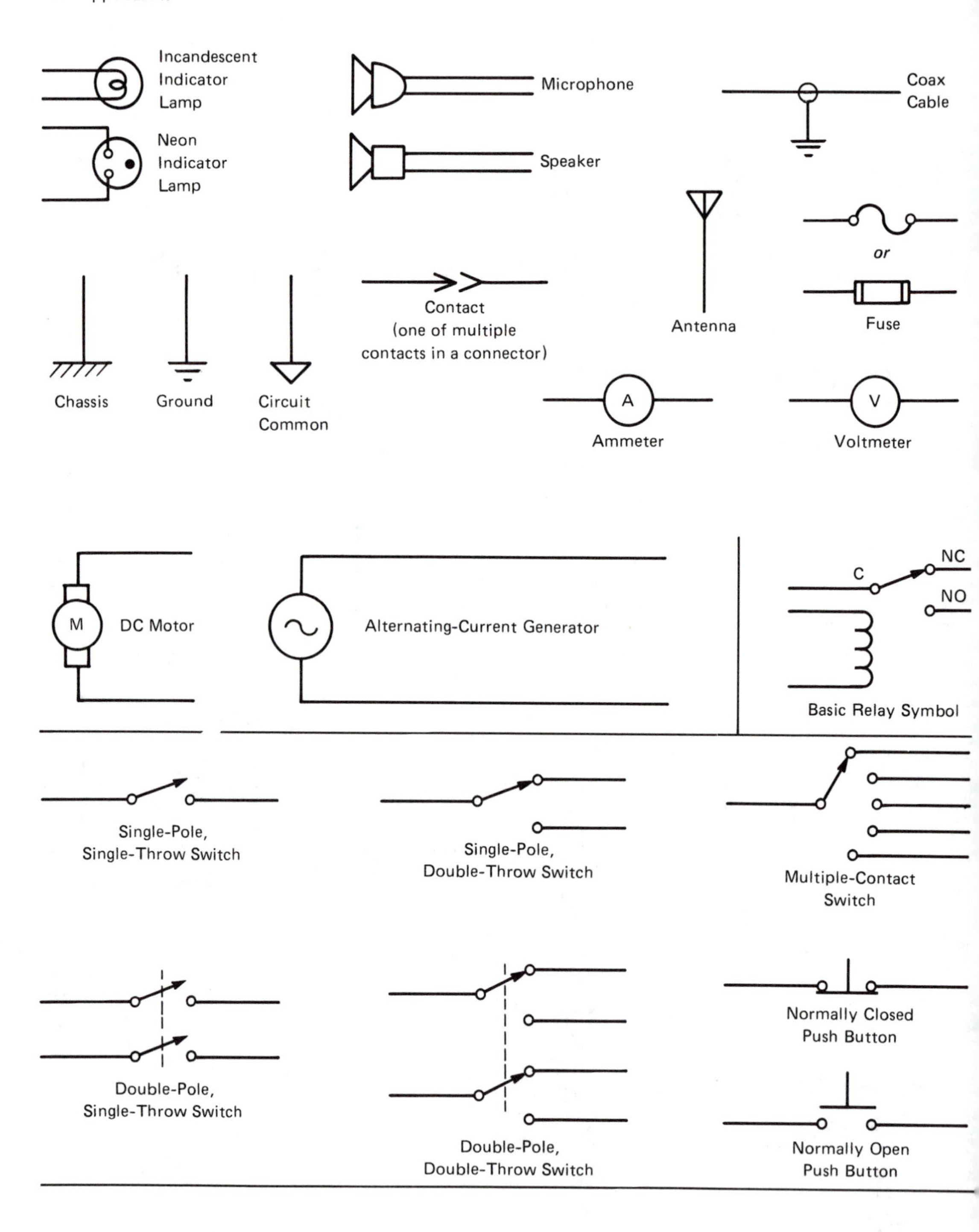
Incandescent
Indicator
Lamp
Neon
Indicator
Lamp
Microphone
Speaker
Coax
Cable
or
Antenna
Fuse
Chassis
Ground
Circuit
Common
Contact
(one of multiple
contacts in a connector)
A
Ammeter
V
Voltmeter
M
DC Motor
Alternating-Current Generator
C
NC
NO
Basic Relay Symbol
Single-Pole,
Single-Throw Switch
Single-Pole,
Double-Throw Switch
Multiple-Contact
Switch
Double-Pole,
Single-Throw Switch
Double-Pole,
Double-Throw Switch
Normally Closed
Push Button
Normally Open
Push Button

SEMICONDUCTOR SYMBOLS

NAME OF DEVICE	CIRCUIT SYMBOL	COMMONLY USED JUNCTION SCHEMATIC	ELECTRICAL CHARACTERISTICS	MAJOR APPLICATIONS
GE-MOV® Varistor		WIRE LEAD, ELECTRODE, INTER-GRANULAR PHASE, ZINC OXIDE GRAINS, ELECTRODE, EPOXY ENCAPSULANT, WIRE LEAD	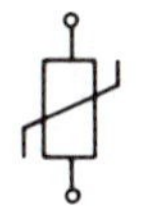 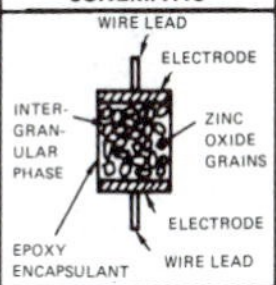 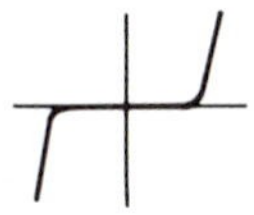 When exposed to high energy transients, the varistor impedance changes from a high standby value to a very low conducting value, thus clamping the transient voltage to a safe level.	Voltage transient protection High voltage sensing Regulation
Diode or Rectifier	ANODE, CATHODE	ANODE, p, n, CATHODE	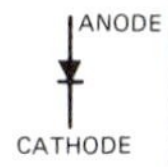 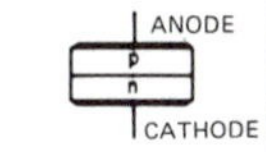 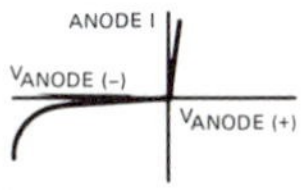 ANODE I; V_{ANODE} (−); V_{ANODE} (+). Conducts easily in one direction, blocks in the other.	Rectification Blocking Detecting Steering
Tunnel Diode	POSITIVE ELECTRODE, NEGATIVE ELECTRODE	POSITIVE ELECTRODE, p, n, NEGATIVE ELECTRODE	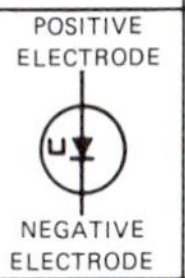 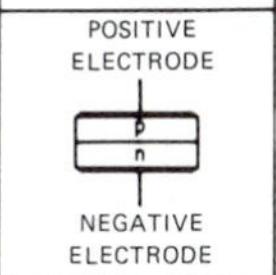 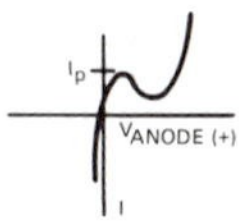 I_p; V_{ANODE} (+); I. Displays negative resistance when current exceeds peak point current I_p.	UHF converter Logic circuits Microwave circuits Level sensing
Back Diode	ANODE, CATHODE	ANODE, n, p, CATHODE	 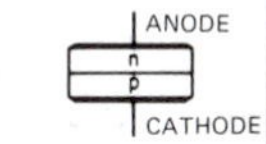 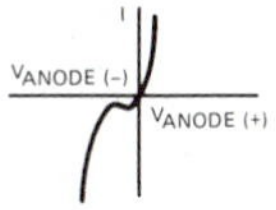 I; V_{ANODE} (−); V_{ANODE} (+). Similar characteristics to conventional diode except very low forward voltage drop.	Microwave mixers and low power oscillators
n-p-n Transistor	COLLECTOR, I_C, BASE, I_B, EMITTER	COLLECTOR, n, p, n, BASE, EMITTER	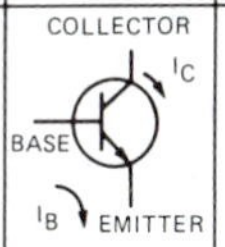 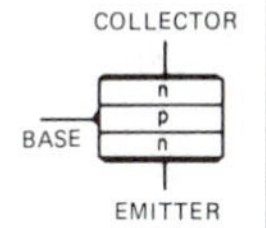 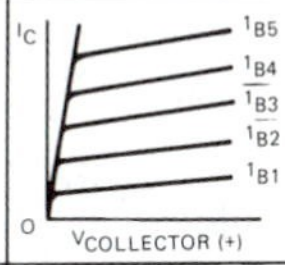 I_C; I_{B5}, I_{B4}, I_{B3}, I_{B2}, I_{B1}; O; $V_{COLLECTOR}$ (+). Constant collector current for given base drive.	Amplification Switching Oscillation
p-n-p Transistor	COLLECTOR, I_C, BASE, I_B, EMITTER	COLLECTOR, p, n, p, BASE, EMITTER	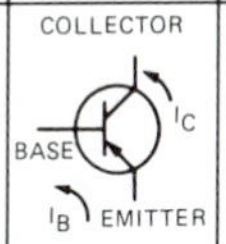 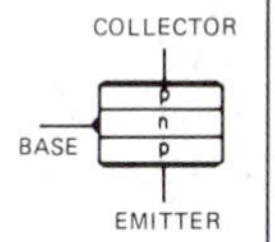 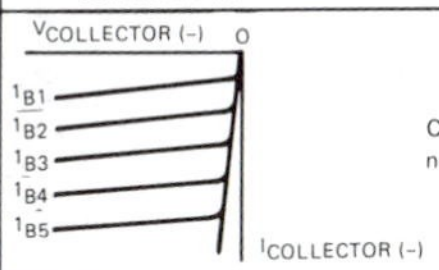 $V_{COLLECTOR}$ (−); O; I_{B1}, I_{B2}, I_{B3}, I_{B4}, I_{B5}; $I_{COLLECTOR}$ (−). Complement to n-p-n transistor.	Amplification Switching Oscillation
Unijunction Transistor (UJT)	BASE 2, EMITTER, BASE 1	BASE 2, EMITTER, p, I_e, n, BASE 1	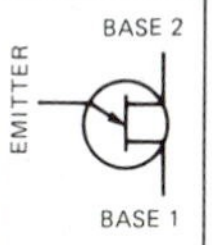 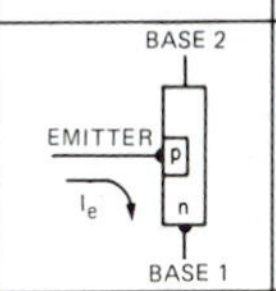 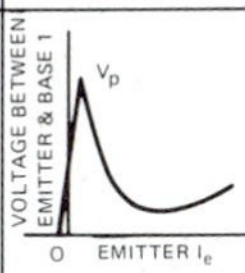 VOLTAGE BETWEEN EMITTER & BASE 1; V_p; O; EMITTER I_e. Unijunction emitter blocks until its voltage reaches V_p; then conducts.	Interval timing Oscillation Level Detector SCR Trigger
Complementary Unijunction Transistor (CUJT)	BASE 1, EMITTER, BASE 2	BASE 1, p, n, p, n, EMITTER, BASE 2	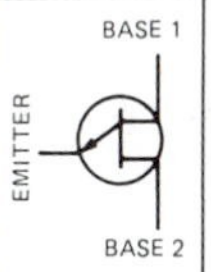 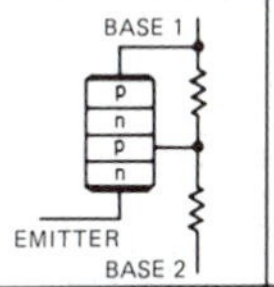 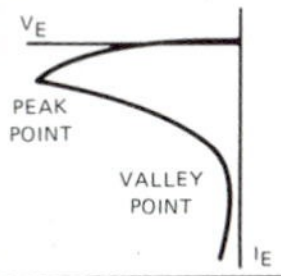 V_E; PEAK POINT; VALLEY POINT; I_E. Functional complement to UJT	High stability timers Oscillators and level detectors
Programmable Unijunction Transistor (PUT)	ANODE, GATE, CATHODE	ANODE, p, n, p, n, GATE, CATHODE	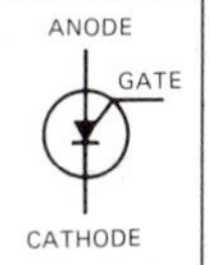 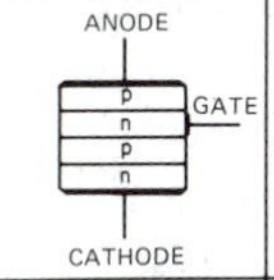 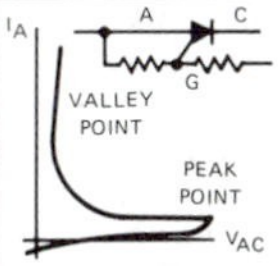 I_A; A; C; G; VALLEY POINT; PEAK POINT; V_{AC}. Programmed by two resistors for V_p, I_p, I_v. Function equivalent to normal UJT.	Low cost timers and oscillators Long period timers SCR trigger Level detector
Photo Transistor	COLLECTOR, BASE, I_B, EMITTER	COLLECTOR, n, p, n, BASE, EMITTER	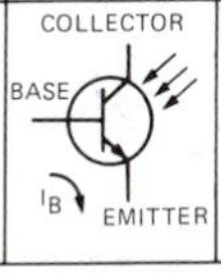 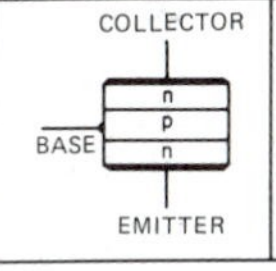 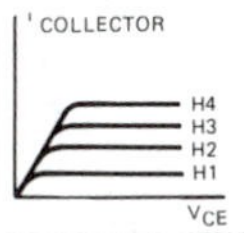 $I_{COLLECTOR}$; H4, H3, H2, H1; V_{CE}. Incident light acts as base current of the photo transistor.	Tape readers Card readers Position sensor Tachometers

Reprinted with permission Microwave Products Department, General Electric Company, Owenburo, Kentucky

NAME OF DEVICE	CIRCUIT SYMBOL	COMMONLY USED JUNCTION SCHEMATIC	ELECTRICAL CHARACTERISTICS	MAJOR APPLICATIONS
Opto Coupler 1) Transistor 2) Darlington Outputs	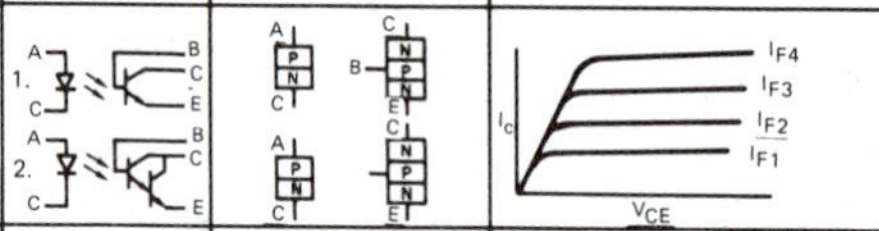 1. A, C, B, C, E 2. A, C, B, C, E	A, P, N, C; C, B, N, P, N, E A, P, N, C; C, N, P, N, E	I_c; I_{F4}, I_{F3}, I_{F2}, I_{F1}; V_{CE} Output characteristics are identical to a normal transistor/Darlington except that the LED current (I_F) replaces the base drive (I_B).	Isolated interfacing of logic systems with other logic systems, power semiconductors and electro-mechanical devices. Solid state relays.
Opto Coupler SCR Output	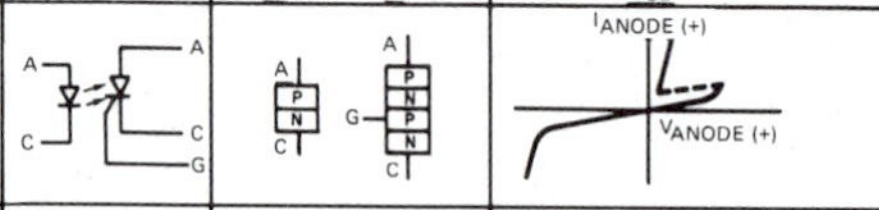 A, C, A, C, G	A, P, N, C; A, G, P, N, P, N, C	I_{ANODE} (+); V_{ANODE} (+) With anode voltage (+) the SCR can be triggered with a forward LED current. (Characteristics identical to a normal SCR except that LED current (I_F) replaces gate trigger current (I_{GT}).	Isolated interfacing of logic systems with AC power switching functions. Replacement of relays; micro-switches.
AC Input Opto Coupler	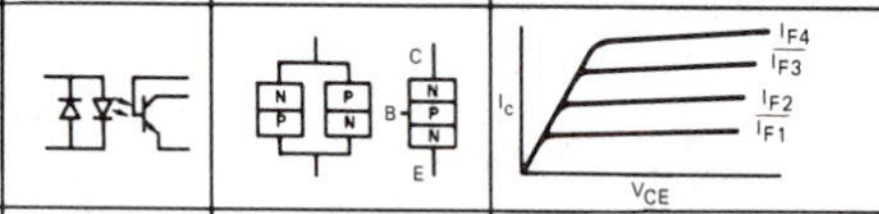	N, P, P, N; C, B, N, P, N, E	I_c; I_{F4}, I_{F3}, I_{F2}, I_{F1}; V_{CE} Identical to a "standard" transistor coupler except that LED current can be of either polarity.	Telecommunications – ring signal detection, monitoring line usage. Polarity insensitive solid state relay. Zero voltage detector.
Silicon Controlled Rectifier (SCR)	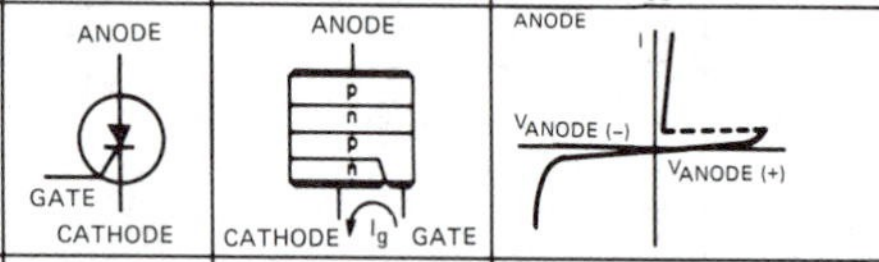 ANODE GATE CATHODE	ANODE p, n, p, n CATHODE I_g GATE	ANODE I; V_{ANODE} (–); V_{ANODE} (+) With anode voltage (+), SCR can be triggered by I_g, remaining in conduction until anode I is reduced to zero.	Power switching Phase control Inverters Choppers
Complementary Silicon Controlled Rectifier (CSCR)	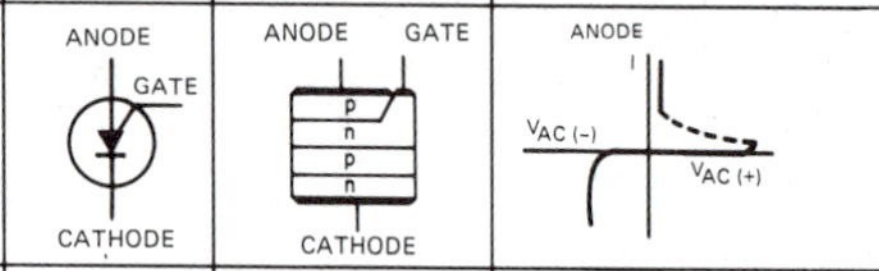 ANODE GATE CATHODE	ANODE GATE p, n, p, n CATHODE	ANODE I; V_{AC} (–); V_{AC} (+) Polarity complement to SCR	Ring counters Low speed logic Lamp driver
Light Activated SCR*	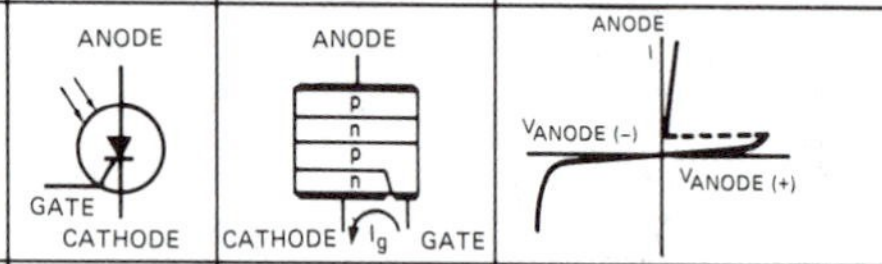 ANODE GATE CATHODE	ANODE p, n, p, n CATHODE I_g GATE	ANODE I; V_{ANODE} (–); V_{ANODE} (+) Operates similar to SCR, except can also be triggered into conduction by light falling on junctions.	Relay Replacement Position controls Photoelectric applications Slave flashes
Silicon Controlled Switch* (SCS)	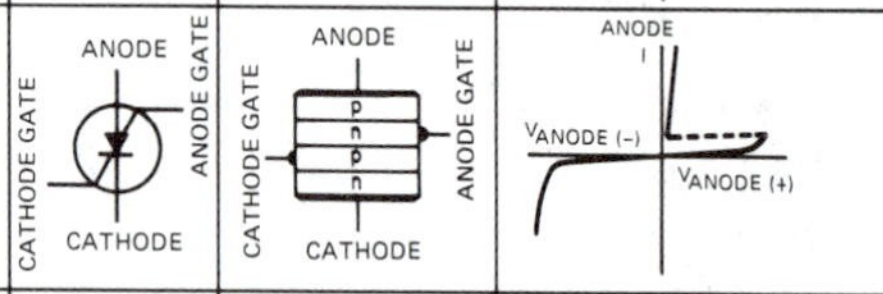 ANODE CATHODE GATE ANODE GATE CATHODE	ANODE CATHODE GATE p, n, p, n ANODE GATE CATHODE	ANODE I; V_{ANODE} (–); V_{ANODE} (+) Operates similar to SCR except can also be triggered on by a negative signal on anode-gate. Also several other specialized modes of operation.	Logic applications Counters Nixie drivers Lamp drivers
Silicon Unilateral Switch (SUS)	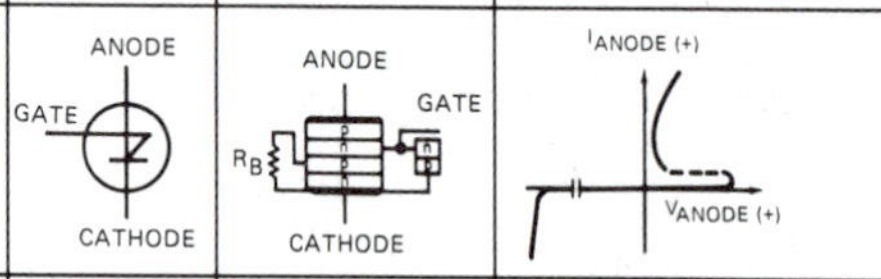 ANODE GATE CATHODE	ANODE GATE R_B CATHODE	I_{ANODE} (+); V_{ANODE} (+) Similar to SCS but zener added to anode gate to trigger device into conduction at ~ 8 volts. Can also be triggered by negative pulse at gate lead.	Switching Circuits Counters SCR Trigger Oscillator
Silicon Bilateral Switch (SBS)	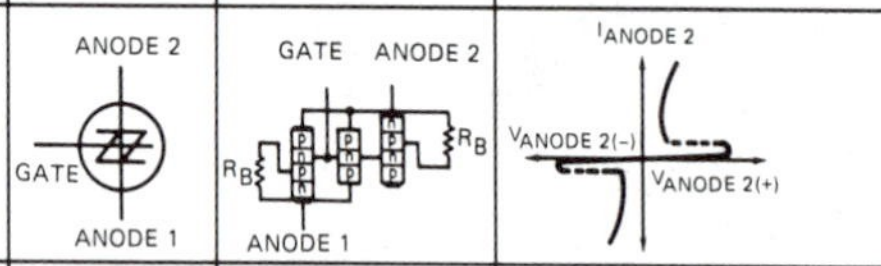 ANODE 2 GATE ANODE 1	GATE ANODE 2 R_B, R_B ANODE 1	$I_{ANODE\ 2}$; $V_{ANODE\ 2}$(–); $V_{ANODE\ 2}$(+) Symmetrical bilateral version of the SUS. Breaks down in both directions as SUS does in forward.	Switching Circuits Counters TRIAC Phase Control
Triac	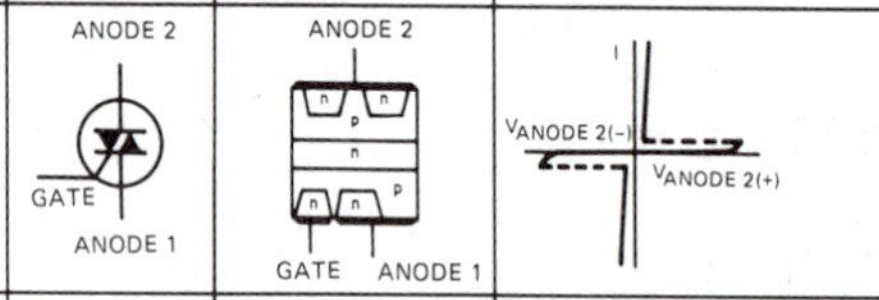 ANODE 2 GATE ANODE 1	ANODE 2 n, n, p, n, n, n, p GATE ANODE 1	I; $V_{ANODE\ 2}$(–); $V_{ANODE\ 2}$(+) Operates similar to SCR except can be triggered into conduction in either direction by (+) or (–) gate signal.	AC switching Phase control Relay replacement
Diac Trigger	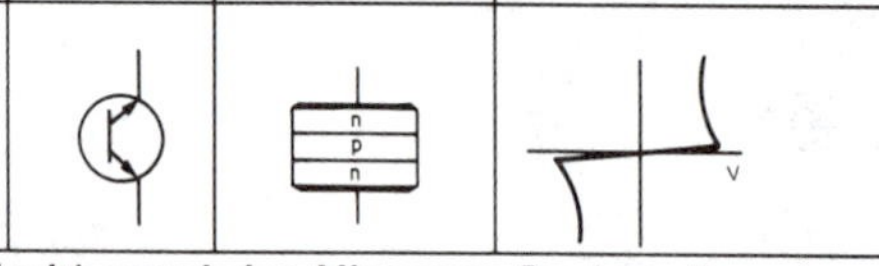	n, p, n	V When voltage reaches trigger level (about 35 volts), abruptly switches down about 10 volts.	Triac and SCR trigger Oscillator

Reprinted with permission Microwave Products Department, General Electric Company, Owenburo, Kentucky

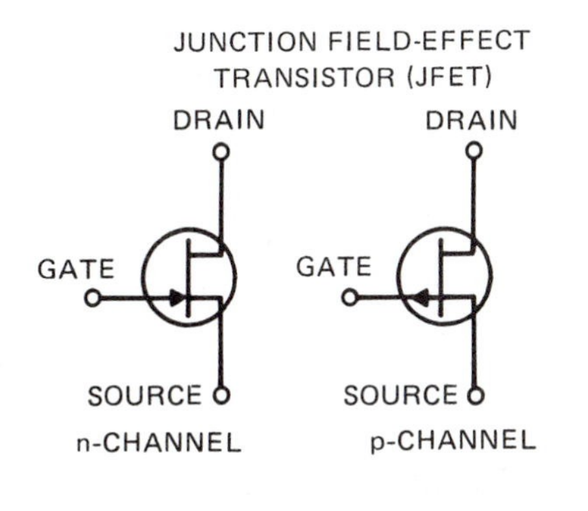

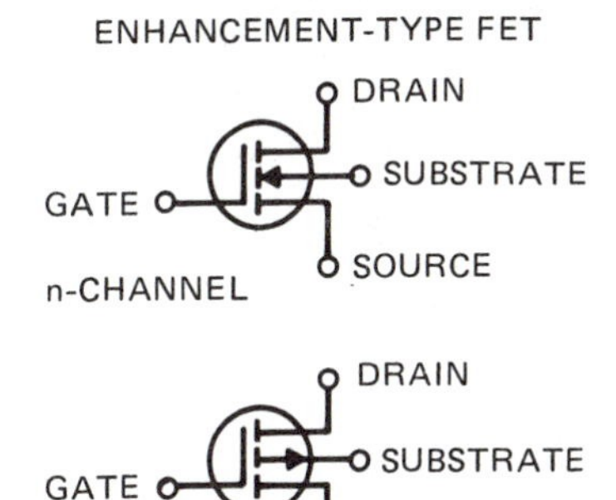

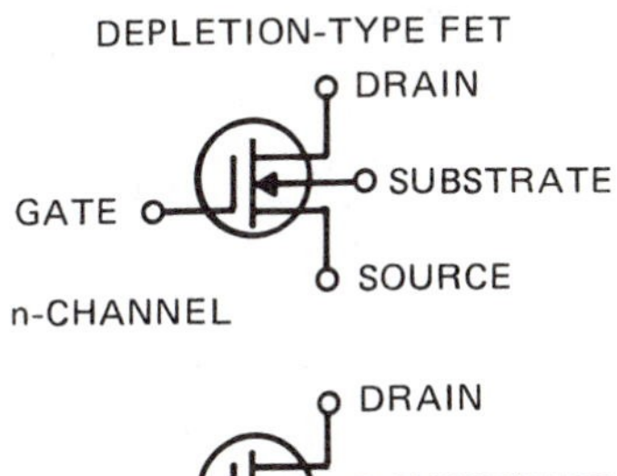

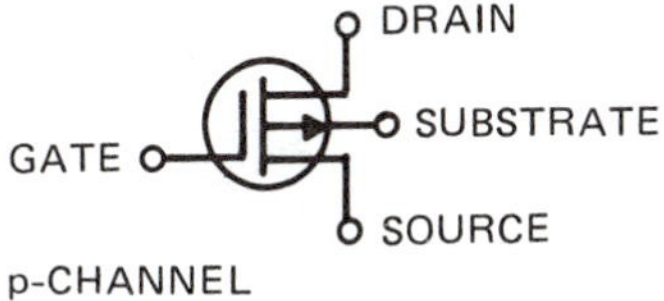

DIGITAL SYMBOLS

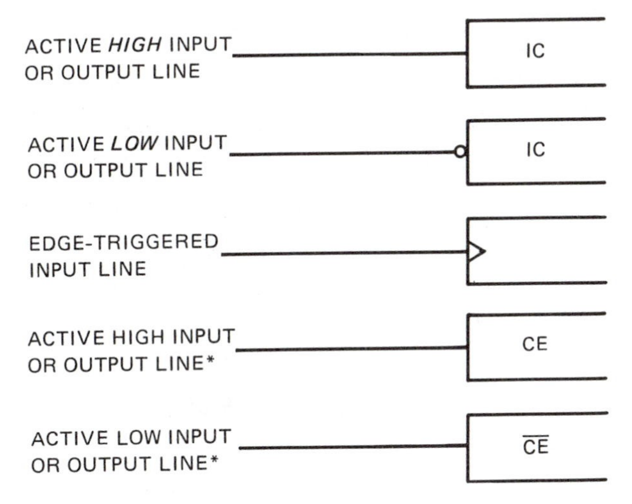

*CE (chip enable) is an example of a lettered IC pin designation. The presence or absence of a bar over the letters determines the active state of the pin.

LOGIC SYMBOLS

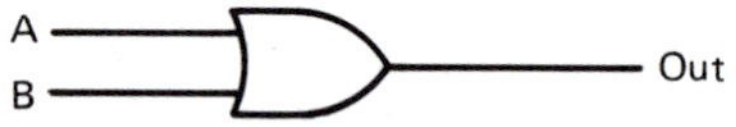

OR GATE

A	B	Out
0	0	0
0	1	1
1	0	1
1	1	1

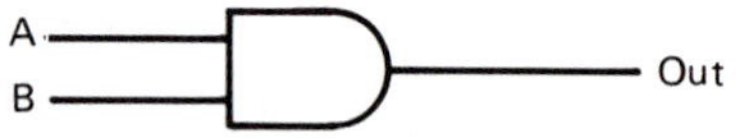

AND GATE

A	B	Out
0	0	0
0	1	0
1	0	0
1	1	1

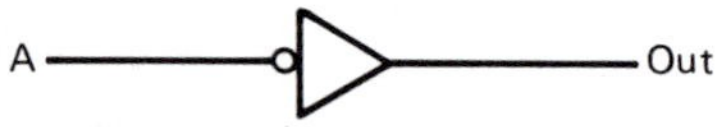

or

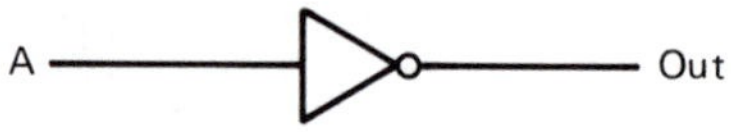

INVERTER

A	Out
0	1
1	0

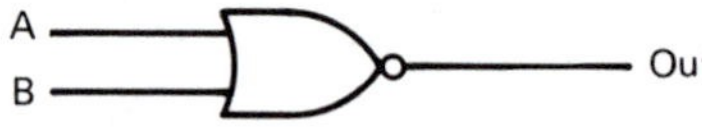

or

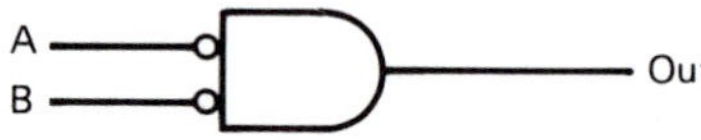

NOR

A	B	Out
0	0	1
0	1	0
1	0	0
1	1	0

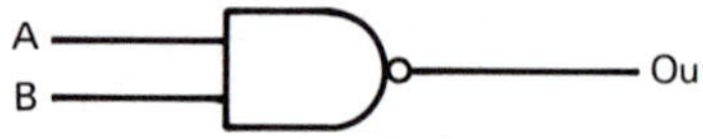

or

NAND

A	B	Out
0	0	1
0	1	1
1	0	1
1	1	0

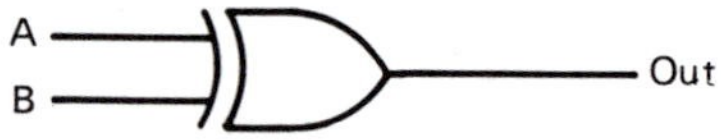

EXCLUSIVE-OR
(X-OR)

A	B	Out
0	0	0
0	1	1
1	0	1
1	1	0

EXCLUSIVE-NOR
(X-NOR)

A	B	Out
0	0	1
0	1	0
1	0	0
1	1	1

COMMON VACUUM-TUBE SYMBOLS

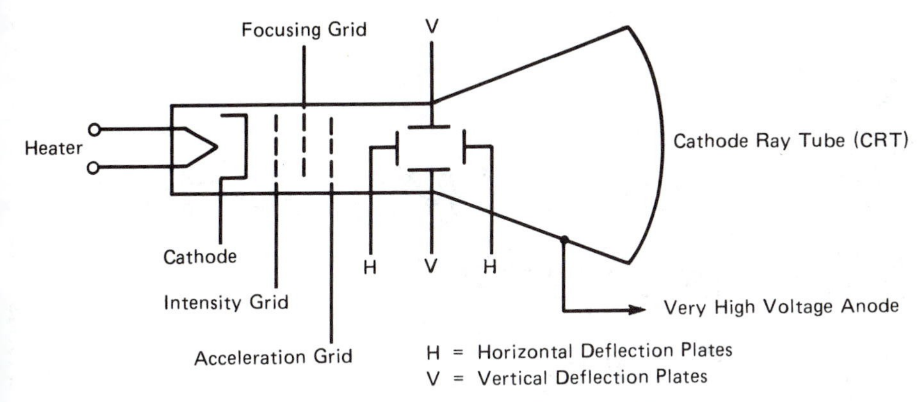

ELECTROSTATIC DEFLECTION CRT

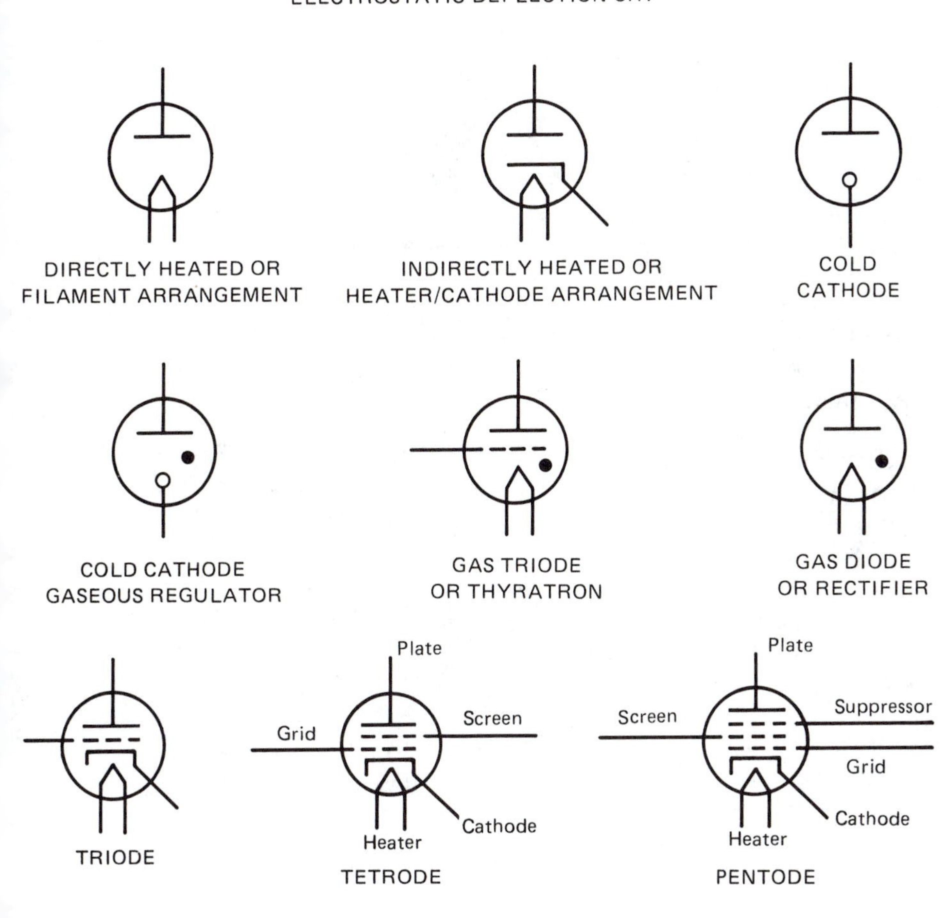

appendix III

Recommended Manufacturers of Test Equipment and Supplies

AP PRODUCTS INC.
9450 Pineneedle Dr.
P.O. Box 603
Mentor, OH 44060
(216) 354-2101
IC Test Clips

BIRD ELECTRONIC CORPORATION
30303 Aurora Rd.
Solon, OH 44139
(216) 248-1200
RF Wattmeters

CREATIVE MICROPROCESSOR SYSTEMS
P.O. Box 1538
Los Gatos, CA 95030
(408) 354-2011
Static Stimulus Tester

DYSAN CORPORATION
5201 Patrick Henry Drive
Santa Clara, CA 95050
(800) 551-9000
(408) 988-3472
Computer Alignment Diskettes

HEWLETT-PACKARD
5301 Stevens Creek Blvd.
Santa Clara, CA 95050-7369
(408) 246-4300
Test Equipment

HOWARD SAMS PUBLISHING COMPANY
4300 West 62nd Street
Indianapolis, IN 46268
Reference Manuals

HUNTRON INSTRUMENTS, INC.
15123 Highway 99 North
Lynwood, WA 98036
(800) 426-9265
(206) 743-3171
Solid-State Testers

JOHN FLUKE MANUFACTURING COMPANY, INC.
P.O. Box C9090
Everett, WA 98206
(206) 347-6100
Digital Multimeters

MOTOROLA SEMICONDUCTOR PRODUCTS INC.
P.O. Box 20912
Phoenix, AZ 85036-0912
Semiconductors, Manuals

TAB BOOKS
Blue Ridge Summit, PA 17214
Technical Books

TEKTRONIX
P.O. Box 500
Beaverton, OR 97077
Test Equipment

UNION CARBIDE CORPORATION
Section A-2, Old Ridgebury Rd.
Danbury, CT 06817
Batteries

VECTOR ELECTRONIC COMPANY INC.
12460 Gladstone Ave.
Slymar, CA 91342
(818) 365-9661
Extender Boards

VIZ MANUFACTURING COMPANY
Test Instruments Group
335 E. Price St.
Philadelphia, PA 19144
(215) 844-2626
Monitored Bench Power Supplies

WELLER PLANT
P.O. Box 868, State Rd.
Cheraw, SC 29520
(803) 537-5167
Soldering Equipment

appendix IV

BASIC Programs for Disk Drive Alignment on IBM and Compatible Computers

The following BASIC programs will enable a technician to use the hookup to the "B" drive of an IBM or compatible computer as a floppy disc drive exerciser. Using an alignment diskette such as that available from Dysan, the technician can perform alignment adjustments on the drive. See Chapter 15 for the details.

Enter all four programs, and begin by loading and running MENU.BAS first. This program will prompt you and will load and run the remaining programs as necessary.

"MENU.BAS"

```
5  CLS
10  PRINT:PRINT:PRINT"Select option:"
15  PRINT
20  PRINT "    (1) Lock on a track"
30  PRINT "    (2) Select various tracks"
35  PRINT "    (3) Cycle between tracks"
37  PRINT "    (4) Exit to DOS
38  PRINT
40  PRINT "Select number of option"
45  A$=INKEY$:IF A$="" THEN 45
50  IF A$="1" THEN 100
60  IF A$="2" THEN 200
65  IF A$="3" THEN 300
67  IF A$="4" THEN 400
70  CLS
80  GOTO 10
100  RUN "locktrac"
```

```
200  RUN "seltrac"
300  RUN "cycle"
400  SYSTEM
```

"LOCKTRAC.BAS"

```
10  REM Following is initialization of drive
20  OUT &H3F2,&H25 : REM select drive
30  OUT &H3F5,&H3 : REM specify
40  OUT &H3F5,&HC0 : REM step rate
50  OUT &H3F5,&H1 : REM non DMA mode
60  OUT &H3F5,&H7 : REM restore
70  OUT &H3F5,&H1 : REM in this drive
80  CLS
90  PRINT:PRINT:PRINT
100  INPUT "Which track do you wish to lock on?";TRAC
105  IF TRAC=99 THEN 300
110  IF TRAC>39 THEN 240
120  PRINT:PRINT "Locked on Track";TRAC
130  PRINT:PRINT:PRINT "To change tracks, hit any key"
140  PRINT "To return to menu, select track 99"
160  REM Following is head positioning command
170  OUT &H3F2,&H25 : REM select drive
180  OUT &H3F5,&HF : REM seek
190  OUT &H3F5,&H1 : REM on head,drive
200  OUT &H3F5,TRAC : REM this track
210  A$=INKEY$:IF A$="" THEN 160
220  GOTO 10
230  RETURN
240  TRAC=39
250  GOTO 120
300  RUN"menu.bas"
```

"SELTRAC.BAS"

```
10  GOSUB 80
20  CLS
25  PRINT:PRINT:PRINT
30  INPUT "Track you wish?";TRAC
31  IF TRAC=99 THEN 350
32  IF TRAC>39 THEN 300
40  PRINT "Running track";TRAC
45  PRINT " to cancel, select track 99"
```

```
50  PRINT
60  GOSUB 160
70  GOTO 30
80  OUT &H3F2,&H25 : REM select drive
90  OUT &H3F5,&H3 : REM specify
100  OUT &H3F5,&HC0 : REM step rate
110  OUT &H3F5,&H1 : REM non DMA mode
120  OUT &H3F5,&H7 : REM restore
130  OUT &H3F5,&H1 : REM in this drive
140  RETURN
150  REM ================================
160  OUT &H3F2,&H25 : REM select drive
170  OUT &H3F5,&HF : REM seek
180  OUT &H3F5,&H1 : REM on head,drive
190  OUT &H3F5,TRAC : REM this track
200  RETURN
300  TRAC=39
310  GOTO 34
350  RUN "menu.bas"
```

"CYCLE.BAS"

```
10  GOSUB 180
20  CLS
30  PRINT:PRINT:PRINT
40  INPUT "LOW TRACK NUMBER?";LOW
50  INPUT "HIGH TRACK NUMBER?";HIGH
55  PRINT "SPEED OF CYCLING IN MILLISECONDS"
56  INPUT " (>250 MS)";M
60  IF TRAC>39 THEN 310
70  INPUT "HOW MANY CYCLES";NR
75  PRINT " To terminate cycling, hit space bar":PRINT
90  PRINT "RUNNING FROM";LOW;"TO";HIGH
100  PRINT
110  A=LOW
120  GOSUB 260
124  FOR N=1 TO M:NEXT N
130  A=HIGH
140  GOSUB 260
150  A$=INKEY$:IF A$=" " THEN 400
160  GOSUB 260
165  FOR N=1 TO M:NEXT N
166  NR=NR-1
167  IF NR=0 THEN 350
170  GOTO 110
```

```
180  OUT &H3F2,&H25 : REM select drive
190  OUT &H3F5,&H3 : REM specify
200  OUT &H3F5,&HC0 : REM step rate
210  OUT &H3F5,&H1 : REM non DMA mode
220  OUT &H3F5,&H7 : REM restore
230  OUT &H3F5,&H1 : REM in this drive
240  RETURN
250  REM ================================
260  OUT &H3F2,&H25 : REM select drive
270  OUT &H3F5,&HF : REM seek
280  OUT &H3F5,&H1 : REM on head,drive
290  OUT &H3F5,A : REM this track
300  RETURN
310  TRAC=39
320  GOTO 80
330  RUN "menu.bas"
350  GOSUB 180
360  GOTO 20
400  PRINT "HIT 1 TO CONTINUE, 0 TO EXIT TO MENU"
410  A$=INKEY$
420  IF A$="1" THEN 20
430  IF A$="0" THEN 330
440  GOTO 410
```

Index

Accuracy, voltmeter, 56
Active, 5
Aligning disk drives, 225
Aligning equipment, 273
Ammeter
  AC clamp-on, 317
  DC clamp-on, 64
  DC clip-on, 317
  multimeters, 62
Analog ICs, see Integrated circuits
Analog switches, 135
Applying power, 76
Arrays
  resistor, 118
  transistor, 118
Audio levels, 37
Automated testing, 31
Average-responding meters, 163
Azimuth alignment, 224

Backup diskettes, 217
Bandpass amplifier sweep adjustment, 277
Batteries, 87
  charging, 95
  dry cells, 87
  gel-cells, 89
  lead-acid, 89
  ni-cad, 88
  testing, 267
Bench, test, 10
Billing the customer, 285
Bipolar transistors, see Transistors, bipolar
Black box, 30
Block diagram, 92
Bridging audio measurement, 36
Burden voltage, 62
Bureau of Standards (WWV), 281
Bus contention, 213

Calibration
  signal generator, 142
  frequency counter, 279
Capacitors, 76
  changing DC levels, 170
  forming, 264
  RC circuits, 172
    differentiators, 173
    integrators, 173
Cathode-ray tube, 42
  voltages, 305
Cautionary notes, 40
Charge rate, nicads, 88
Chip enable input, 194
Circuit-card extenders, 42
Circuit common, 7
Clamp-on ammeter, 317
Classes of operation, 7
Cleaning equipment, 278
Clocked logic, 204
CMOS, 184
  static precautions, 186
Coax, 139
Color codes
  audio transformers, 296
  power transformers, 296
  resistor, 329—30
  RF/IF transformers, 296
  schematic, 45
  wiring, 295
Communications receiver as detector, 275
Complementary metal oxide semiconductor (CMOS), 184
Complementary-symmetry circuit troubleshooting, 116
Complete failures, 14
Computer
  crashes, 217

diagnostics, 214
disk, 215
RAM, 214
option cards, 219
problems, 212
hardware, 213
RAM, 214
setup, 219
software, 217
setup switches, 215
Conductance, 292
Consumer ICs, 133
Crowbar circuit, 77
CRT, *see* Cathode-ray tube
Current limiting, 85
Current measuring, 62
Current tracer, 199

Darlington opto-isolators, 135
d'Arsonval movement, 54
DBs, *see* decibels
DC blocking, 99
DC offset, 127
DC-to-DC power inverter, 77
Dead circuits, 6
testing, 230
Decibels (DBs)
audio, 36
RF, 279
Demodulator probe, 278
Desoldering, 247
Differential amplifier, 128
Differentiators, 173
Digital multimeter (DMM)
average responding, 168
high-voltage probe, 168
LF signal detector, 103
measuring
current, 62
resistance, 233
voltage, 57
true vs RMS readings, 163
Digital schematics, 187
Digital troubleshooting, 181
Dipping adjustments, 217
DIP switches, 215
Discrete semiconductors, 108
Disk, *see* Floppy diskette)
Distortion, 26
tracing, 106

DMM, *see* Digital multimeter
Drip loop, 293
Drive level, 170
Dropped equipment, 19
Dummy load, 147
Duty cycle, 157

ECL logic, 186
Edge-triggered devices, 192
Electric motors
AC
split phase, 315
3-phase, 317
DC
alternators, 314
compound wound, 314
pancake, 312
permanent magnet, 311
series wound, 313
shunt wound, 313
Enabling inputs, 192
Erratic intermittent, 16
Errors, RF, 7
Extenders, *see* circuit-card extenders

Failures, 14
batteries, 87
complete, 14
distortion, 26
dropped equipment, 19
erratic intermittents, 16
filter capacitors, 76
hum, 24
intermittents, 16
lightning, 21
massive traumas, 18
dropped equipment, 19
fire/smoke, 18
lightning, 21
polarity reversal, 18
water immersed, 19
wrong voltage, 20
microphonics, 27
motorboating, 18
multiple problems, 27
noisy controls, 27
operator problems, 28, 30
overheating, 22
poor performance, 14
rectifiers, 74

Failures *(continued)*
- regulators,
  - series, 84
  - shunt, 87
- tampered equipment, 15
- thermal intermittent, 16
- transformers, 74
- transients, 21
- water immersion, 19
- wrong voltage, 20

Feedback circuits, 91
Field-effect transistor voltmeter, 56
Field strength meter, 275
Filter capacitors, 76
Floppy diskette, 222
- backups, 217
- damage, 217
- drives, repair, 221

Foldback power supply, 85
Forming capacitors, 264
Frequency counters, 102
- in pulsed circuits, 163
- in RF circuits, 149

Frustration, *see* Tough problems
Functional block diagram, 43
Function generator, 159
Fuses,
- failure analysis, 73
- finding bad, 72
- testing, 72

Gates. *See also* Logic gates
- pulse, 168

Grid dipper, 141
Ground, 8

Hardwire short, 236
Heat sink, 243
Hum, 24
- tracing unwanted, 99

Huntron
- switcher, 235
- tracker, 234

ICs, *see* Integrated circuits
Impedance, 9
In-circuit testing
- defined, 7
- with schematic, 239
- without schematic, 241

Inductors
- RL circuits, 176
  - differentiators, 173
  - integrators, 173
- testing, 266
- polarity reversal, 157

Insulated gate field-effect transistor (IGFET)
- meters, 56
- troubleshooting circuits, 120

Integrated circuits
- analog (linear), 125
- digital, 180
- open collector, 131
- pin ID, 2
- testing, 263
- troubleshooting, 125

Integrated test systems, 31
Integrators, 173
Intermittents, 15
Inverters
- digital, 190
- power, 77

Inverting, 125

Junction FET (JFET)
- testing, 255
- troubleshooting circuits, 118

Lead dress, 244
Levels, *see* Decibels
Linear ICs, *see* Integrated circuits
Line drivers, 164
Lissajous pattern, 106
Live circuits, 6
Logic analyzer, 201
Logic clip, 198
Logic Comparator, 200
Logic gates,
- AND, 188
- Drivers, 191
- Inverters, 190
- logic pulser, 194
- NAND, 189
- NOR, 189
- OR, 189
- X-NOR, 190
- X-OR, 190

Logic probe, 199

Low-power TTL, 183
Lubricating equipment, 278
Lubrication, 290

Maintenance
  batteries, 88
  preventive, 287
Mechanical intermittent, 16
Megohmmeter, 291
Meter, *see* Volt-ohm-milliammeter, Digital multimeter
Milking panel, 97
Mixing circuits, 91
Mockups, 10
Motors, electric, *see* Electric motors

Non-inverting, 127
No-signal test, 7
Nulling adjustment, 277

Ohmmeter
  DMM
    autoranging, 233
    continuity range, 234
    diode range, 233
  lead polarity, 233
  negative reading, 231
  VOM
    battery, 231
    cautions, 230
Open collector, 131
Opens, 9
Operation, classes of, 7
Operational amplifiers
  DC offset adjustment, 127
  gain, 126
  troubleshooting, 125
Operator problems, 30
Optical isolators, 135
Originating circuits, 90
Oscilloscope
  alternate sweep, 197
  bandwidth, 161
  chopped sweep, 197
  DC balance, 61
  demodulator probe, 278
  determining frequency, 146
  distortion tracing, 106
  frequency determination, 146
  frequency response, 144
  holdoff control, 162
  LF signals, 102
  probes, 61
    calibration, 145
    loading, 145
    voltage limits, 160
  rise time, 158
  signal isolator, 318
  storage, 162
  timing analysis, 196
  trigger, setting, 196
  triggering, 161
  use on
    AC line, 318
    DC circuits, 59
    digital circuits, 195
    pulse circuits, 160
    RF circuits, 143
Outlets, 309
Out-of-circuit, 7

Packing equipment, 285
Paper work, 284
Peaking adjustments, 276
Power outlets, 309
Power supplies
  battery chargers, 93
  bench, 92
  troubleshooting, 72
  types, 93
Predicting failures, 288
Preventive maintenance, 287
Printed circuit board
  dead-circuit testing, 230
  extenders, 42
  repairs, 244
  shorts, finding, 238
Probes, voltmeter, 53
Processing circuits, 90
PRR, *see* Pulse repetition rate
Pulse
  amplifiers, 164
  distortion by scope, 161
  generator, 159
  oscillators, 164
Pulse circuits, troubleshooting, 157
Pulse repetition rate (PRR), 160
Push-pull circuit, 115

Radial alignment, diskette, 222
Radio frequency
  Amplifiers
    large signal, 152
    small signal, 151
  detector circuits, 151
  loads, dummy, 153
  mixers, 154
  oscillators, 150
  probe, 143
  synthesizers, 154
  troubleshooting, 139
  wattmeter, 147
Rails, 129
Reassembling equipment, 277
Receiver as RF detector, 149
Rectifiers, 74
Regulators
  protection diode, 132
  series-type, 83
  shunt-type, 87
  switching, 86
  vacuum tube, 303
Removing parts, 243
Repairing
  computers, 213
  Floppy disk drives, 221
  PC boards, 244
  power supplies, 72
Replacing components, 244
Reset input, 194
Resistance measuring, 4, 230
Resistor
  color code, 329-30
  multiplier, 55
Reversing electric motors, *see* Electric motors
RF, *see* Radio frequency
Rivets, removing, 243

Safety rules, 40
Schematics
  diagrams, 42
  drawing circuit, 47
  easel, 43
  how to read, 45
  symbols, 331
Schmitt trigger, 181
Schottky TTL, 184
SCR, *see* Silicon controlled rectifier
Semiconductor short, 237
Set input, 192
Setup switches, computer, 215
Shorting signals, 99
Shorts
  defined, 9
  finding on PC boards, 238
  on multi-level boards, 239
Shunt, 63
Siemens (Mhos), 292
Signal generator
  audio, 101
  calibration, 142, 281
  function, 159
  pulse, 159
  radio frequency, 142
Signal rectification, 170
Signal tracing, 98
Signature analysis, 201
Silicon-controlled rectifier (SCR)
  testing, 259
  troubleshooting circuits, 123
Silicon grease, 244
Single-phase power, 309
Single-stepping logic, 209
Skew check, diskette drive, 225
"Soft fuse," 80
Solder removal, 247
Solid-state tester, 234
Solvents, 290
Standing wave ratio (SWR) meter, 148
Static precautions, 186
Static stimulus testing, 216
Stuck levels, 207
Substituting parts
  batteries, 324
  capacitors, 323
  diodes, 320
  fuses, 327
  ICs, 321
  meters, 327
  microphones, 324
  relays, 326
  resistors, 321
  speakers, 324
  switches, 325
  transformers, 325
  transistors, 320
  tubes, 328
Substituting PC boards, 213

Sweep generator, 277
SWR meter, 148
Symbols
  component reference, 46
  digital, 5
  ground, 46
  schematic, 331
System trigger, 160

Terminating circuits, 91
Terminating measurement, *see* Bridging audio measurement
Test equipment, list of, 12
Test leads, 53
Testing parts, out of circuit
  batteries, 267
  bipolar transistors, 253
  Darlington transistors, 254
  diodes, 251
  FETs, 255
  fuses, 270
  gas displays, 261
  ICs, 263
  inductors, 266
  LEDs, 260
  meters, 270
  microphones, 265
  optical isolators, 255
  photo diodes, 261
  photo transistors, 262
  relays, 270
  resistors, 263
  SCRs, 259
  speakers, 265
  switches, 269
  transformers, 267
  tubes, 271
  UJTs, 258
  varactors, 263
Thermal intermittent, 16
Three-phase power, 311
Three-state logic, 212
Time constants
  RC, 172
  RL, 176
Time domain
  reflectometry (TDR), 288
Tools
  minimum, 12
  PC board, 244
  RF, 140
Totem pole output, 184
Tough problems, 28
Tracing signals, 98
Transformers
  flyback, 167
  polarity, 158
  power, 74
  pulse, 158
Transistors, bipolar
  checker circuit, 254
  SCRs, 123
  troubleshooting, 109
  typical base diagram, 2
  UJTs, 122
  wire bundles, 33
  zener diodes, 109
Transit, VOM, 57
Transmitter tests, 279
Trigger, system, 160
Trouble reports, 30
Troubleshooting, live circuits
  analog ICs, 125
  audio lines, 36
  charts, 32
  coax, 33
  complementary-symmetry, 116
  digital circuits, 180
  diodes, 108
  if too high, 68
  if too low, 67
  IGFETs, 120
  JFETs, 119
  LF circuits, 101
  motor circuits, 311
  power supplies, 72
    filters, 76
    rectifiers, 74
    regulators, 83
    transformers, 74
Truth tables, 5
TTL logic, 183

UJTs, troubleshooting, 122
Unclocked logic, 204
Unsoldering, 247

Vacuum tubes
  amplifiers, 302
    push-pull, 303

Vacuum tubes *(continued)*
RC-coupled, 302
RF/IF, 304
safety, 295
typical voltages, 301
Variable transformer, 79
Vibration, 244
Virtual ground, 126
Visual checks, 70
Voltage, measuring
DC, 53
estimating, 65
Voltage comparator, 130
Voltage-controlled oscillator (VCO), 155
Voltage regulator ICs, 132
Volt-ohm-milliammeter (VOM)
average responding, 163
high-voltage probe, 168
LF signals, 102
loading circuits, 55
measuring
current, 62
resistance in-circuit, 239
resistance out-of-circuit, 263
voltage, 55
VOM, *see* Volt-ohm-milliammeter

Wiring diagrams, 43
WWV, 281

Zener diode
troubleshooting circuits, 109